集成电路科学与工程丛书

半导体先进封装技术

[美] 刘汉诚（John H. Lau） 著

蔡 坚 王 谦 俞杰勋 徐柘淇 译

机械工业出版社

本书作者在半导体封装领域拥有40多年的研发和制造经验。本书共分为11章，重点介绍了先进封装，系统级封装，扇入型晶圆级／板级芯片尺寸封装，扇出型晶圆级／板级封装，2D、2.1D和2.3D IC 集成，2.5D IC 集成，3D IC 集成和 3D IC 封装，混合键合，芯粒异质集成，低损耗介电材料和先进封装未来趋势等内容。通过对这些内容的学习，能够让读者快速学会解决先进封装问题的方法。

本书可作为高等院校微电子学与固体电子学、电子科学与技术、集成电路科学与工程等专业的高年级本科生和研究生的教材和参考书，也可供相关领域的工程技术人员参考。

First published in English under the title:
Semiconductor Advanced Packaging
by John H. Lau, edition:1
Copyright © The Editor(s) (if applicable) and The Author(s), under exclusive license to Springer Nature Singapore Pte Ltd. 2021

This edition has been translated and published under licence from Springer Nature Singapore Pte Ltd..

Springer Nature Singapore Pte Ltd. takes no responsibility and shall not be made liable for the accuracy of the translation.

此版本仅限在中国大陆地区（不包括香港、澳门特别行政区及台湾地区）销售。未经出版者书面许可，不得以任何方式抄袭、复制或节录本书中的任何部分。

北京市版权局著作权合同登记　图字：01-2021-5495 号。

图书在版编目（CIP）数据

半导体先进封装技术／（美）刘汉诚著；蔡坚等译 . — 北京：机械工业出版社，2023.6（2023.11重印）
（集成电路科学与工程丛书）
书名原文：Semiconductor Advanced Packaging
ISBN 978-7-111-73094-1

Ⅰ.①半… Ⅱ.①刘…②蔡… Ⅲ.①半导体器件 – 封装工艺 Ⅳ.① TN305.94

中国国家版本馆 CIP 数据核字（2023）第 074207 号

机械工业出版社（北京市百万庄大街 22 号　邮政编码 100037）
策划编辑：刘星宁　　　　　　　责任编辑：刘星宁　闫洪庆
责任校对：郑　婕　梁　静　　　封面设计：马精明
责任印制：李　昂
北京捷迅佳彩印刷有限公司印刷
2023 年 11 月第 1 版第 2 次印刷
184mm×240mm · 26.75 印张 · 645 千字
标准书号：ISBN 978-7-111-73094-1
定价：189.00 元

电话服务　　　　　　　　　　网络服务
客服电话：010-88361066　　　机　工　官　网：www.cmpbook.com
　　　　　010-88379833　　　机　工　官　博：weibo.com/cmp1952
　　　　　010-68326294　　　金　书　网：www.golden-book.com
封底无防伪标均为盗版　　　　机工教育服务网：www.cmpedu.com

前言

当前半导体产业有五个确定的增长引擎,它们分别是:①移动终端,如智能手机、智能手表、可穿戴设备、笔记本电脑和平板电脑;②高性能计算(HPC),也被称为超级计算,它能够在超级计算机上高速处理数据和进行复杂的计算;③自动驾驶汽车;④物联网(internet of things, IoT),如智慧工厂和智慧医疗;⑤用于云计算的大数据和用于边缘计算的实时数据处理。

封装技术专家正在使用各种先进的封装方法如倒装芯片、晶圆级/板级芯片尺寸封装;扇出型晶圆级/板级封装;封装堆叠(PoP);硅通孔;2.1D、2.3D、2.5D 以及 3D IC 集成;高带宽存储器(HBM);多芯片模组;系统级封装(SiP);异质集成;芯粒技术;互连桥等,以容纳(封装)面向这五类主要应用的半导体器件。

系统技术的驱动力,如 5G(第五代标准宽带蜂窝网络技术)和 AI(人工智能,是指任何能让计算机模拟人类智力的技术),也正在持续推动这五类半导体应用的增长。由于 5G 和 AI 的推动,半导体器件的速度不断提高、密度不断增加、焊盘节距不断减小、芯片尺寸不断增大,同时功耗也随之增加。所有这些变化都为半导体先进封装技术提供了新的机遇和挑战。

可是,对于大多数从业的工程师和管理人员以及科研工作人员而言,先进封装仍没有得到很好理解。目前无论是工业界还是学术界,都亟需一本能对当前先进封装技术进行全面讲解的书籍。本书写作的目的就是为了让读者能快速学会解决先进封装问题的方法;通过阅读本书,还可以学习到在做系统级决策时所必须具备的折中意识。

本书共分为 11 章,它们分别是:①先进封装;②系统级封装;③扇入型晶圆级/板级芯片尺寸封装;④扇出型晶圆级/板级封装;⑤ 2D、2.1D 和 2.3D IC 集成;⑥ 2.5D IC 集成;⑦ 3D IC 集成和 3D IC 封装;⑧混合键合;⑨芯粒异质集成;⑩低损耗介电材料;⑪先进封装未来趋势。

第 1 章先简要介绍了什么是先进封装。随后列出了 16 种不同的先进封装技术,并针对每一种先进封装技术给出一个案例。该章还简要讨论了技术驱动、半导体器件和封装三者之间的关系。

第 2 章介绍了系统级封装(SiP)技术及其组装工艺,如表面安装技术(SMT)和倒装芯片(FC)技术。该章首先还介绍了片上系统(SoC)的概念和系统级封装的概念,并对两者做出区分。

第 3 章详细介绍了扇入型晶圆级/板级芯片尺寸封装。该章分为 4 个部分:①扇入型晶圆级芯片尺寸封装;②扇入型板级芯片尺寸封装;③ 6 面模塑晶圆级芯片尺寸封装;④ 6 面模塑板级芯片尺寸封装。

第 4 章介绍了扇出型晶圆级/板级封装。该章分为 6 个部分：①扇出型（先上晶且面朝下）晶圆级封装；②扇出型（先上晶且面朝下）板级封装；③扇出型（先上晶且面朝上）晶圆级封装；④扇出型（先上晶且面朝上）板级封装；⑤扇出型（后上晶或先 RDL）晶圆级封装；⑥扇出型（后上晶或先 RDL）板级封装。

第 5 章简单介绍了 2D、2.1D 和 2.3D IC 集成，并就每一种封装技术给出相应案例。该章还简要介绍了再布线层（RDL）的概念，包括有机 RDL、无机 RDL 和混合 RDL。

第 6 章简单介绍了 2.5D IC 集成（或无源 TSV 转接板），并给出几个相应案例。该章还提及了 2.5D IC 集成的开端以及最新进展。

第 7 章介绍了 3D IC 封装（无 TSV）和 3D IC 集成（含 TSV）或有源 TSV 转接板以及相应的案例。该章还提及了高带宽存储器（HBM）的概念。

第 8 章讨论了混合键合，也介绍了 Cu-Cu 热压键合（TCB）、SiO_2-SiO_2 热压键合以及室温 Cu-Cu 热压键合。该章还简要提及了一些混合键合技术的新进展。

第 9 章简单介绍了芯粒异质集成及其优缺点，还简要介绍了 DARPA 的 COSMOS、DAHI、CHIPS 和 SHIP 项目。

第 10 章系统介绍了近几年文献报道的高速高频应用中介电材料的 Df 和 Dk 性质，并首先说明了为什么 5G 应用中需要用到低 Df、Dk 以及低热膨胀系数的介电材料。

第 11 章介绍了先进封装技术未来趋势及其组装工艺，主要是未来半导体行业 SoC 和芯粒发展的趋势。该章还简要讨论了 COVID-19 对半导体行业的影响。

本书服务的主要对象是以下三类专业人员：①对于诸如 2D 扇出型（先上晶）IC 集成、2D 倒装 IC 集成、封装堆叠、系统级封装或异质集成、2D 扇出型（后上晶）IC 集成、2.1D 倒装 IC 集成、2.1D 含互连桥倒装 IC 集成、2.1D 含互连桥扇出 IC 集成、2.3D 扇出型（先上晶）IC 集成、2.3D 倒装 IC 集成、2.3D 扇出型（后上晶）IC 集成、2.5D（焊料凸点）IC 集成、2.5D（微凸点）IC 集成、微凸点 3D IC 集成、微凸点芯粒 3D IC 集成、无凸点 3D IC 集成以及无凸点芯粒 3D IC 集成等先进封装感兴趣的专业人员；②在实际生产中遭遇封装问题并想要理解和学习更多解决问题方法的技术人员；③希望为产品选择一个可靠的、创新的、高性能、高密度、低功耗以及性价比高的先进封装方法的专业人士。本书同样也可以作为有志成为微电子、光电子行业未来的管理人员、科学家以及工程师的大学本科生和研究生的教科书。

我希望在先进封装技术发展前所未有的今天，当各位在面临挑战性难题的时候，本书可以为各位提供有价值的参考。我也希望它有助于进一步推动先进封装有关的研发工作，为我们提供更多技术全面的产品。当机构或企业掌握了如何为他们的产品设计并制造先进封装技术的方法时，他们将有望在微电子、光电子行业尽享性能、功能、密度、功率、带宽、品质、尺寸以及重量多方面提升所带来的效益。我十分憧憬本书所提供的内容可以帮助先进封装的发展破除障碍，避免无效的投入，缩短设计、材料、工艺和制造的研发周期。

John H. Lau
美国加利福尼亚州帕罗奥图

目　　录

前言
第 1 章　先进封装 ··· 1
1.1　引言 ··· 1
1.2　半导体的应用 ·· 1
1.3　系统技术的驱动力 ·· 1
　　1.3.1　AI ·· 1
　　1.3.2　5G ·· 2
1.4　先进封装概述 ·· 3
　　1.4.1　先进封装种类 ··· 3
　　1.4.2　先进封装层级 ··· 3
1.5　2D 扇出型（先上晶）IC 集成 ··· 5
1.6　2D 倒装芯片 IC 集成 ·· 5
1.7　PoP、SiP 和异质集成 ··· 6
1.8　2D 扇出型（后上晶）IC 集成 ··· 8
1.9　2.1D 倒装芯片 IC 集成 ··· 8
1.10　含互连桥的 2.1D 倒装芯片 IC 集成 ···································· 9
1.11　含互连桥的 2.1D 扇出型 IC 集成 ······································· 9
1.12　2.3D 扇出型（先上晶）IC 集成 ······································· 10
1.13　2.3D 倒装芯片 IC 集成 ·· 10
1.14　2.3D 扇出型（后上晶）IC 集成 ······································· 11
1.15　2.5D（C4 凸点）IC 集成 ·· 12
1.16　2.5D（C2 凸点）IC 集成 ·· 12
1.17　微凸点 3D IC 集成 ·· 13
1.18　微凸点芯粒 3D IC 集成 ·· 14
1.19　无凸点 3D IC 集成 ·· 14
1.20　无凸点芯粒 3D IC 集成 ·· 15

1.21 总结和建议 ·· 15
参考文献 ·· 16

第 2 章　系统级封装 ·· 22

2.1 引言 ··· 22
2.2 片上系统 ··· 22
2.3 SiP 概述 ·· 23
2.4 SiP 的使用目的 ·· 23
2.5 SiP 的实际应用 ·· 23
2.6 SiP 举例 ·· 24
2.7 SMT ·· 25
　2.7.1 PCB ·· 26
　2.7.2 SMD ··· 28
　2.7.3 焊膏 ·· 29
　2.7.4 模板印刷焊膏和自动光学检测 ·· 30
　2.7.5 SMD 的拾取和放置 ·· 32
　2.7.6 对 PCB 上的 SMD 的 AOI ·· 33
　2.7.7 SMT 焊料回流 ··· 33
　2.7.8 缺陷的 AOI 和 X 射线检测 ··· 34
　2.7.9 返修 ·· 35
　2.7.10 总结和建议 ··· 36
2.8 倒装芯片技术 ··· 36
　2.8.1 基于模板印刷的晶圆凸点成型技术 ·· 37
　2.8.2 C4 晶圆凸点成型技术 ·· 38
　2.8.3 C2 晶圆凸点成型技术 ·· 40
　2.8.4 倒装芯片组装——C4/C2 凸点批量回流（CUF） ····················· 40
　2.8.5 底部填充提升可靠性 ·· 42
　2.8.6 倒装芯片组装——C4/C2 凸点的小压力热压键合（CUF） ······ 44
　2.8.7 倒装芯片组装——C2 凸点的大压力热压键合（NCP） ··········· 45
　2.8.8 倒装芯片组装——C2 凸点的大压力热压键合（NCF） ··········· 45
　2.8.9 一种先进的倒装芯片组装——C2 凸点液相接触热压键合 ······· 47
　2.8.10 总结和建议 ··· 53
参考文献 ·· 54

第 3 章　扇入型晶圆级 / 板级芯片尺寸封装 ·· 63

3.1 引言 ··· 63
3.2 扇入型晶圆级芯片尺寸封装（WLCSP） ··· 65
　3.2.1 封装结构 ·· 65

3.2.2　WLCSP的关键工艺步骤 ·· 67
　　3.2.3　WLCSP在PCB上的组装 ·· 68
　　3.2.4　WLCSP在PCB上组装的热仿真 ·· 68
　　3.2.5　总结和建议 ··· 74
3.3　扇入型板级芯片尺寸封装（PLCSP） ·· 75
　　3.3.1　测试芯片 ··· 75
　　3.3.2　测试封装体 ··· 76
　　3.3.3　PLCSP工艺流程 ··· 77
　　3.3.4　PLCSP的PCB组装 ··· 83
　　3.3.5　PLCSP PCB组装的跌落试验 ·· 84
　　3.3.6　PLCSP PCB组装的热循环试验 ·· 86
　　3.3.7　PLCSP PCB组装的热循环仿真 ·· 92
　　3.3.8　总结和建议 ··· 95
3.4　6面模塑晶圆级芯片尺寸封装 ·· 96
　　3.4.1　星科金朋的eWLCSP ·· 97
　　3.4.2　联合科技的WLCSP ··· 97
　　3.4.3　矽品科技的mWLCSP ··· 97
　　3.4.4　华天科技的WLCSP ··· 99
　　3.4.5　矽品科技和联发科的mWLCSP ·· 99
　　3.4.6　总结和建议 ··· 102
3.5　6面模塑板级芯片尺寸封装 ·· 102
　　3.5.1　6面模塑PLCSP的结构 ·· 102
　　3.5.2　晶圆正面切割和EMC模塑 ··· 104
　　3.5.3　背面减薄和晶圆背面模塑 ··· 104
　　3.5.4　等离子体刻蚀和划片 ··· 106
　　3.5.5　测试的PCB ··· 106
　　3.5.6　6面模塑PLCSP在PCB上的SMT组装 ·· 106
　　3.5.7　6面模塑PLCSP的热循环试验 ··· 108
　　3.5.8　6面模塑PLCSP的PCB组装热循环仿真 ·· 111
　　3.5.9　总结和建议 ··· 115
参考文献 ··· 115

第4章　扇出型晶圆级/板级封装 ··· 124
4.1　引言 ·· 124
4.2　扇出型（先上晶且面朝下）晶圆级封装（FOWLP） ······································ 125
　　4.2.1　测试芯片 ··· 125
　　4.2.2　测试封装体 ··· 126

- 4.2.3 传统的先上晶（面朝下）晶圆级工艺 …… 127
- 4.2.4 异质集成封装的新工艺 …… 128
- 4.2.5 干膜 EMC 层压 …… 128
- 4.2.6 临时键合另一块玻璃支撑片 …… 128
- 4.2.7 再布线层 …… 130
- 4.2.8 焊球植球 …… 131
- 4.2.9 最终解键合 …… 131
- 4.2.10 PCB 组装 …… 134
- 4.2.11 异质集成的可靠性（跌落试验） …… 135
- 4.2.12 总结和建议 …… 137
- 4.3 扇出型（先上晶且面朝下）板级封装（FOPLP） …… 137
  - 4.3.1 测试封装体的异质集成 …… 138
  - 4.3.2 一种新的 Uni-SIP 工艺 …… 140
  - 4.3.3 ECM 面板的干膜层压 …… 140
  - 4.3.4 Uni-SIP 结构的层压 …… 141
  - 4.3.5 新 ABF 的层压、激光钻孔、去胶渣 …… 141
  - 4.3.6 激光直写图形和 PCB 镀铜 …… 144
  - 4.3.7 总结和建议 …… 145
- 4.4 扇出型（先上晶且面朝上）晶圆级封装 …… 146
  - 4.4.1 测试芯片 …… 146
  - 4.4.2 工艺流程 …… 146
- 4.5 扇出型（先上晶且面朝上）板级封装 …… 148
  - 4.5.1 封装结构 …… 148
  - 4.5.2 工艺流程 …… 148
- 4.6 扇出型（后上晶或先 RDL）晶圆级封装 …… 150
  - 4.6.1 IME 的先 RDL FOWLP …… 151
  - 4.6.2 测试结构 …… 151
  - 4.6.3 先 RDL 关键工艺步骤 …… 152
  - 4.6.4 先 RDL FOWLP 的 PCB 组装 …… 154
- 4.7 扇出型（后上晶或先 RDL）板级封装 …… 154
  - 4.7.1 测试芯片 …… 154
  - 4.7.2 测试封装体 …… 154
  - 4.7.3 异质集成用先 RDL 板级封装 …… 157
  - 4.7.4 RDL 基板的制作 …… 157
  - 4.7.5 晶圆凸点成型 …… 160
  - 4.7.6 芯片 - 基板键合 …… 160

	4.7.7 底部填充和 EMC 模塑	162
	4.7.8 面板/条带转移	163
	4.7.9 阻焊层开窗和表面处理	163
	4.7.10 植球、解键合和条带切割	163
	4.7.11 先 RDL 板级封装的 PCB 组装	165
	4.7.12 跌落试验结果和失效分析	165
	4.7.13 热循环试验结果和失效分析	169
	4.7.14 热循环仿真	174
	4.7.15 总结和建议	175
4.8	Mini-LED RGB 显示器的扇出型板级封装	176
	4.8.1 测试 mini-LED	177
	4.8.2 测试 mini-LED RGB 显示器的 SMD 封装	178
	4.8.3 RDL 和 mini-LED RGB SMD 制造	179
	4.8.4 PCB 组装	182
	4.8.5 跌落试验	185
	4.8.6 热循环仿真	185
	4.8.7 总结和建议	191
参考文献		191

第 5 章 2D、2.1D 和 2.3D IC 集成 200

5.1	引言	200
5.2	2D IC 集成——引线键合	200
5.3	2D IC 集成——倒装芯片	201
5.4	2D IC 集成——引线键合和倒装芯片	201
5.5	RDL	202
	5.5.1 有机 RDL	202
	5.5.2 无机 RDL	202
	5.5.3 混合 RDL	202
5.6	2D IC 集成——扇出型（先上晶）	203
	5.6.1 HTC 的 Desire 606W	203
	5.6.2 4 颗芯片异质集成	203
5.7	2D IC 集成——扇出型（后上晶）	205
	5.7.1 IME 的后上晶扇出型封装	205
	5.7.2 Amkor 的 SWIFT	206
	5.7.3 Amkor 的 SLIM	207
	5.7.4 矽品科技的混合 RDL 扇出	208
	5.7.5 欣兴电子的扇出型后上晶工艺	209

- 5.8 2.1D IC 集成 ········· 210
  - 5.8.1 Shinko 的 i-THOP ········· 210
  - 5.8.2 日立的 2.1D 有机转接板 ········· 212
  - 5.8.3 日月光的 2.1D 有机转接板 ········· 212
  - 5.8.4 矽品科技的 2.1D 有机转接板 ········· 213
  - 5.8.5 长电科技的 uFOS ········· 215
  - 5.8.6 英特尔的 EMIB ········· 216
  - 5.8.7 应用材料的互连桥 ········· 217
  - 5.8.8 台积电的 LSI ········· 217
- 5.9 2.3D IC 集成 ········· 217
- 5.10 采用 SAP/PCB 法的 2.3D IC 集成 ········· 218
  - 5.10.1 Shinko 的无芯板有机转接板 ········· 218
  - 5.10.2 思科的有机转接板 ········· 219
- 5.11 采用扇出型（先上晶）技术的 2.3D IC 集成 ········· 220
  - 5.11.1 星科金朋的 2.3D eWLB ········· 220
  - 5.11.2 联发科的扇出型（先上晶）技术 ········· 222
  - 5.11.3 日月光的 FOCoS（先上晶） ········· 223
  - 5.11.4 台积电的 InFO_oS 和 InFO_MS ········· 224
- 5.12 采用扇出型（后上晶）技术的 2.3D IC 集成 ········· 225
  - 5.12.1 矽品科技的 NTI ········· 225
  - 5.12.2 三星的无硅 RDL 转接板 ········· 225
  - 5.12.3 日月光的 FOCoS（后上晶） ········· 228
  - 5.12.4 台积电的多层 RDL 转接板 ········· 229
  - 5.12.5 Shinko 的 2.3D 有机转接板 ········· 229
  - 5.12.6 欣兴电子的 2.3D RDL 转接板 ········· 232
- 5.13 总结和建议 ········· 247
- 参考文献 ········· 247

第 6 章 2.5D IC 集成 ········· 251
- 6.1 引言 ········· 251
- 6.2 Leti 的 SoW 技术（2.5D IC 集成技术的起源） ········· 251
- 6.3 IME 的 2.5D IC 集成技术 ········· 252
  - 6.3.1 2.5D IC 集成的三维非线性局部及全局分析 ········· 252
  - 6.3.2 用于电气和流体互连的 2.5D IC 集成技术 ········· 254
  - 6.3.3 双面堆叠无源 TSV 转接板 ········· 256
  - 6.3.4 作为应力（可靠性）缓冲的 TSV 转接板 ········· 257
- 6.4 中国香港科技大学双面集成芯片的 TSV 转接板技术 ········· 258

- 6.5 中国台湾"工业技术研究院"的 2.5D IC 集成 259
  - 6.5.1 双面集成芯片 TSV 转接板的热管理 259
  - 6.5.2 应用于 LED 含嵌入式流体微通道的 TSV 转接板 260
  - 6.5.3 集成有片上系统和存储立方的 TSV 转接板 262
  - 6.5.4 半嵌入式 TSV 转接板 263
  - 6.5.5 双面粘接芯片的 TSV 转接板 264
  - 6.5.6 双面集成芯片的 TSV 转接板 266
  - 6.5.7 TSH 转接板 268
- 6.6 台积电的 CoWoS 技术 270
- 6.7 赛灵思 / 台积电的 2.5D IC 集成 270
- 6.8 Altera/ 台积电的 2.5D IC 集成 273
- 6.9 AMD/ 联电的 2.5D IC 集成 273
- 6.10 英伟达 / 台积电的 2.5D IC 集成 274
- 6.11 台积电 CoWoS 路线图 275
- 6.12 2.5D IC 集成的近期进展 276
  - 6.12.1 台积电的集成有深槽电容 CoWoS 276
  - 6.12.2 IME 2.5D IC 集成的非破坏性失效定位方法 277
  - 6.12.3 Fraunhofer 的光电转接板 277
  - 6.12.4 Dai Nippon/AGC 的玻璃转接板 278
  - 6.12.5 富士通的多层玻璃转接板 280
- 6.13 总结和建议 280
- 参考文献 281

第 7 章 3D IC 集成和 3D IC 封装 287
- 7.1 引言 287
- 7.2 3D IC 封装 287
  - 7.2.1 3D IC 封装——引线键合式存储芯片堆叠 287
  - 7.2.2 3D IC 封装——面对面键合后引线键合到基板 291
  - 7.2.3 3D IC 封装——背对背键合后引线键合到基板 292
  - 7.2.4 3D IC 封装——面对面键合后通过凸点 / 焊球到基板上 293
  - 7.2.5 3D IC 封装——面对背 296
  - 7.2.6 3D IC 封装——SiP 中的埋入式芯片（面对面） 296
  - 7.2.7 3D IC 封装——采用倒装芯片技术的 PoP 298
  - 7.2.8 3D IC 封装——采用扇出技术的 PoP 300
  - 7.2.9 总结和建议 303
- 7.3 3D IC 集成 303
  - 7.3.1 3D IC 集成——HBM 标准 303

XII 半导体先进封装技术

- 7.3.2 3D IC 集成——HBM 组装 ·········· 305
- 7.3.3 3D IC 集成——采用 TSV 的芯片堆叠 ·········· 307
- 7.3.4 3D IC 集成——采用 TSV 的无凸点混合键合芯片堆叠 ·········· 311
- 7.3.5 3D IC 集成——无 TSV 的无凸点混合键合芯片堆叠 ·········· 313
- 7.3.6 总结和建议 ·········· 313
- 参考文献 ·········· 314

第 8 章 混合键合 ·········· 319
- 8.1 引言 ·········· 319
- 8.2 Cu-Cu TCB ·········· 319
  - 8.2.1 Cu-Cu TCB 的一些基本原理 ·········· 319
  - 8.2.2 IBM/RPI 的 Cu-Cu TCB ·········· 321
- 8.3 室温 Cu-Cu TCB ·········· 321
  - 8.3.1 室温 Cu-Cu TCB 的一些基本原理 ·········· 321
  - 8.3.2 NIMS/AIST/ 东芝 / 东京大学的室温 Cu-Cu TCB ·········· 322
- 8.4 SiO_2-SiO_2 TCB ·········· 322
  - 8.4.1 SiO_2-SiO_2 TCB 的一些基本原理 ·········· 322
  - 8.4.2 麻省理工学院的 SiO_2-SiO_2 TCB ·········· 324
  - 8.4.3 Leti/ 飞思卡尔 / 意法半导体的 SiO_2-SiO_2 TCB ·········· 325
- 8.5 低温 DBI ·········· 326
  - 8.5.1 低温 DBI 的一些基本原理 ·········· 326
  - 8.5.2 有 TSV 的索尼 CMOS 图像传感器 ·········· 328
  - 8.5.3 无 TSV（混合键合）的索尼 CMOS 图像传感器 ·········· 329
- 8.6 低温混合键合的近期发展 ·········· 332
  - 8.6.1 IME 混合键合的热机械性能 ·········· 332
  - 8.6.2 台积电的混合键合 ·········· 335
  - 8.6.3 IMEC 的混合键合 ·········· 338
  - 8.6.4 格罗方德的混合键合 ·········· 339
  - 8.6.5 三菱的混合键合 ·········· 340
  - 8.6.6 Leti 的混合键合 ·········· 341
  - 8.6.7 英特尔的混合键合 ·········· 343
- 8.7 总结和建议 ·········· 343
- 参考文献 ·········· 344

第 9 章 芯粒异质集成 ·········· 347
- 9.1 引言 ·········· 347
- 9.2 DARPA 在芯粒异质集成方面的工作 ·········· 347
- 9.3 SoC（片上系统） ·········· 348

9.4	芯粒异质集成	349
9.5	芯粒异质集成的优缺点	350
9.6	应用于芯粒异质集成的先进封装	351
	9.6.1 有机基板上的 2D 芯粒异质集成	351
	9.6.2 有机基板上的 2.1D 芯粒异质集成	352
	9.6.3 有机基板上的 2.3D 芯粒异质集成	353
	9.6.4 硅基板（无源 TSV 转接板）上的 2.5D 芯粒异质集成	354
	9.6.5 硅基板（有源 TSV 转接板）上的 3D 芯粒异质集成	355
	9.6.6 带互连桥的有机基板上的芯粒异质集成	356
	9.6.7 PoP 芯粒异质集成	357
	9.6.8 扇出型 RDL 基板上的芯粒异质集成	358
9.7	AMD 的芯粒异质集成	359
9.8	英特尔的芯粒异质集成	362
9.9	台积电的芯粒异质集成	364
9.10	总结和建议	367
参考文献		367

第 10 章 低损耗介电材料 … 371

10.1	引言	371
10.2	为什么需要低 Dk 和 Df 的介电材料	372
10.3	为什么需要低热膨胀系数的介电材料	372
10.4	NAMICS 材料的 Dk 和 Df	372
10.5	Arakawa 材料的 Dk 和 Df	375
10.6	杜邦材料的 Dk 和 Df	376
10.7	日立/杜邦微系统材料的 Dk 和 Df	377
10.8	JSR 材料的 Dk 和 Df	378
10.9	Toray 材料的 Dk 和 Df	381
10.10	富士通材料的 Dk 和 Df	381
10.11	Kayaku 材料的 Dk 和 Df	382
10.12	三菱材料的 Dk 和 Df	385
10.13	TAITO INK 材料的 Dk 和 Df	386
10.14	浙江大学材料的 Dk 和 Df	388
10.15	总结和建议	389
参考文献		391

第 11 章 先进封装未来趋势 … 392

11.1	引言	392
11.2	COVID-19 对半导体产业的影响	392
11.3	COVID-19 对晶圆代工行业的影响	393

11.4　COVID-19 对半导体客户的影响 ·········· 393
11.5　COVID-19 对封测行业的影响 ·········· 394
11.6　驱动端、半导体和先进封装 ·········· 395
11.7　先进封装的组装工艺 ·········· 397
　　11.7.1　引线键合 ·········· 398
　　11.7.2　SMT ·········· 399
　　11.7.3　倒装芯片技术的晶圆凸点成型 ·········· 400
　　11.7.4　有机基板上的倒装芯片技术 ·········· 400
　　11.7.5　CoC、CoW 和 WoW TCB 以及混合键合 ·········· 401
11.8　扇出型先上晶（芯片面朝上）、先上晶（芯片面朝下）以及后上晶技术 ·········· 402
11.9　互连桥与 TSV 转接板 ·········· 405
11.10　SoC 与芯粒 ·········· 407
11.11　高速/高频器件对材料的需求 ·········· 410
11.12　总结和建议 ·········· 411
参考文献 ·········· 412

第 1 章

先进封装

1.1 引言

首先明确，本书讨论的范围并非半导体技术，而是半导体先进封装技术。本章会先定义什么是"先进封装"，列出所有先进封装的种类并给出一个相应的案例。本章还会简要提及驱动端、半导体和封装三者之间的关系。

1.2 半导体的应用

当前半导体产业有五个确定的增长引擎（应用），它们分别是：①移动终端，如智能手机、智能手表、可穿戴设备、笔记本电脑和平板电脑；②高性能计算（high-performance computing，HPC），也被称为超级计算，它能够在超级计算机上高速处理数据和进行复杂的计算；③自动驾驶汽车；④物联网（internet of things，IoT），如智慧工厂和智慧医疗；⑤用于云计算的大数据和用于边缘计算的实时数据处理。

1.3 系统技术的驱动力

系统技术驱动端有很多，本书只讨论人工智能（artificial intelligence，AI）和第五代标准宽带蜂窝网络技术（5th generation technology standard for broadband cellular networks，5G），它们正在持续推动这五类半导体应用的增长。

1.3.1 AI

任何使得计算机能够模拟人脑工作的技术均可称为 AI。比如，AI 需要 HPC，HPC 的实现需要的基础设施包括存储数据的数据中心和用于操作运算的超级计算机。如图 1.1 所示，它的硬件设施（半导体和封装）就包括中央处理器（central processing unit，CPU）、图形处理器（graphics processing unit，GPU）、可编程门阵列（field programmable gate array，FPGA）、存储器、服务器和交换机。

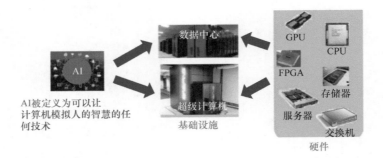

> AI正在推动半导体在HPC中的应用。
> AI需要HPC，HPC的基础设施包括存储数据的数据中心和用于操作运算的超级计算机。
> 这些基础设施所需要的硬件（半导体/封装）包括如CPU、GPU、FPGA、存储器、服务器和交换机。

图1.1 AI、HPC、基础设施和硬件之间的关系

1.3.2 5G

根据美国联邦通信委员会（US Federal Communications Commission）制定的标准：①中频频谱定义为频率处于900MHz～6GHz之间（因此也被称作亚6GHz 5G，sub-6GHz 5G）、数据传输速度≤1Gbit/s的通信频段；②高频频谱定义为频率处于24～100GHz之间（也被称作5G毫米波波段，5G mmWave）、数据传输速度处于1～10Gbit/s之间的通信频段（见图1.2）。在天线和多模射频收发器之间的距离很远的情况下，亚6GHz 5G和4G的应用依旧可以并存。28/39GHz频段可用于5G移动技术的天线，60GHz频段可用于高速无线数据连接，77GHz频段可用于汽车雷达，而94GHz频段可用于雷达成像（见图1.3）。为了快速满足信号传输速率提升的需求，且同时能够管理大数据流，先进封装技术的发展势在必行。

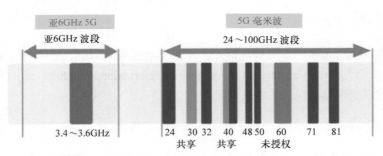

① 中频频段：通常指的是亚6GHz 5G
> 频率为2.5GHz、3.5GHz和3.7~4.2GHz（<6GHz），但是>900MHz
> 峰值数据传输速度≤1Gbit/s

② 高频频段：通常指的是5G毫米波
> 频率≥24GHz，比如24GHz、28GHz、37GHz、47GHz、…、≤100GHz
> 峰值数据传输速度>1Gbit/s但≤10Gbit/s

图1.2 美国联邦通信委员会关于5G的标准

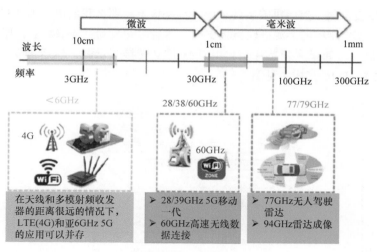

图 1.3　5G 的应用

1.4　先进封装概述

1.4.1　先进封装种类

目前有很多类型的先进封装技术，如 2D 扇出型（先上晶）IC 集成、2D 倒装芯片 IC 集成、封装堆叠（package-on-package，PoP）、系统级封装（system-in-package，SiP）或异质集成、2D 扇出型（后上晶）IC 集成、2.1D 倒装芯片 IC 集成、2.1D 含互连桥倒装芯片 IC 集成、2.1D 含互连桥扇出型 IC 集成、2.3D 扇出型（先上晶）IC 集成、2.3D 倒装芯片 IC 集成、2.3D 扇出型（后上晶）IC 集成、2.5D（焊料凸点）IC 集成、2.5D（微凸点）IC 集成、微凸点 3D IC 集成、微凸点芯粒 3D IC 集成、无凸点 3D IC 集成、无凸点芯粒 3D IC 集成。图 1.4 是各种先进封装技术的性能和密度对应图。图 1.5 描述了这些先进封装的层级。

1.4.2　先进封装层级

最简单的封装方法是直接将半导体芯片安装到印制电路板（printed circuit board，PCB）上，比如板上芯片（chip-on-board，COB）、直接芯片粘接（direct chip attach，DCA）[1-3]。引线框架类封装，如塑料方形扁平封装（plastic quad flat pack，PQFP）、小外形集成电路（small outline integrated circuit，SOIC）均为普通封装[4]。甚至单芯片的塑料焊球阵列封装（plastic ball grid array，PBGA）和倒装芯片级尺寸封装（flip chip-chip scale package，fcCSP）也都只能算是传统封装[5]。本书定义的先进封装，至少是在封装基板上进行多颗芯片的 2D 集成（这是最低的要求）。如果积层（build-up）封装基板顶部有薄膜布线层，我们就称之为 2.1D IC 集成。如果在积层封装基板中，或者在环氧树脂模塑料（epoxy molding compound，EMC）中含有嵌入式互连桥，那么我们就称之为含互连桥的 2.1D IC 集成。如果多颗芯片先是由无芯板的无机 / 有机无TSV的转接板承载，然后再安装到积层封装基板上，那么我们称之为2.3D IC集成。如果多颗芯片由含有TSV结构的无源硅转接板支撑，然后再安装到封装基板上，那么就可以称之为2.5D IC集成。最后，如果多颗芯片是先由有源TSV转接板承载，然后再安装到封装基板上，那么我们就称之为3D IC集成（见图1.5）。

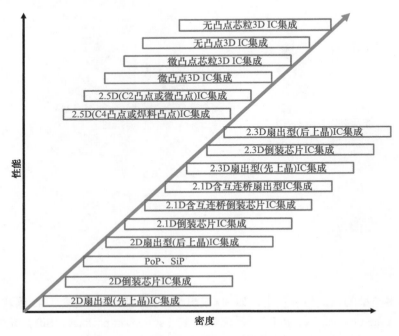

图1.4 各种先进封装技术的密度和性能对应图

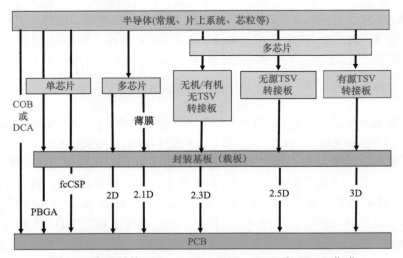

图1.5 先进封装：2D、2.1D、2.3D、2.5D和3D IC集成

纵观全书，图 1.4 所示的各类先进封装技术都会被讨论。本书还会讨论到表面安装技术（surface mount technology，SMT）、引线键合技术、倒装芯片技术等组装技术，以及芯片 - 芯片（chip-on-chip，CoC）、芯片 - 圆片（chip-on-wafer，CoW）、圆片 - 圆片（wafer-on-wafer，WoW）、热压键合（thermocompression bonding，TCB）技术和混合键合（hybrid bonding）技术。在本章中，我们还会针对图 1.4 的各项封装技术简要介绍一个案例。

1.5　2D 扇出型（先上晶）IC 集成

图 1.6 为一个 2D 扇出型（先上晶且面朝下）IC 集成案例[6-21]。从图中可以看出，四颗芯片先被埋入 EMC 中，然后通过再布线层（redistribution layer，RDL）进行扇出，最终连接到焊球上。这些焊球将直接连接到 PCB 上。关于扇出型封装的更多信息，请阅读第 4 章。

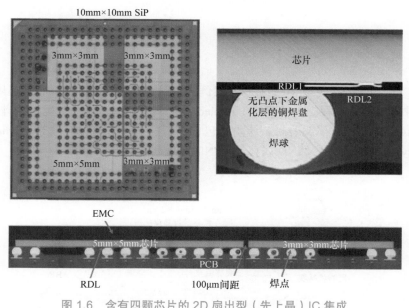

图 1.6　含有四颗芯片的 2D 扇出型（先上晶）IC 集成

1.6　2D 倒装芯片 IC 集成

图 1.7 为一个 2D 倒装芯片 IC 集成的案例。从图中可以看到，一颗芯片通过倒装技术连接到带可控塌陷芯片互连（controlled collapse chip connection，C4）凸点或芯片互连（chip connection，C2）凸点的积层封装基板上。在芯片和封装基板之间一般需要施加底部填充料保护，再把封装基板组装到 PCB 上。关于 2D 倒装芯片 IC 集成的更多信息，请阅读第 2 章和第 5 章。

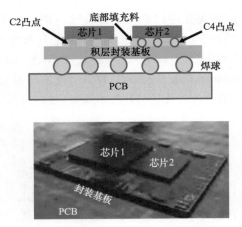

图 1.7　2D 倒装芯片 IC 集成

1.7　PoP、SiP 和异质集成

图 1.8 是一个三星生产的智能手表 PoP 案例。从图中可以看到，底部的封装体通过扇出和先上晶工艺并排封装了一颗应用处理器（applied processor，AP）和一颗电源管理芯片（power management IC，PMIC）。顶部的封装体内则包含控制器、动态随机存储器（dynamic random-access memory，DRAM）和 NAND 闪存（NAND Flash）。

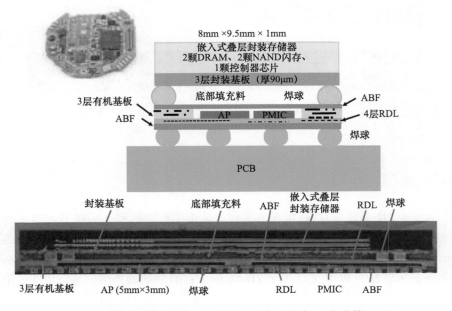

图 1.8　底部封装体采用 2D 扇出型（先上晶）IC 集成的 PoP

图 1.9 为一个苹果生产的智能手表 SiP 案例。从图中可以看到,所有芯片和分立元器件(系统)都在同一个封装基板上。

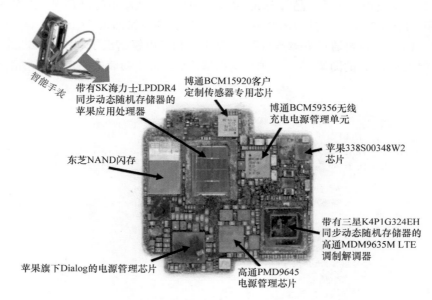

图 1.9 采用 2D IC 集成的 SiP

图 1.10 为一个 IBM 9121 热导模块(thermal conduction module,TCM)的异质集成案例。在一个 63 层的陶瓷基板上涵盖了 121 颗芯片(面积为 8~10mm²)。它的热性能表现十分惊人,每颗芯片的散热功率达到了 10W,每个 TCM 的散热功率达到了 600W。

IBM 9121 TCM
- TCM 重2.2kg
- 内含多达121颗芯片,面积为8~10mm²
- 每颗芯片都有一个弹簧式铜质活塞,以导出热量
- 每颗芯片散热功率高达10W
- 每个TCM散热功率高达600W
  陶瓷基板有
  - 63层
  - 布线长度达400m
  - 多达200万个孔
- 5kg的风冷热沉用于导出TCM的热量

图 1.10 陶瓷基板上 121 颗芯片的 2D 异质集成

1.8 2D 扇出型（后上晶）IC 集成

图 1.11 为一个扇出型（后上晶）的 IC 集成案例[22-40]。从图中可以看到，先在晶圆上制作了线宽/线距（L/S）2μm/2μm 的扇出 RDL。然后，芯片通过微凸点（Cu 柱 + 焊料帽）键合到 RDL 基板上，RDL 基板再通过焊球连接到 PCB 上。扫描电镜给出了其中一颗芯片、微凸点、RDL 基板、焊点和 PCB 的图像[22, 23]。关于扇出型（后上晶）封装的更多信息，请阅读第 4 章和第 5 章。

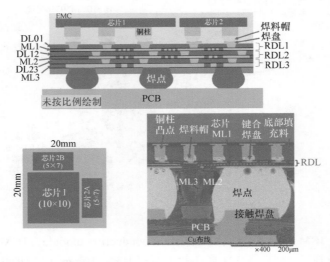

图 1.11 含三颗芯片的 2D 扇出型（后上晶）IC 集成

1.9 2.1D 倒装芯片 IC 集成

图 1.12 为一个 2.1D 倒装芯片 IC 集成的案例[41-46]。从图中可以看到，积层封装基板上有薄膜布线层。薄膜布线层的金属线宽/线距（L/S）可做到 2μm/2μm，通过微凸点支撑倒装芯片[41, 42]。关于 2.1D 倒装芯片 IC 集成的更多信息，请阅读第 5 章。

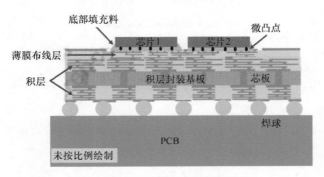

图 1.12 2.1D 倒装芯片 IC 集成

1.10 含互连桥的 2.1D 倒装芯片 IC 集成

图 1.13 为一个英特尔（Intel）含互连桥的 2.1D 倒装芯片 IC 集成案例[47, 48]。从图中可以看到，嵌入式多芯片互连桥（embedded multi-die interconnect bridge，EMIB）嵌入在积层封装基板的顶层，支持两侧两颗倒装芯片的横向通信。这种封装技术试图替代硅通孔（through silicon via，TSV）转接板技术。关于含互连桥的 2.1D 倒装芯片 IC 集成的更多信息，请阅读第 5 章。

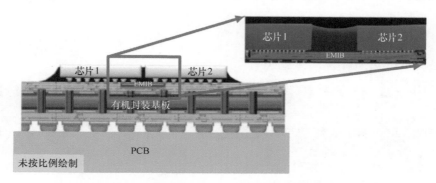

图 1.13　含互连桥的 2.1D 倒装芯片 IC 集成 [47, 48]

1.11 含互连桥的 2.1D 扇出型 IC 集成

图 1.14 为应用材料公司（Applied Materials）提供的含互连桥的 2.1D 扇出型 IC 集成案例[49]。从图中可以看到，互连桥不是被嵌入到积层封装基板中，而是被嵌入到 EMC 中。关于含互连桥的 2.1D 扇出型芯片 IC 集成的更多信息，请阅读第 5 章。

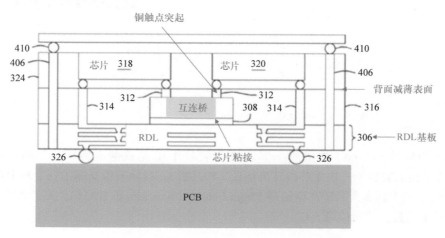

图 1.14　含互连桥的 2.1D 扇出型 IC 集成 [49]

1.12　2.3D 扇出型（先上晶）IC 集成

图 1.15 为一个 2.3D 扇出型（先上晶）IC 集成的案例[50-54]。从图中可以看到，TSV 转接板、微凸点和底部填充料被扇出型 RDL 转接板代替。日月光（ASE）计划将这项技术在 2021 年大规模量产（high volume manufacturing，HVM）。关于 2.3D 扇出型（先上晶）IC 集成的更多信息，请阅读第 5 章。

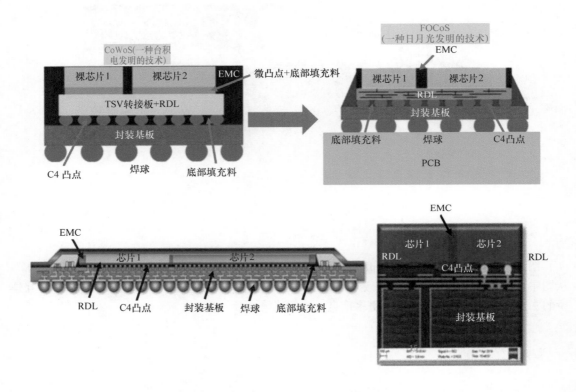

图 1.15　2.3D 扇出型（先上晶）IC 集成[50]

1.13　2.3D 倒装芯片 IC 集成

图 1.16 为思科（Cisco）提供的一个 2.3D 倒装芯片 IC 集成的案例[55]。从图中可以看到，在积层封装基板上安装了一块无芯板有机基板（转接板），这块基板支撑了一个片上系统（system-on-chip，SoC）和多个高带宽存储器（high-bandwidth memory，HBM）。关于 2.3D 倒装芯片 IC 集成的更多信息，请阅读第 5 章。

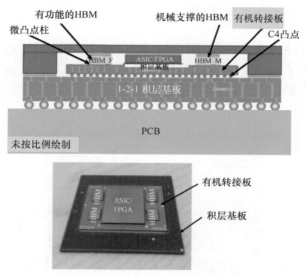

图 1.16 2.3D 倒装芯片 IC 集成[55]

1.14 2.3D 扇出型（后上晶）IC 集成

图 1.17 为一个 2.3D 扇出型（后上晶）IC 集成的案例[56-65]。从图中可以看出，先通过扇出型封装技术制作了一个有机转接板。接下来芯片通过微凸点实现与有机转接板的键合并完成底部填充。然后整个模组再通过 C4 凸点安装到积层封装基板。关于 2.3D 扇出型（后上晶）IC 集成的更多信息，请阅读第 5 章。

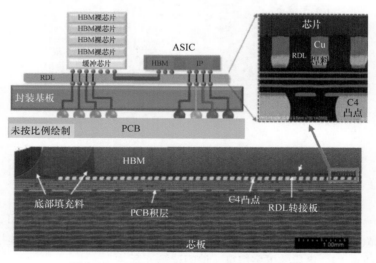

图 1.17 2.3D 扇出型（后上晶）IC 集成[56]

1.15　2.5D（C4凸点）IC集成

图1.18为一个2.5D倒装芯片（C4凸点）IC集成的案例[66-77]。从图中可以看出，射频（RF）芯片和逻辑芯片通过C4凸点连接在无源TSV转接板上（硅基板1、2）。关于2.5D（C4凸点）IC集成的更多信息，请阅读第6章。

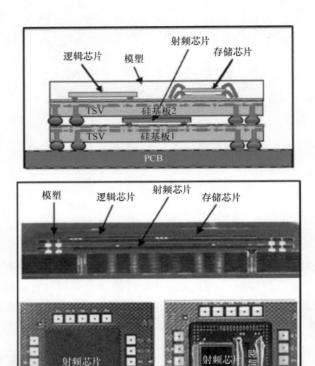

图1.18　2.5D倒装芯片（C4凸点）IC集成[78]

1.16　2.5D（C2凸点）IC集成

图1.19为一个2.5D倒装芯片（C2凸点）IC集成的案例[78-93]。从图中可以看出，GPU和HBM2通过C2微凸点连接到无源TSV转接板上。然后整个模组通过C4凸点连接到封装基板上。关于2.5D（C2凸点）IC集成的更多信息，请阅读第6章。

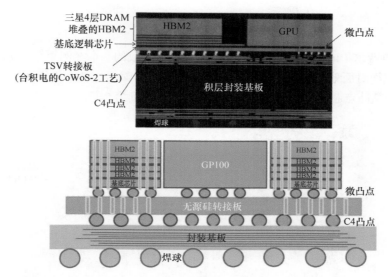

图 1.19　2.5D 含微凸点的倒装芯片 IC 集成

1.17　微凸点 3D IC 集成

图 1.20 为新加坡微电子研究所（IME）提供的微凸点 3D IC 集成的案例[94]。从图中可以看出，顶部芯片通过微凸点与含 TSV 结构的底部芯片连接。然后，整个模组通过 C4 凸点连接到封装基板上。关于微凸点 3D IC 集成的更多信息，请阅读第 7 章。

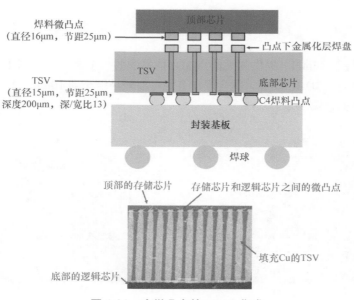

图 1.20　含微凸点的 3D IC 集成

1.18 微凸点芯粒 3D IC 集成

图 1.21 为英特尔（Intel）提供的微凸点芯粒 3D IC 集成的一个案例[95-97]。从图中可以看出，芯粒通过微凸点面对面地连接到一个含 TSV 的基底芯片。然后，整个模组通过 C4 凸点连接到封装基板。关于微凸点芯粒 3D IC 集成的更多信息，请阅读第 8 章。

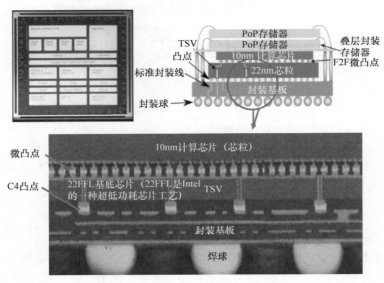

图 1.21 微凸点芯粒 3D IC 集成[96]

1.19 无凸点 3D IC 集成

图 1.22 为英特尔（Intel）提供的无凸点 3D IC 集成的一个案例。从图 1.22b 可以看出，通过无凸点（混合键合）3D IC 集成，键合焊盘节距可以轻易缩小到 10μm。关于无凸点 3D IC 集成的更多信息，请阅读第 8 章。

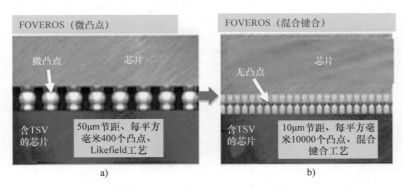

图 1.22 无凸点 3D IC 集成[97]

1.20 无凸点芯粒 3D IC 集成

图 1.23 为台积电（TSMC）发布的集成片上系统（system on integrated chip，SoIC）无凸点芯粒 3D IC 集成[98-101]。从图中可以看到，芯粒（带 TSV 的 SoC1 和 SoC2）采用 CoW 或是 WoW 实现无凸点混合键合。台积电计划将该技术在 2021 年大规模量产。关于无凸点芯粒 3D IC 集成的更多信息，请阅读第 8 章。

1.21 总结和建议

重要的结论和建议总结如下。

1）当前半导体产业有五个确定的增长引擎（应用）：

① 移动终端；
② 高性能计算；
③ 无人驾驶汽车；
④ 物联网；
⑤ 大数据（云计算）和实时数据（边缘计算）。

2）以下两种系统技术驱动端在促进五个应用快速发展：

① AI；
② 5G。

3）先进封装技术种类包括：

① 2D 扇出型（先上晶）IC 集成；
② 2D 倒装芯片 IC 集成；
③ PoP（封装堆叠）；
④ SiP 或异质集成；
⑤ 2D 扇出型（后上晶）IC 集成；
⑥ 2.1D 倒装芯片 IC 集成；
⑦ 含互连桥的 2.1D 倒装芯片 IC 集成；
⑧ 含互连桥的 2.1D 扇出型 IC 集成；
⑨ 2.3D 扇出型（先上晶）IC 集成；
⑩ 2.3D 倒装芯片 IC 集成；
⑪ 2.3D 扇出型（后上晶）IC 集成；
⑫ 2.5D（C4 焊料凸点）IC 集成；
⑬ 2.5D（C2 微凸点）IC 集成；
⑭ 微凸点 3D IC 集成；

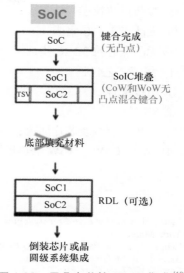

图 1.23 无凸点芯粒 3D IC 集成[100]

⑮ 微凸点芯粒 3D IC 集成；

⑯ 无凸点 3D IC 集成；

⑰ 无凸点芯粒 3D IC 集成。

4）组装工艺包括：

① 引线键合；

② SMT；

③ 有机基板倒装芯片回流；

④ CoC、CoW、WoW 的热压键合以及混合键合。

参 考 文 献

1. Lau, J. H., *Chip On Board Technologies for Multichip Modules*, Van Nostrand Reinhold, New York, March 1994.
2. Lau, J. H., and Y. Pao, *Solder Joint Reliability of BGA, CSP, Flip Chip, and Fine Pitch SMT Assemblies*, McGraw-Hill, New York, 1997.
3. Lau, J. H., *Low Cost Flip Chip Technologies for DCA, WLCSP, and PBGA Assemblies*, McGraw-Hill, New York, 2000.
4. Lau, J. H., and N. C. Lee, *Assembly and Reliability of Lead-Free Solder Joints*, Springer, New York, 2020.
5. Lau, J. H., C. P. Wong, J. Prince, and W. Nakayama, *Electronic Packaging: Design, Materials, Process, and Reliability*, McGraw-Hill, New York, 1998.
6. Lau, J. H., M. Li, M. Li, T. Chen, I. Xu, X. Qing, Z. Cheng, N. Fan, E. Kuah, Z. Li, K. Tan, Y. Cheung, E. Ng, P. Lo, K. Wu, J. Hao, S. Koh, R. Jiang, X. Cao, R. Beica, S. Lim, N. Lee, C. Ko, H. Yang, Y. Chen, M. Tao, J. Lo, and R. Lee, "Fan-Out Wafer-Level Packaging for Heterogeneous Integration", *IEEE Transactions on CPMT*, 2018, September 2018, pp. 1544–1560.
7. Lau, J. H., M. Li, Y. Lei, M. Li, I. Xu, T. Chen, Q. Yong, Z. Cheng, K. Wu, P. Lo, Z. Li, K. Tan, Y. Cheung, N. Fan, E. Kuah, C. Xi, J. Ran, R. Beica, S. Lim, N. Lee, C. Ko, H. Yang, Y. Chen, M. Tao, J. Lo, and R. Lee, "Reliability of Fan-Out Wafer-Level Heterogeneous Integration", *IMAPS Transactions, Journal of Microelectronics and Electronic Packaging*, Vol. 15, Issue: 4, October 2018, pp. 148–162.
8. Ko, CT, H. Yang, J. H. Lau, M. Li, M. Li, C. Lin, J. W. Lin, T. Chen, I. Xu, C. Chang, J. Pan, H. Wu, Q. Yong, N. Fan, E. Kuah, Z. Li, K. Tan, Y. Cheung, E. Ng, K. Wu, J. Hao, R. Beica, M. Lin, Y. Chen, Z. Cheng, S. Koh, R. Jiang, X. Cao, S. Lim, N. Lee, M. Tao, J. Lo, and R. Lee, "Chip-First Fan-Out Panel-Level Packaging for Heterogeneous Integration", *IEEE Transactions on CPMT*, September 2018, pp. 1561–1572.
9. Ko, C. T., H. Yang, J. H. Lau, M. Li, M. Li, C. Lin, J. Lin, C. Chang, J. Pan, H. Wu, Y. Chen, T. Chen, I. Xu, P. Lo, N. Fan, E. Kuah, Z. Li, K. Tan, C. Lin, R. Beica, M. Lin, C. Xi, S. Lim, N. Lee, M. Tao, J. Lo, and R. Lee, "Design, Materials, Process, and Fabrication of Fan-Out Panel-Level Heterogeneous Integration", *IMAPS Transactions, Journal of Microelectronics and Electronic Packaging*, Vol. 15, Issue: 4, October 2018, pp. 141–147.
10. Lau, J. H., "Recent Advances and Trends in Fan-Out Wafer/Panel-Level Packaging", *ASME Transactions, Journal of Electronic Packaging*, Vol. 141, December 2019, pp. 1–27.
11. Lau, J. H., "Recent Advances and Trends in Heterogeneous Integrations", *IMAPS Transactions, Journal of Microelectronics and Electronic Packaging*, Vol. 16, April 2019, pp. 45–77.
12. Hedler, H., T. Meyer, and B. Vasquez, "Transfer wafer level packaging," *US Patent 6,727,576*, filed on Oct. 31, 2001; patented on April 27, 2004.
13. Lau, J. H., "Patent Issues of Fan-Out Wafer/Panel-Level Packaging", *Chip Scale Review*, Vol. 19, November/December 2015, pp. 42–46.
14. Brunnbauer, M., E. Furgut, G. Beer, T. Meyer, H. Hedler, J. Belonio, E. Nomura, K. Kiuchi, and K. Kobayashi, "An Embedded Device Technology Based on a Molded Reconfigured Wafer", *IEEE/EPTC Proceedings*, May 2006, pp. 547–551.

15. Brunnbauer, M., E. Furgut, G. Beer, and T. Meyer, "Embedded Wafer Level Ball Grid Array (eWLB)", *IEEE/EPTC Proceedings,* May 2006, pp. 1–5.
16. Keser, B., C. Amrine, T. Duong, O. Fay, S. Hayes, G. Leal, W. Lytle, D. Mitchell, and R. Wenzel, "The Redistributed Chip Package: A Breakthrough for Advanced Packaging", *Proceedings of IEEE/ECTC*, May 2007, pp. 286–291.
17. Kripesh, V., V. Rao, A. Kumar, G. Sharma, K. Houe, X. Zhang, K. Mong, N. Khan, and J. H. Lau, "Design and Development of a Multi-Die Embedded Micro Wafer Level Package", *IEEE/ECTC Proceedings*, May 2008, pp. 1544–1549.
18. Khong, C., A. Kumar, X. Zhang, S. Gaurav, S. Vempati, V. Kripesh, J. H. Lau, and D. Kwong, "A Novel Method to Predict Die Shift During Compression Molding in Embedded Wafer Level Package", *IEEE/ECTC Proceedings*, May 2009, pp. 535–541.
19. Sharma, G., S. Vempati, A. Kumar, N. Su, Y. Lim, K. Houe, S. Lim, V. Sekhar, R. Rajoo, V. Kripesh, and J. H. Lau, "Embedded Wafer Level Packages with Laterally Placed and Vertically Stacked Thin Dies", *IEEE/ECTC Proceedings*, 2009, pp. 1537–1543. Also, *IEEE Transactions on CPMT*, Vol. 1, No. 5, May 2011, pp. 52–59.
20. Kumar, A., D. Xia, V. Sekhar, S. Lim, C. Keng, S. Gaurav, S. Vempati, V. Kripesh, J. H. Lau, and D. Kwong, "Wafer Level Embedding Technology for 3D Wafer Level Embedded Package", *IEEE/ECTC Proceedings*, May 2009, pp. 1289–1296.
21. Lim, Y., S. Vempati, N. Su, X. Xiao, J. Zhou, A. Kumar, P. Thaw, S. Gaurav, T. Lim, S. Liu, V. Kripesh, and J. H. Lau, "Demonstration of High Quality and Low Loss Millimeter Wave Passives on Embedded Wafer Level Packaging Platform (EMWLP)", *IEEE/ECTC Proceedings*, 2009, pp. 508–515. Also, *IEEE Transactions on Advanced Packaging*, Vol. 33, 2010, pp. 1061–1071.
22. Lau, J. H., C. Ko, T. Peng, K. Yang, T. Xia, P. Lin, J. Chen, P. Huang, T. Tseng, E. Lin, L. Chang, C. Lin, and W. Lu, "Chip-Last (RDL-First) Fan-Out Panel-Level Packaging (FOPLP) for Heterogeneous Integration", *IMAPS Transactions, Journal of Microelectronics and Electronic Packaging*, Vol. 17, No. 3, October 2020, pp. 89–98.
23. Lau, J. H., C. Ko, K. Yang, C. Peng, T. Xia, P. Lin, J. Chen, P. Huang, H. Liu, T. Tseng, E. Lin, and L. Chang, "Panel-Level Fan-Out RDL-first Packaging for Heterogeneous Integration", *IEEE Transactions on CPMT*, Vol. 10, No. 7, July 2020, pp. 1125–1137.
24. Bu, L., F. Che, M. Ding, S. Chong, and X. Zhang, "Mechanism of Moldable Underfill (MUF) Process for Fan-Out Wafer Level Packaging", *IEEE/EPTC Proceedings*, 2015, pp. 1–7.
25. Che, F., D. Ho, M. Ding, and D. Woo, "Study on Process Induced Wafer Level Warpage of Fan-Out Wafer Level Packaging", *IEEE/ECTC Proceedings*, 2016, pp. 1879–1885.
26. Rao, V., C. Chong, D. Ho, D. Zhi, C. Choong, S. Lim, D. Ismael, and Y. Liang, "Development of High Density Fan Out Wafer Level Package (HD FOWLP) with Multilayer Fine Pitch RDL for Mobile Applications", *IEEE/ECTC Proceedings*, 2016, pp. 1522–1529.
27. Chen, Z., F. Che, M. Ding, D. Ho, T. Chai, V. Rao, "Drop Impact Reliability Test and Failure Analysis for Large Size High Density FOWLP Package on Package", *IEEE/ECTC Proceedings*, 2017, pp. 1196–1203.
28. Lim, T., and D. Ho, "Electrical design for the development of FOWLP for HBM integration", *IEEE/ECTC Proceedings*, 2018, pp. 2136–2142.
29. Ho, S., H. Hsiao, S. Lim, C. Choong, S. Lim, and C. Chong, "High Density RDL build-up on FO-WLP using RDL-first Approach", *IEEE/EPTC Proceedings*, 2019, pp. 23–27.
30. Boon, S., D. Wee, R. Salahuddin, and R. Singh, "Magnetic Inductor Integration in FO-WLP using RDLfirst Approach", *IEEE/EPTC Proceedings*, 2019, pp. 18–22.
31. Hsiao, H., S. Ho, S. S. Lim, W. Ching, C. Choong, S. Lim, H. Hong, and C. Chong, "Ultra-thin FO Packageon-Package for Mobile Application", *IEEE/ECTC Proceedings*, 2019, pp. 21–27.
32. Lin, B., F. Che, V. Rao, and X. Zhang, "Mechanism of Moldable Underfill (MUF) Process for RDL-1st Fan-Out Panel Level Packaging (FOPLP)", *IEEE/ECTC Proceedings*, 2019, pp. 1152–1158.
33. Sekhar, V., V. Rao, F. Che, C. Choong, and K. Yamamoto, "RDL-1st Fan-Out Panel Level Packaging (FOPLP) for Heterogeneous and Economical Packaging", *IEEE/ECTC Proceedings*, 2019, pp. 2126–2133.
34. Huemoeller, R. and C. Zwenger, "Silicon wafer integrated fan-out technology," *Chip Scale Review*, Mar/Apr 2015, pp. 34–37.

35. Hiner, D., M. Kelly, R. Huemoeller, and R. Reed, "Silicon interposer-less integrated module-SLIM," *IMAPS/Device Packaging*, March 2015.
36. Hiner, D., M. Kolbehdari, M. Kelly, Y. Kim, W. Do, J. Bae, "SLIM™ advanced fan-out packaging for high performance multi-die solutions," *IEEE/ECTC Proceedings*, May 2017, pp. 575–580.
37. Kim, Y., J. Bae, M. Chang, A. Jo, J. Kim, S. Park, et al., "SLIM™, high density wafer-level fan-out package development with sub-micron RDL," *IEEE/ECTC Proceedings*, May 2017, pp. 18–13.
38. Zwenger, C., G. Scott, B. Baloglu, M. Kelly, W. Do, W. Lee, and J. Yi, "Electrical and Thermal Simulation of SWIFT™ High-density Fan-out PoP Technology", *IEEE/ECTC Proceedings*, May 2017, pp. 1962–1967.
39. Scott, G., J. Bae, K. Yang, W. Ki, N. Whitchurch, M. Kelly, C. Zwenger, J. Jeon, and T. Hwang, "Heterogeneous Integration Using Organic Interposer Technology", *IEEE/ECTC Proceedings*, May 2020, pp. 885–892.
40. Ma, M., S. Chen, P. I. Wu, A. Huang, C. H. Lu, A. Chen, C. Liu, and S. Peng, "The development and the integration of the 5 μm to 1 μm half pitches wafer level Cu redistribution layers", *IEEE/ECTC Proceedings*, May 2016, pp. 1509–1514.
41. Shimizu, N., W. Kaneda, H. Arisaka, N. Koizumi, S. Sunohara, A. Rokugawa, and T. Koyama, "Development of Organic Multi Chip Package for High Performance Application", *IMAPS Proceedings of International Symposium on Microelectronics*, October 2013, pp. 414–419.
42. Oi, K., S. Otake, N. Shimizu, S. Watanabe, Y. Kunimoto, T. Kurihara, T. Koyama, M. Tanaka, L. Aryasomayajula, and Z. Kutlu, "Development of New 2.5D Package with Novel Integrated Organic Interposer Substrate with Ultra-fine Wiring and High Density Bumps", *IEEE/ECTC Proceedings*, May 2014, pp. 348–353.
43. Uematsu, Y., N. Ushifusa, and H. Onozeki, "Electrical Transmission Properties of HBM Interface on 2.1-D System in Package using Organic Interposer", *IEEE/ECTC Proceedings*, May 2017, pp. 1943–1949.
44. Chen, W., C. Lee, M. Chung, C. Wang, S. Huang, Y. Liao, H. Kuo, C. Wang, and D. Tarng, "Development of novel fine line 2.1 D package with organic interposer using advanced substrate-based process", *IEEE/ECTC Proceedings*, May 2018, pp. 601–606.
45. Huang, C., Y. Xu, Y. Lu, K. Yu, W. Tsai, C. Lin, C. Chung, "Analysis of Warpage and Stress Behavior in a Fine Pitch Multi-Chip Interconnection with Ultrafine-Line Organic Substrate (2.1D)", *IEEE/ECTC Proceedings*, May 2018, pp. 631–637.
46. Islam, N., S. Yoon, K. Tan, and T. Chen, "High Density Ultra-Thin Organic Substrate for Advanced Flip Chip Packages", *IEEE/ECTC Proceedings*, May 2019, pp. 325–329.
47. Chiu, C., Z. Qian, and M. Manusharow, "Bridge interconnect with air gap in package assembly," *US Patent No. 8,872,349*, 2014.
48. Mahajan, R., R. Sankman, N. Patel, D. Kim, K. Aygun, Z. Qian, et al., "Embedded multi-die interconnect bridge (EMIB) – a high-density, high-bandwidth packaging interconnect," *IEEE/ECTC Proceedings*, May 2016, pp. 557–565.
49. Hsiung, C., and a. Sundarrajan, "Methods and Apparatus for Wafer-Level Die Bridge", US 10,651,126 B2, Filed on December 8, 2017, Granted on May 12, 2020.
50. Lin, Y., W. Lai, C. Kao, J. Lou, P. Yang, C. Wang, and C. Hseih, "Wafer warpage experiments and simulation for fan-out chip on substrate," *IEEE/ECTC Proceedings*, May 2016, pp. 13–18.
51. Pendse, R., "Semiconductor Device and Method of Forming Extended Semiconductor Device with Fan-Out Interconnect Structure to Reduce Complexity of Substrate", *US 9,484,319 B2*, Filed: December 23, 2011, Granted: November 1, 2016.
52. Yoon, S., P. Tang, R. Emigh, Y. Lin, P. Marimuthu, and R. Pendse, "Fanout Flipchip eWLB (Embedded Wafer Level Ball Grid Array) Technology as 2.5D Packaging Solutions", *IEEE/ECTC Proceedings*, 2013, pp. 1855–1860.
53. Chen, N., "Flip-Chip Package with Fan-Out WLCSP", *US 7,838,975 B2*, Filed: February 12, 2009, Granted: November 23, 2010.
54. Chen, N. C., T. Hsieh, J. Jinn, P. Chang, F. Huang, J. Xiao, A. Chou, B. Lin, "A Novel System in Package with Fan-out WLP for high speed SERDES application", IEEE/ECTC Proceedings, May 2016, pp. 1496–1501.

55. Li, L., P. Chia, P. Ton, M. Nagar, S. Patil, J. Xue, J. DeLaCruz, M. Voicu, J. Hellings, B. Isaacson, M. Coor, and R. Havens, "3D SiP with Organic Interposer for ASIC and Memory Integration", *IEEE/ECTC Proceedings,* May 2016, pp. 1445–1450.
56. Suk, K., S. Lee, J. Kim, S. Lee, H. Kim, S. Lee, P. Kim, D. Kim, D. Oh, and J. Byun, "Low Cost Si-less RDL Interposer Package for High Performance Computing Applications", *IEEE/ECTC Proceedings*, May 2018, pp. 64–69.
57. You, S., S. Jeon, D. Oh, K. Kim, J. Kim, S. Cha, G. Kim, "Advanced Fan-Out Package SI/PI/Thermal Performance Analysis of Novel RDL Packages", *IEEE/ECTC Proceedings*, May 2018, pp. 1295–1301.
58. Kwon, W., S. Ramalingam, X. Wu, L. Madden, C. Huang, H. Chang, et al., "Cost-effective and high-performance 28 nm FPGA with new disruptive silicon-less interconnect technology (SLIT)," *Proc. of Inter. Symp. on Micro.*, October 2014, pp. 599–605.
59. Liang, F., H. Chang, W. Tseng, J. Lai, S. Cheng, M. Ma, et al., "Development of non-TSV interposer (NTI) for high electrical performance package," *IEEE/ECTC Proceedings,* May 2016, pp. 31–36.
60. Lin, Y., M. Yew, M. Liu, S. Chen, T. Lai, P. Kavle, C. Lin, T. Fang, C. Chen, C. Yu, K. Lee, C. Hsu, P. Lin, F. Hsu, and S. Jeng, "Multilayer RDL Interposer for Heterogeneous Device and Module Integration", *IEEE/ECTC Proceedings*, May 2019, pp. 931–936.
61. Chang, K., C. Huang, H. Kuo, M. Jhong, T. Hsieh, M. Hung, C. Wang, "Ultra High Density IO Fan-Out Design Optimization with Signal Integrity and Power Integrity", *IEEE/ECTC Proceedings*, May 2019, pp. 41–46.
62. Lai, W., P. Yang, I. Hu, T. Liao, K. Chen, D. Tarng, and C. Hung, "A Comparative Study of 2.5D and Fan-out Chip on Substrate: Chip First and Chip Last", *IEEE/ECTC Proceedings*, May 2020, pp. 354–360.
63. Fang, J., M. Huang, H. Tu, W. Lu, P. Yang, "A Production-worthy Fan-Out Solution – ASE FOCoS Chip Last", *IEEE/ECTC Proceedings*, May 2020, pp. 290–295.
64. Miki, S., H. Taneda, N. Kobayashi, K. Oi, K. Nagai, T. Koyama, "Development of 2.3D High Density Organic Package using Low Temperature Bonding Process with Sn-Bi Solder", *IEEE/ECTC Proceedings*, May 2019, pp. 1599–1604.
65. Murayama, K., S. Miki, H. Sugahara, and K. Oi, "Electro-migration evaluation between organic interposer and build-up substrate on 2.3D organic package", *IEEE/ECTC Proceedings*, May 2020, pp. 716–722.
66. Khan, N., V. Rao, S. Lim, H. We, V. Lee, X. Zhang, E. Liao, R. Nagarajan, T. C. Chai, V. Kripesh, and J. H. Lau, "Development of 3-D Silicon Module With TSV for System in Packaging", *IEEE/ECTC Proceedings*, May 2008, pp. 550–555.
67. Khan, N., V. Rao, S. Lim, H. We, V. Lee, X. Zhang, E. Liao, R. Nagarajan, T. C. Chai, V. Kripesh, and J. H. Lau, "Development of 3-D Silicon Module With TSV for System in Packaging", *IEEE Transactions on CPMT*, Vol. 33, No. 1, March 2010, pp. 3–9.
68. Selvanayagam, C., J. H. Lau, X. Zhang, S. Seah, K. Vaidyanathan, and T. Chai, "Nonlinear Thermal Stress/Strain Analyses of Copper Filled TSV (Through Silicon Via) and Their Flip-Chip Microbumps", *IEEE Transactions on Advanced Packaging*, Vol. 32, No. 4, November 2009, pp. 720–728.
69. Khan, N., L. Yu, P. Tan, S. Ho, N. Su, H. Wai, K. Vaidyanathan, D. Pinjala, J. H. Lau, T. Chuan, "3D Packaging with Through Silicon Via (TSV) for Electrical and Fluidic Interconnections", *IEEE/ECTC* Proceedings, May, 2009, pp. 1153–1158.
70. Yu, A., N. Khan, G. Archit, D. Pinjala, K. Toh, V. Kripesh, S. Yoon, and J. H. Lau, "Fabrication of Silicon Carriers With TSV Electrical Interconnections and Embedded Thermal Solutions for High Power 3-D Packages", *IEEE Transactions on CPMT*, Vol. 32, No. 3, September 2009, pp. 566–571.
71. Tang, G. Y., S. Tan, N. Khan, D. Pinjala, J. H. Lau, A. Yu, V. Kripesh, and K. Toh, "Integrated Liquid Cooling Systems for 3-D Stacked TSV Modules", *IEEE Transactions on CPMT*, Vol. 33, No. 1, March 2010, pp. 184–195.
72. Khan, N., H. Li, S. Tan, S. Ho, V. Kripesh, D. Pinjala, J. H. Lau, and T. Chuan, "3-D Packaging With Through-Silicon Via (TSV) for Electrical and Fluidic Interconnections", *IEEE Transactions on CPMT*, Vol. 3, No. 2, February 2013, pp. 221–228.

73. Zhang, X., T. Chai, J. H. Lau, C. Selvanayagam, K. Biswas, S. Liu, D. Pinjala, et al., "Development of Through Silicon Via (TSV) Interposer Technology for Large Die (21x21mm) Fine-pitch Cu/low-k FCBGA Package", *IEEE/ECTC Proceedings*, May 2009, pp. 305–312.
74. Chai, T. C., X. Zhang, J. H. Lau, C. S. Selvanayagam, D. Pinjala, et al., "Development of Large Die Fine-Pitch Cu/low-*k* FCBGA Package with through Silicon via (TSV) Interposer", *IEEE Transactions on CPMT*, Vol. 1, No. 5, May 2011, pp. 660–672.
75. Lau, J. H., S. Lee, M. Yuen, J. Wu, C. Lo, H. Fan, and H. Chen, "Apparatus having thermal-enhanced and cost-effective 3D IC integration structure with through silicon via interposer". US Patent No: 8,604,603, Filed Date: February 19, 2010, Date of Patent: December 10, 2013.
76. Lau, J. H., Y. S. Chan, and R. S. W. Lee, "3D IC Integration with TSV Interposers for High-Performance Applications", *Chip Scale Review*, Vol. 14, No. 5, September/October, 2010, pp. 26–29.
77. Chien, H. C., J. H. Lau, Y. Chao, R. Tain, M. Dai, S. T. Wu, W. Lo, and M. J. Kao, "Thermal Performance of 3D IC Integration with Through-Silicon Via (TSV)", *IMAPS Transactions, Journal of Microelectronic Packaging*, Vol. 9, 2012, pp. 97–103.
78. Hou, S., W. Chen, C. Hu, C. Chiu, K. Ting, T. Lin, W. Wei, W. Chiou, V. Lin, V. Chang, C. Wang, C. Wu, and D. Yu, "Wafer-Level Integration of an Advanced Logic-Memory System Through the Second-Generation CoWoS Technology", *IEEE Transactions on Electron Devices*, October 2017, pp. 4071–4077.
79. Banijamali, B., S. Ramalingam, K. Nagarajan, and R. Chaware, "Advanced Reliability Study of TSV Interposers and Interconnects for the 28 nm Technology FPGA", *Proceedings of IEEE/ECTC*, May 2011, pp. 285–290.
80. Kim, N., D. Wu, D. Kim, A. Rahman, and P. Wu, "Interposer Design Optimization for High Frequency Signal Transmission in Passive and Active Interposer using Through Silicon Via (TSV)", *IEEE/ECTC Proceedings*, May 2011, pp. 1160–1167.
81. Banijamali, B., S. Ramalingam, N. Kim, C. Wyland, N. Kim, D. Wu, J. Carrel, J. Kim, and Paul Wu, "Ceramics vs. low-CTE Organic packaging of TSV Silicon Interposers", *IEEE/ECTC Proceedings*, May 2011, pp. 573–576.
82. Chaware, R., K. Nagarajan, and S. Ramalingam, "Assembly and Reliability Challenges in 3D Integration of 28 nm FPGA Die on a Large High Density 65 nm Passive Interposer", *Proceedings of IEEE/ECTC*, May 2012, San Diego, CA, pp. 279–283.
83. Banijamali, B., S. Ramalingam, H. Liu and M. Kim, "Outstanding and Innovative Reliability Study of 3D TSV Interposer and Fine Pitch Solder Micro-bumps", *Proceedings of IEEE/ECTC*, San Diego, CA, May 2012, pp. 309–314.
84. Kim, N., D. Wu, J. Carrel, J. Kim, and P. Wu, "Channel Design Methodology for 28 Gb/s SerDes FPGA Applications with Stacked Silicon Interconnect Technology", *IEEE/ECTC Proceedings*, May 2012, pp. 1786–1793.
85. Banijamali, B., C. Chiu, C. Hsieh, T. Lin, C. Hu, S. Hou, et al., "Reliability evaluation of a CoWoS-enabled 3D IC package," *IEEE/ECTC Proceedings*, May 2013, pp. 35–40.
86. Hariharan, G., R. Chaware, L. Yip, I. Singh, K. Ng, S. Pai, M. Kim, H. Liu, and S. Ramalingam, "Assembly Process Qualification and Reliability Evaluations for Heterogeneous 2.5D FPGA with HiCTE Ceramic", *IEEE/ECTC Proceedings*, May 2013, pp. 904–908.
87. Kwon, W., M. Kim, J. Chang, S. Ramalingam, L. Madden, G. Tsai, S. Tseng, J. Lai, T. Lu, and S. Chin, "Enabling a Manufacturable 3D Technologies and Ecosystem using 28 nm FPGA with Stack Silicon Interconnect Technology", *IMAPS Proceedings of International Symposium on Microelectronics*, Orlando, FL, October 2013, pp. 217–222.
88. Banijamali, B., T. Lee, H. Liu, S. Ramalingam, I. Barber, J. Chang and M. Kim, and L. Yip, "Reliability Evaluation of an Extreme TSV Interposer and Interconnects for the 20 nm Technology CoWoS IC Package", *IEEE/ECTC Proceedings*, May 2015, pp. 276–280.
89. Hariharan, G., R. Chaware, I. Singh, J. Lin, L. Yip, K. Ng, and S. Pai, "A Comprehensive Reliability Study on a CoWoS 3D IC Package", *IEEE/ECTC Proceedings*, May 2015, pp. 573–577.
90. Chaware, R., G. Hariharan, J. Lin, I. Singh, G. O'Rourke, K. Ng, S. Pai, C. Li, Z. Huang, and S. Cheng, "Assembly Challenges in Developing 3D IC Package with Ultra High Yield and High Reliability", *IEEE/ECTC Proceedings*, May 2015, pp. 1447–1451.

91. Xu, J., Y. Niu, S. Cain, S. McCann, H. Lee, G.Ahmed, and S. Park, "The Experimental and Numerical Study of Electromigration in 2.5D Packaging", *IEEE/ECTC Proceedings,* May 2018, pp. 483–489.
92. McCann, S., H. Lee, G. Ahmed, T. Lee, S. Ramalingam, "Warpage and Reliability Challenges for Stacked Silicon Interconnect Technology in Large Packages", *IEEE/ECTC Proceedings,* May 2018, pp. 2339–2344.
93. Wang, H., J. Wang, J. Xu , V. Pham, K. Pan, S. Park, H. Lee, and G. Ahmed, "Product Level Design Optimization for 2.5D Package Pad Cratering Reliability during Drop Impact", *IEEE/ECTC Proceedings,* May 2019, pp. 2343–2348.
94. Yu, A. B., J. H. Lau, S. Ho, A. Kumar, W. Hnin, W. Lee, M. Jong, et al., "Fabrication of High Aspect Ratio TSV and Assembly with Fine-Pitch Low-Cost Solder Microbump for Si Interposer Technology with High-Density Interconnects", *IEEE Transactions on CPMT*, Vol. 1, No. 9, September 2011, pp. 1336–1344.
95. Ingerly, D., S. Amin, L. Aryasomayajula, A. Balankutty, D. Borst, A. Chandra, K. Cheemalapati, C. Cook, R. Criss, K. Enamul1, W. Gomes, D. Jones, K. Kolluru, A. Kandas, G. Kim, H. Ma, D. Pantuso, C. Petersburg, M. Phen-givoni, A. Pillai, A. Sairam, P. Shekhar, P. Sinha, P. Stover, A. Telang, and Z. Zell, "Foveros: 3D Integration and the use of Face-to-Face Chip Stacking for Logic Devices", *IEEE/IEDM Proceedings*, December 2019, pp. 19.6.1–19.6.4.
96. Gomes, W., S. Khushu, D. Ingerly, P. Stover, N. Chowdhury, F. O'Mahony, etc., "Lakefield and Mobility Computer: A 3D Stacked 10 nm and 2FFL Hybrid Processor System in 12 × 12 mm^2, 1 mm Package-on-Package", *IEEE/ISSCC Proceedings*, February 2020, pp. 40–41.
97. WikiChip, "A Look at Intel Lakefield: A 3D-Stacked Single-ISA Heterogeneous Penta-Core SoC", https://en.wikichip.org/wiki/chiplet, May 27, 2020.
98. Chen, M. F., C. S. Lin, E. B. Liao, W. C. Chiou, C. C. Kuo, C. C. Hu, C. H. Tsai, C. T. Wang and D. Yu, "SoIC for Low-Temperature, Multi-Layer 3D Memory Integration", *IEEE/ECTC Proceedings,* May 2020, pp. 855–860.
99. Chen, Y. H., C. A. Yang, C. C. Kuo, M. F. Chen, C. H. Tung, W. C. Chiou, and D. Yu, "Ultra High Density SoIC with Sub-micron Bond Pitch", *IEEE/ECTC Proceedings,* May 2020, pp. 576–581.
100. Chen, F., M. Chen, W. Chiou, D. Yu, "System on Integrated Chips (SoIC^{TM}) for 3D Heterogeneous Integration", *IEEE/ECTC Proceedings,* May 2019, pp. 594–599.
101. Chen, M., C. Lin, E. Liao, W. Chiou, C. Kuo, C. Hu, C. Tsai, C. Wang and D. Yu, "SoIC for Low-Temperature, Multi-Layer 3D Memory Integration", *IEEE/ECTC Proceedings*, May 2020, pp. 855–860.

第 2 章

系统级封装

2.1 引言

系统级封装（system-in-package，SiP）技术已经在消费类电子产品中广泛使用，比如智能手表、智能手机、平板电脑、笔记本电脑、真无线立体声耳机（true wireless stereo，TWS）等。SiP 技术的核心组装工艺是表面组装技术（surface mount technology，SMT）和倒装芯片技术（flip chip technology），本章会重点介绍这两项技术。除此之外，本章还会简要提及 SiP 和片上系统（system-on-chip，SoC）的区别、SiP 的使用目的和应用场景，并给出一些利用 SiP 技术制作消费类电子产品的案例。

2.2 片上系统

由于摩尔定律的推动和移动产品（如智能手机、桌上电脑、笔记本电脑、可穿戴设备等）需求的增长，在过去十多年间 SoC 变得非常普遍。SoC 将不同功能的集成电路（integrated circuit，IC），如中央处理器（central processing unit，CPU）、图形处理器（graphic processing unit，GPU）、存储器等都集成在同一个芯片上，形成一个系统或子系统。图 2.1 为高通的骁龙（Snapdragon）888 5G 处理器芯片示意图。图 2.2 为三星的 Exynos 990 5G 处理器芯片示意图。但是当下摩尔定律正在快速走向终点，通过缩小特征尺寸来制造 SoC 变得越来越困难，相应的成本也越来越高。

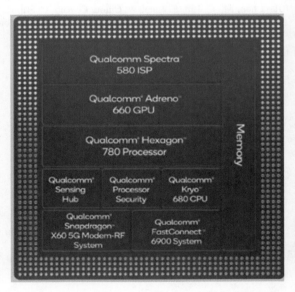

图 2.1 SoC：高通（Qualcomm）的应用处理器

图 2.2　三星（Samsung）的应用处理器

2.3　SiP 概述

SiP[1-109] 与 SoC 形成对比。SiP 使用封装技术将 CPU、GPU、存储器等不同的芯片集成到一个系统或子系统中；SiP 除了可以集成 IC 芯片外，还可以对光子器件，以及不同材料、不同功能的，甚至来自不同设计公司、不同晶圆厂、不同晶圆尺寸、不同特征尺寸的各种电子元器件（肩并肩、堆叠或两者兼有的方式）进行集成。SiP 与异质集成的概念非常接近[1-109]，只不过异质集成适用于更窄节距、更高输入 / 输出（I/O）、更高密度和更高性能的应用场景。

2.4　SiP 的使用目的

十多年前 SiP 的目标就是将不同的芯片和分立元件以及其他封装芯片 / 裸片的三维芯片堆叠结构 [如高带宽存储立方、用硅通孔（through-silicon vias，TSV）连接的存储芯片和逻辑芯片等] 通过一块通用的基板（硅、陶瓷或有机基板）以"肩并肩"的方式集成在一起，共同形成一个系统或子系统，并将该系统应用到智能手机、平板电脑、笔记本电脑等产品中。人们希望通过 SiP 技术同时实现水平方向和垂直方向的集成，即所谓的垂直多芯片模组或三维多芯片模组（multichip module，MCM）。

2.5　SiP 的实际应用

可惜的是，由于 TSV 技术对于智能手机、平板电脑等产品而言成本较高[110, 111]，这种设想没有成为现实。过去 10 年间大部分大规模生产（high volume manufacturing，HVM）的 SiP 产品实际上是 MCM-L（MCM on laminated substrate）的形式。这些产品多用于智能手机、平板电脑、智能手表、医药、可穿戴电子、游戏机、消费类产品以及物联网（internet of things，IoT）相关产品（如智能家居、智能能源和智能工业自动化）。大部分外包半导体封测厂商（outsourced semiconductor assembly and test，OSAT）的 SiP 产品都是在一块有机基板上集成 2 颗或更多不同的芯片、元器件和分立元件。

2.6 SiP 举例

例如，主营消费类电子的生产厂商苹果（Apple）正在大力推进更轻、更薄、更小以及更低成本的先进半导体封装技术，SiP 技术就在其中发挥着无可替代的作用。自苹果在 2015 年发布了采用 SiP 技术的苹果智能手表以来，该产品系列的全球累积出货量已经直逼 1 亿件。图 2.3 为 Apple Watch Series 4 中的 SiP。2019 年，苹果的 AirPods Pro 也使用了 SiP 技术（见图 2.4）。图 2.5 为 iPhone 12（2020 年）中的 5G mmWave 天线封装（antenna-in-package，AiP）模组，这个集成了多阵列天线、波束成形电路以及多个 RF 器件的 AiP 模组（见图 2.5）本质上就是基于 SiP 的概念。

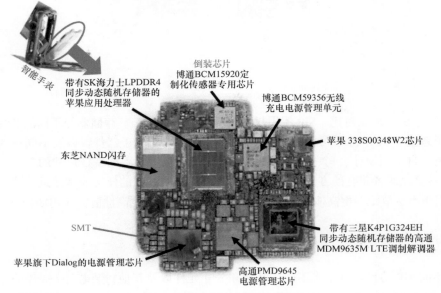

图 2.3　苹果的智能手表：SiP

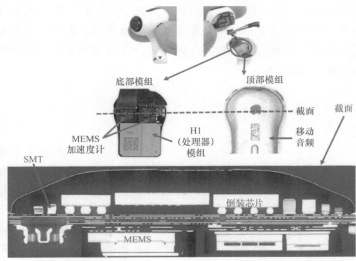

图 2.4　苹果的真无线立体声（true wireless stereo，TWS）产品 AirPods Pro

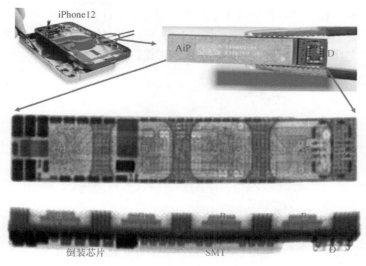

图 2.5　SiP：iPhone 12 内的 AiP

除了苹果以外还有很多使用 SiP 的案例。2019 年，高通（Qualcomm）的 QSiP（Qualcomm SiP）出货。该技术将一些在骁龙移动平台上常见的器件，如处理器芯片、电源管理芯片、RF 前端芯片和音频转码器芯片统统集成在一个封装体内，这为其他的内置器件如镜头、电池等腾出了很多空间。

因为 SiP 成功帮助 Airpods Pro 等真无线立体声（TWS）产品达到市场所需的多功能、小尺寸、长续航目标，它得到了其他非苹果用户/制造商的关注。第一个使用 SiP 技术的非苹果 TWS 产品将在 2021 年上半年问世。

SiP 技术的主要组装工艺为 SMT[112, 113] 和倒装芯片技术[115-118]，如图 2.3、图 2.4、图 2.5 和图 2.6 所示。因此本章后半部分将介绍 SMT 和倒装芯片技术。

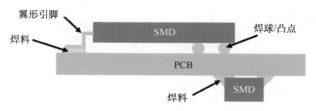

图 2.6　SMT 组装的示意图（双面组装）

2.7　SMT

如前文所述，SMT 是 SiP 的一项主要的组装技术。SMT 是一项非常成熟的技术[112, 113]。SMT 可以实现将更多的表面安装器件（surface mount device，SMD）放到更小、密度更高的印制电路板（printed circuit board，PCB）上。元器件密度的提升意味着在一个更小的封装体内来提升性能和功效，也有助于在较低的成本下实现外形更小、性能更好的产品。SMT 还支持双面组装。图 2.6 为一个 SMT 组装的示意图；图 2.7 为一个实际的 SMT PCB 组装。图 2.8 为 SMT 中的关键元素和工艺步骤，以下各节对此进行详细讨论。

图 2.7　典型的 PCB 组装案例图

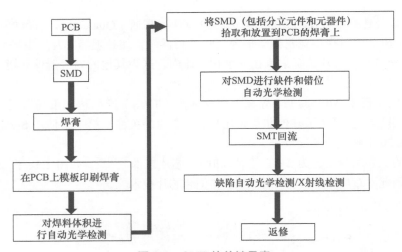

图 2.8　SMT 的关键元素

2.7.1　PCB

图 2.9a 为一个典型的 PCB。大多数 PCB 都是用 FR-4 板（FR 代表 flame retardant，即阻燃材料）制得，FR-4 板是一种用环氧树脂粘合物包裹交织的玻璃纤维形成的复合材料（图 2.9b 所示的玻璃纤维增强的层压环氧树脂）。如图 2.10 所示，一块 PCB 包含导线、走线、轨道、布线等。在 PCB 表层有微带线（microstrip line）分布，PCB 内层有带状线（strip line）分布。图 2.11 为不同结构的 PCB，比如单面板、双面板、多层板等。目前，制作超过 100 层的 PCB 并不稀奇；但是一般 30~60 层是 PCB 最多的层数了。

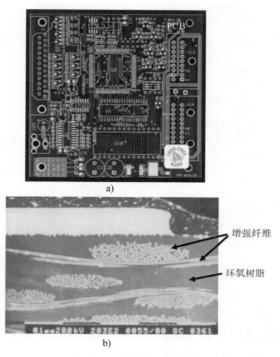

图 2.9 PCB

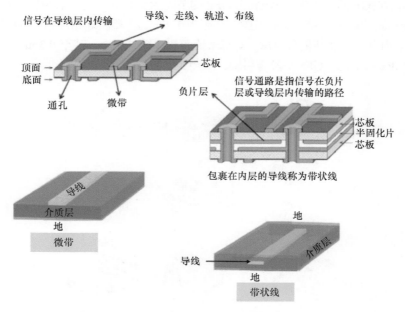

图 2.10 PCB 上的导线、走线、轨道、布线等

形式	层数	示意图
单面板	一层信号层	
双面板	两层信号层	
多层板	两层信号层和两层电源层	
多层板	四层信号层和两层电源层	
多层板	两层信号层和两层电源层	

图 2.11　PCB 的截面图

2.7.2　SMD

SMD 种类有很多。图 2.12 给出了部分常见的 SMD。在电子产品中像电容、电阻等的分立元件都通常采用 SMD 的形式。今天，最小的电容尺寸仅 0.25mm × 0.125mm × 0.125mm。小外形集成电路（small outline integrated circuit，SOIC）、薄型小尺寸封装（thin small outline package，TSOP）、塑料四方扁平封装（plastic quad flat package，PQFP）都带有翼形引脚。SOIC 的节距为 1.27mm，TSOP 的节距为 0.65mm，PQFP 的节距为 0.4mm。塑料有引脚芯片载体（plastic leaded chip carrier，PLCC）带有 J 形引脚，节距为 1.27mm。塑料球栅阵列（plastic ball grid array，PBGA）采用有机封装基板和焊球，焊球节距可以小到 0.35mm 甚至 0.3mm。四面扁平无引脚封装（quad-flat no-lead，QFN）没有引脚，节距为 0.5mm。

图 2.12　SMD 实例

2.7.3 焊膏

如图 2.13 所示，锡膏由混合在被称为助焊剂的黏性介质材料中的金属焊料粉末组成。添加的助焊剂类似一种临时黏合剂，焊接过程中在焊料熔化并形成更为牢固的物理连接前起固定元件的作用。

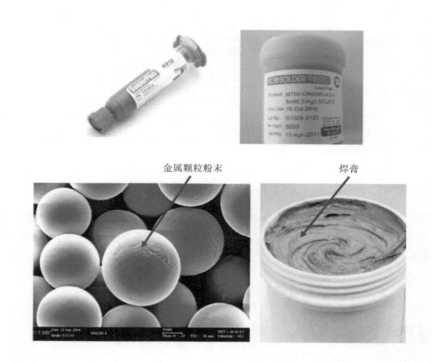

图 2.13 焊膏

根据 JEDEC 标准 J-STD-004 "焊接助焊剂要求"，焊膏可以根据助焊剂种类分为三种。松香基焊膏由松香制成，松香是松树的天然提取物。这种助焊剂在焊接后需要使用溶剂（可能包括氯氟烃）清洗。目前松香助焊剂已不再占主导地位。水溶性助焊剂由有机材料和乙二醇碱组成。有多种清洁剂可用于清洗这类助焊剂。免清洗助焊剂由树脂和不同程度的固体残留物制成。免清洗焊膏不仅节省了清洁成本，还节省了相应的固定资产投资和设备占地面积。但是，这些焊膏需要非常干净的装配环境，并且可能需要惰性回流环境。

根据 IPC 标准 J-TSD-005，焊膏的种类可以根据焊料金属粉末（颗粒）粒径区分，见表2.1。例如，4 号焊膏中粒径大于 50μm 的颗粒占比小于 0.5%，粒径位于 38～50μm 的颗粒占比小于 10%，粒径位于 20～38μm 的颗粒占比大于 80%，粒径小于 20μm 的颗粒占比小于 10%。目前普遍使用的焊膏是 3、4、5 号焊膏（尤其是 4 号焊膏）。

表 2.1 焊膏的种类

类型	占比小于 0.5% 的颗粒粒径大于	占比小于 10% 的颗粒粒径位于	占比大于 80% 的颗粒粒径位于	占比小于 10% 的颗粒粒径小于
1	160	150～160	75～150	75
2	80	75～80	45～75	45
3	60	45～60	25～45	25
4	50	38～50	20～38	20
5	40	25～40	15～25	15
6	25	15～25	5～15	5
7	15	11～15	2～11	2

注：第 2～4 列数据的单位为 μm。

2.7.4 模板印刷焊膏和自动光学检测

SMT 工艺中的关键一环是在 PCB 上模板印刷焊膏。图 2.14 为一个包含焊膏、模板和刮刀的模板印刷系统。图 2.15a、b 为模板，它由带有开窗的不锈钢制成。图 2.16a 为印刷在 PCB 上的焊盘节距 0.3mm PQFP 组装的焊膏；图 2.16b 为印刷在 PCB 上的焊盘节距 0.4mm PBGA 组装的焊膏。

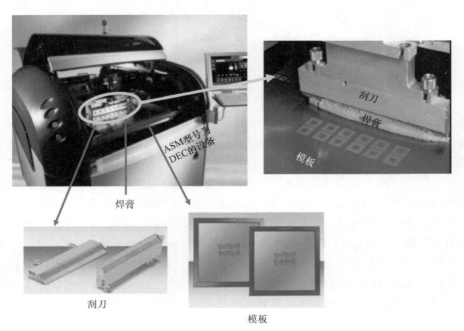

图 2.14 典型的模板印刷系统

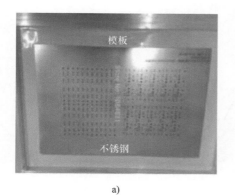
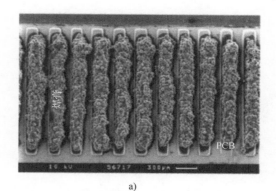

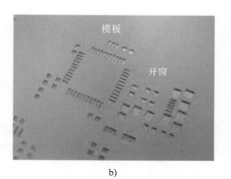
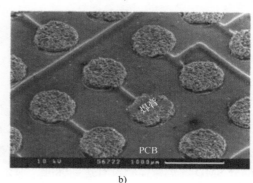

图 2.15　典型的模板

图 2.16　a）PCB 上焊盘节距为 0.3mm 的 PQFP 上的焊膏；b）PCB 上焊盘节距为 0.4mm 的 PBGA 上的焊膏

英特尔（Intel）计算了模板厚度（高度）对不同焊盘节距 PBGA 的焊膏体积[114]的影响，见表 2.2。从表中可以看到，对于焊盘节距为 0.4mm 的 PBGA，若要得到的焊膏体积（300）> 250mil³，需要模板的厚度为 100μm。这些结论是英特尔从焊点质量 / 可靠性的角度出发考虑的。因为随着 PBGA 焊球节距的减小，焊膏对于焊点质量 / 可靠性的影响越发明显。在进行焊点质量 / 可靠性的设计之外，还需利用自动光学检测（automated optical inspection，AOI）设备来控制焊膏的体积。图 2.17 为用于焊料体积检测的 AOI 设备。

表 2.2　不同节距 PBGA 焊盘所用的焊膏体积

节距 /mm	焊点 / 焊盘尺寸 / μm（mil）	体积 /mil³ 模板 / 高度			
		2mil（50μm）	3mil（75μm）	4mil（100μm）	5mil（125μm）
0.5	300（11.8）	218	327	436	545
0.5	275（10.8）	184	276	368	460
0.4	250（9.8）	150	225	300	450
0.35	200（7.8）	96	144	192	288
0.3	188（7.5）	89	133.2	177.6	222
0.25	150（5.9）	55	82	109	137

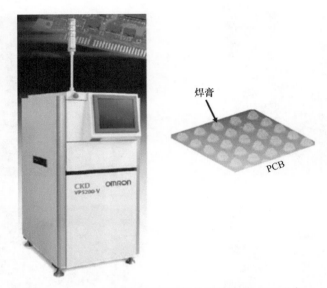

图 2.17 一台典型的用于焊料体积检测的 AOI 设备

AOI 有多项针对焊膏体积的功能。如果大部分的印刷焊膏体积没有达到 AOI 的检测标准，那么整块 PCB 上的焊膏都会被移除、清洗、烘干，然后再次进行模板印刷焊膏操作。如果只有少部分 SMD 的焊膏体积没有达到标准，那么设备会告诉计算机系统（后续的 AOI 设备或 X 射线设备旁边），在拾取和放置（pick & place）及回流操作后对这些 SMD 给予更多关注。

2.7.5 SMD 的拾取和放置

一般来说，SMD 可以分为 2 类，一类是像电容、电阻这样的分立 SMD，另一类是像 PQFP、PBGA、QFN 等封装的 IC 元件。通常，通过一个高速射片机（见图 2.18a）对分立 SMD 进行拾取和放置，通过一个高速拾放机器（见图 2.18b）对 IC 元件进行拾取和放置。

a)

b)

图 2.18 a) 一台芯片射片机；b) 一台 SMD 拾放机器

2.7.6 对 PCB 上的 SMD 的 AOI

如图 2.19 所示，将 SMD 拾放到 PCB 焊盘的焊膏上后，下一步的 AOI 操作必须进行。这次 AOI 是检测 SMD 有无遗漏、误放、偏心。这些数据对回流后的 AOI 和 X 射线检测非常重要。

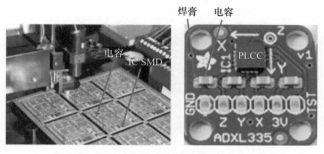

图 2.19 典型的带有焊膏和 SMD 的 PCB

2.7.7 SMT 焊料回流

SMT 工艺中的另一关键环节是 SMT 焊料回流。图 2.20 为一个典型的焊料回流炉，回流炉可能有 6 个加热区或者 8 个加热区、10 个加热区、12 个加热区。根据 IPC/JEDEC J-STD-020C 无铅回流曲线，焊料回流炉的作用包括：①升温并浸润 SMD、焊膏、PCB；②预热并浸润 SMD、焊膏、PCB；③升温至峰值温度；④焊膏回流；⑤焊点、SMD、PCB 冷却。图 2.21 为回流炉中 PCB 组装随时间变化的温度曲线。从图中可以看到，温度先从室温升高到 150℃以使 SMD、焊膏和 PCB 浸润；然后，温度又从 150℃升高到 200℃以对 SMD、焊膏和 PCB 进行预热和浸润；接下来温度又以最高 3℃/s 的升温速度快速从 200℃升高到顶峰（260℃），焊料开始回流。随后温度以最高 6℃/s 的降温速度快速从 260℃下降到 217℃，此时焊点发生固化。回流炉中 PCB 组装从室温升高到峰值温度的总时间不超过 8min。图 2.22 为一些 PCB 组装的实物图。

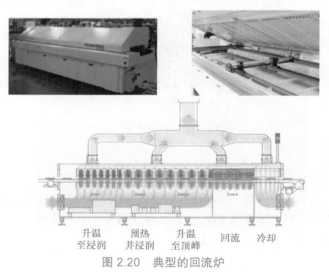

图 2.20 典型的回流炉

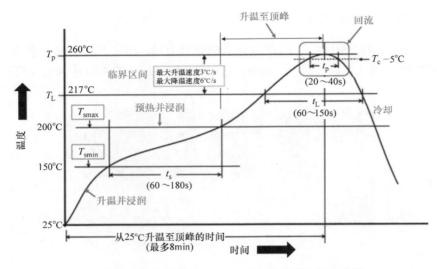

图 2.21　IPC/JEDEC J-STD-020C 无铅回流温度曲线

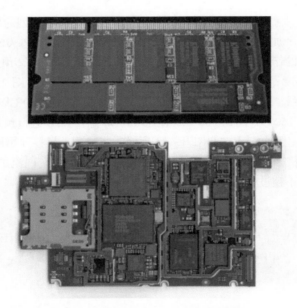

图 2.22　回流后的 PCB 组装典型案例

2.7.8　缺陷的 AOI 和 X 射线检测

焊料回流后,需对 SMD 遗漏、偏心进行快速的 AOI,并对 SMD 内部缺陷(见图 2.24)进行 X 射线检测(见图 2.23)。通常 X 射线检测可以发现至少 6 种不同的缺陷,分别是开路、焊料过多、回流不充分、短路、空洞和未对准。

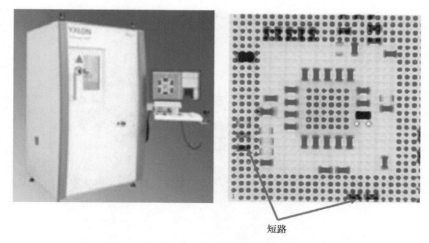

图 2.23　X 射线检测

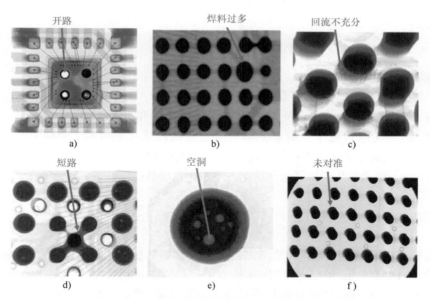

图 2.24　X 射线设备可能检测到的焊点缺陷

2.7.9　返修

PCB 组装缺陷需要进入返修站进行返修。图 2.25 所示是 SRT Summit 2100RD 返修站。图中底部和顶部的加热器可以将焊料回流熔化，方便移除元件。该设备适用于 PBGA、翼形引脚 PQFP 和底部端接元件。此外，返修站内的焊膏点胶器可以实现精准的焊膏放置。

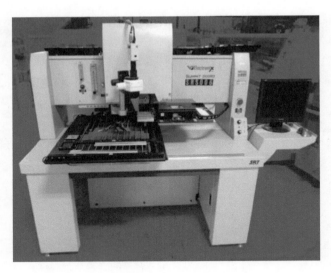

图 2.25 常见的返修站

2.7.10 总结和建议

重要的结论和建议总结如下。

1）本节介绍了 PCB、SMD、焊膏、模板印刷焊膏和 AOI、SMD 拾取和放置、SMT 焊料回流、缺陷 AOI 和 X 射线检测以及 SMT 返修等主要要素和工艺步骤。

2）SMT 给现代消费电子、商业、军事电子系统提供了更灵活/便利的互连电子元件。SMT 极大促进了板级组装工艺的选择，推动了应用创新和技术进步。

3）SMT 未来的趋势是变得更加自动化、小型化、高性能、高可靠性、高效率和环境友好。

2.8 倒装芯片技术

如早前所述，倒装芯片技术也是 SiP 的一项关键组装技术。倒装芯片技术是一项很成熟的技术[115-120]。IBM 在 20 世纪 60 年代初提出了用于固态逻辑技术的倒装芯片技术，此后该技术发展成为 IBM System/360 计算机产品线的坚实基础[119]。所谓的 C4（controlled-collapse chip connection，可控塌陷芯片互连）技术[120]，实质是利用在芯片可润湿金属端子上沉积高铅焊料凸点，以及相应的基板上的可润湿焊盘。如图 2.26 所示，将凸点成型之后的倒装芯片与基板（陶瓷、硅或有机）对准，同时对凸点进行加热回流形成焊点。本书中，即使倒装芯片焊料凸点不采用高铅焊料，我们仍然称其为 C4 技术。本节将简要介绍 4 种不同的倒装芯片组装工艺，另外还将介绍一种高产量的热压键合倒装芯片工艺。由于晶圆凸点成型技术是倒装芯片技术的前提，本节首先介绍模板印刷和电镀实现晶圆凸点成型的工艺。

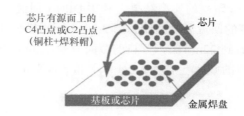

图 2.26　采用 C4 凸点或 C2 凸点的倒装芯片组装

2.8.1　基于模板印刷的晶圆凸点成型技术

用模板印刷进行晶圆凸点成型技术最为方便，本节对此做简要介绍[121]。图 2.27 为本节实验所用的 8in⊖ 晶圆（直径约 200mm）。从图中可以看到，每颗芯片有 48 个焊盘，焊盘的节距为 0.75mm。焊盘直径为 0.33mm，焊盘上有高度约 60μm 的铜凸台。模板为不锈钢材质，厚度为 8mil⊖（约 0.2mm），采用 LPKF 的四倍频 Nd：YAG 激光系统进行钻孔。然后再用 30% 的氢氧化钾溶液对模板进行电抛光，如图 2.27 所示。从图中可以看到，第一象限为 20mil 的方形开口，第二象限为 20mil 的圆形开口，第三象限为 22mil 的方形开口，第四象限为 22mil 的圆形开口。晶圆上的凸点成型的目标高度为 13mil（0.33mm）。

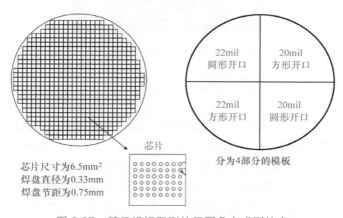

图 2.27　基于模板印刷的晶圆凸点成型技术

⊖　1in = 0.0254m。
⊖　1mil = 25.4 × 10⁻⁶m。

图 2.28 为每个象限焊膏印刷的典型结果。利用 DIMA 回流炉对晶圆上的焊膏进行回流，回流过程采用与 SMT 兼容的回流温度曲线。图 2.29 为不同模板开口的焊点；图 2.30 为这些焊点的截面。从图中可以看到 22mil 方形开口的截面最大，20mil 圆形开口的截面最小；22mil 圆形开口和 20mil 方形开口的截面基本一致。4 水平和 8 水平正交试验的数据分析请见参考文献 [121]。

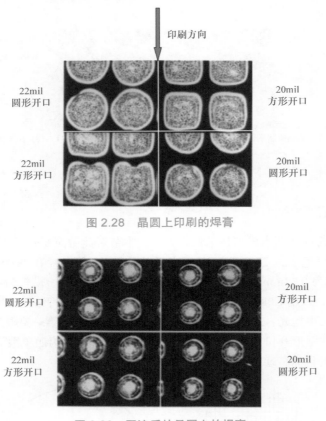

图 2.28 晶圆上印刷的焊膏

图 2.29 回流后的晶圆上的焊膏

2.8.2 C4 晶圆凸点成型技术

图 2.31 为采用电镀的 C4 晶圆凸点成型技术。通常焊盘尺寸为 100μm，凸点目标高度为 100μm。在重新定义阻焊层开窗后（通常不需要这步），先在整片晶圆表面溅射一层 Ti 层或 TiW 层（0.1~0.2μm），然后再溅射一层 Cu 层（0.3~0.8μm）。这里 Ti-Cu 和 TiW-Cu 称为凸点下金属化层（under bump metallization，UBM）。为了得到高 100μm 的凸点，在 Ti-Cu 或 TiW-Cu 层上旋涂 40μm 厚的光刻胶，然后用焊料凸点掩模版定义（紫外曝光）焊球凸点图形，如图 2.31（1）~（4）所示。光刻胶开窗要比钝化层焊盘开窗宽 7~10μm。然后，在 UBM 上电镀 5μm 厚的 Cu 层，并电镀上焊料。电镀过程中晶圆位于阴极，可以采用直流或者脉冲电流。为

确保得到厚度达到目标值（100μm）的凸点，在电镀过程中控制凸点的电镀高度超出旋涂的光刻层 15μm，让凸点形成一个蘑菇状。然后剥离光刻胶，并用过氧化氢或等离子体刻蚀去除 Ti-Cu 或 TiW-Cu 层。最终在晶圆上添加助焊剂，并回流，得到如图 2.31 所示的光滑无顶的球形 C4 焊料凸点。

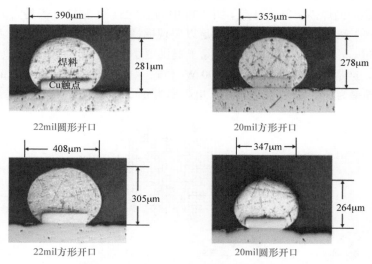

图 2.30　基于模板印刷的回流焊点截面图

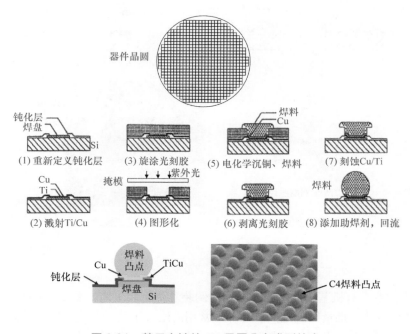

图 2.31　基于电镀的 C4 晶圆凸点成型技术

2.8.3　C2晶圆凸点成型技术

随着芯片引脚数增多、焊盘节距缩小，芯片上相邻的C4焊球容易发生桥接短路的现象。针对该问题，可以采用金属引线互连[122]和带焊料帽的铜柱[123, 124]的结构代替C4焊球。该结构制作的过程基本上与C4凸点一致，只是在图2.32（5）中直接电镀纯铜而不是电镀焊料。之后再电镀焊料并添加助焊剂回流就得到了图2.32中的C2凸点（铜柱＋焊料帽）。由于该凸点结构中焊料体积比C4凸点更小，表面张力不足以实现铜柱和焊料帽的自对准，我们将该凸点称为芯片互连（chip connection，C2）凸点。除了提供更小节距以外，C2凸点具有相比C4凸点更好的热电性能。这是因为Cu的热导率[400 W/（m·K）]和电阻率（0.0172μΩ·m）都比焊料优异 [焊料热导率为55～60W/（m·K），电阻率为0.12～0.14μΩ·m]。

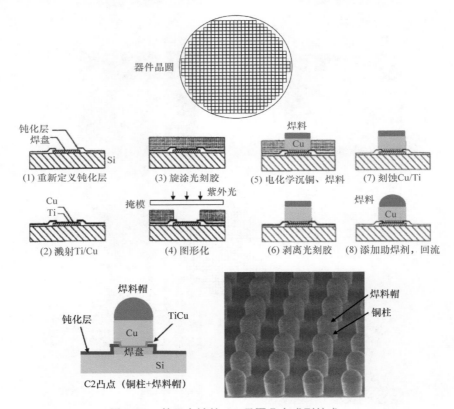

图2.32　基于电镀的C2晶圆凸点成型技术

2.8.4　倒装芯片组装——C4/C2凸点批量回流（CUF）

图2.33为主要在有机基板上的四种不同的倒装芯片组装工艺。图2.33a为C4/C2凸点的批量回流倒装芯片组装。批量回流是倒装芯片最常用的组装工艺。首先，用一对向上向下成像的摄像机识别出芯片上凸点和基板上焊盘的位置并进行对准；然后，在C4焊球或基板或两者同

时添加助焊剂；最后，将带有 C4 焊球的芯片拾取和放置在基板上，然后采用如图 2.21 所示的温度曲线进行批量回流。如图 2.34 所示，由于 C4 焊球在回流过程中存在表面张力，整个工艺过程具有非常好的鲁棒性（自对准）。

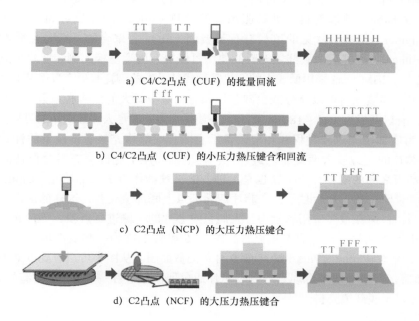

图 2.33 倒装芯片组装工艺（主要在有机基板上）：a) C4/C2 凸点（CUF）的批量回流；b) C4/C2 凸点（CUF）的小压力热压键合和回流；c) C2 凸点（NCP）的大压力热压键合；d) C2 凸点（NCF）的大压力热压键合

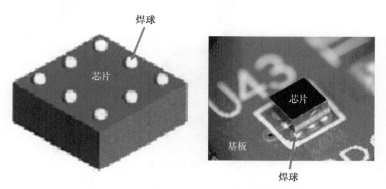

图 2.34 含 C4 凸点的倒装芯片批量回流

过去几年里，在有机封装基板上批量回流 C2 凸点（铜柱 + 焊料帽）的方法已被大量运用在高引脚数和窄节距的倒装芯片组装中。它的组装工艺（见图 2.33a）与 C4 凸点的完全一致，只是 C2 凸点熔化时的自对准能力比较差。

2.8.5 底部填充提升可靠性

通过使用底部填充（underfill）可提升倒装芯片焊点，对于有机基板上焊点的可靠性更是如此[125-148]。大部分底部填充料由低膨胀系数的填料（filler）和加热可固化的液态预聚合物（liquid prepolymer）组成。常见的填料有石英玻璃（SiO_2），聚合物一般是热固性树脂。1987年，日立（Hitachi）研究表明底部填充料的使用可以提升陶瓷基板上倒装芯片焊点的热疲劳寿命[125]。1992年，IBM Yasu 采用低成本的有机基板代替昂贵的陶瓷基板实现倒装芯片组装[126-128]。它们表明采用底部填充料可以充分减少硅芯片和有机基板间大的热膨胀系数不匹配 [硅芯片为 $2.5 \times 10^{-6}/℃$，有机基板为 $(15 \sim 18) \times 10^{-6}/℃$]，使得焊点能满足大多数应用的可靠性要求。

图 2.35 为底部填充点胶过程和相应设备示意图。底部填充料通过设备的针管挤出，点胶过程可以在芯片的一边或者两边、三边进行，在毛细作用下底部填充料会保持流动直到完全填满芯片和基板之间的缝隙为止。这也是这种方法又被称作毛细底部填充（capillary underfill，CUF）的原因。请注意，开始时不要在芯片四周都点上底部填充料，这样的话在芯片和基板中心会产生空洞。目前在很多应用场景中，人们只对芯片的一侧进行点胶，等待底部填充料自动从另一侧流出后再从另一侧点底部填充料。图 2.36 和图 2.37 为一些底部填充并固化后的芯片实物图。图 2.38 为 iPhone 6 Plus（2015年9月）的截面图。从图中可以看到以 PoP 形式封装的 A7 应用处理器，其中一颗带有焊点的倒装芯片通过回流组装在一块 2-2-2 的有机封装基板上，芯片的四周包裹了底部填充料。

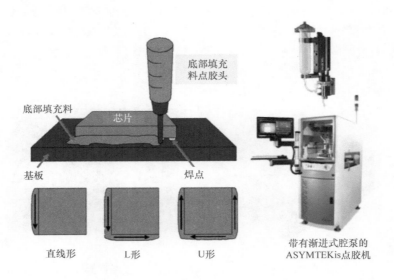

图 2.35 底部填充点胶过程及相应设备

图 2.36 带有 C4 凸点和底部填充料的倒装芯片组装批量回流

图 2.37 带有 C4 凸点和底部填充料的倒装芯片组装

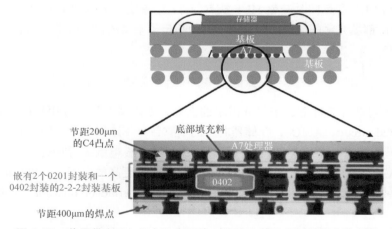

图 2.38 苹果带有 C4 凸点和底部填充料的应用处理器倒装芯片组装

除了 CUF，还有模塑底部填充（molded underfill，MUF），最早由确信电子（Cookson Electronics）[149]在 2000 年提出，此后被 Dexter[150]、英特尔（Intel）[151]、安靠（Amkor）[152]、星科金朋（STATSChipPAC）[153]和 Leti/ 意法半导体（STMicroelectronics）[154]等公司采用。对于 MUF，如图 2.39 所示，改进后的环氧模塑料（epoxy molding compound，EMC）既可完成芯片的模塑，同时还能填充倒装芯片组装体中芯片、焊点和基板之间的间隙。芯片的模塑和底部填充同时完成，可以极大地提高生产效率。但是，MUF 也存在以下挑战：① MUF 在芯片和基板间的流动往往需要借助真空条件；②为了保证流动性，EMC 中的二氧化硅填料必须足够小；③ MUF 采用的 EMC 成本要远高于普通芯片模塑；④ EMC、芯片和基板之间的热膨胀系数不匹配会引起严重的翘曲；⑤模塑的温度受到焊点熔点限制；⑥焊点的站高（standoff-height）和节距不能太小。

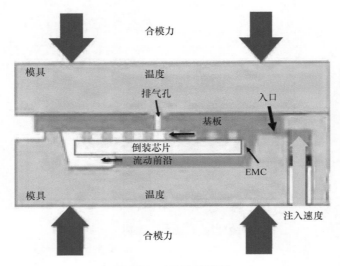

图 2.39　MUF 示意图

有一种提高 CUF 产量同时避免 MUF 缺点的方法，就是为倒装芯片组装设计一块用于印刷（而不是点胶）底部填充料的模板。但是这个话题超出了本书的讨论范围，若想获得相关信息可见参考文献 [155]。

2.8.6　倒装芯片组装——C4/C2 凸点的小压力热压键合（CUF）

最近，由于对芯片高性能、小尺寸需求的提升，处理器芯片、专用集成电路（application specific integrated circuit，ASIC）、存储器芯片的引脚数目不断增多，节距（或引脚焊盘之间的间距）不断减小。另外，由于移动设备（如智能手机和平板电脑）、便携设备（如笔记本电脑）追求小型化，芯片和封装基板的厚度也必须尽可能变薄。更高的引脚数、更窄的节距、更薄的芯片和更小的封装基板，使得倒装芯片组装中热压键合（thermocompression bonding，TCB）的使用势在必行。

图 2.33b 为基于 CUF 的含 C4/C2 凸点小压力热压键合的倒装芯片组装。首先，用一对向上

和向下成像的摄像头对准芯片上 C4/C2 焊球和相应基板上的焊盘；然后，在焊球（或焊料帽）上或基板上或两者都添加助焊剂；再将芯片拾取和放置到基板上并施加温度（T）使焊料熔化，期间用一个小压力（f）使芯片与基板维持一定距离。最后，进行 CUF 点胶并固化。以上过程一次只能处理一颗芯片，因此相比于 C4/C2 凸点的批量回流，该方法的效率较低。图 2.40 为一个典型的通过 C2 凸点小压力热压键合形成的倒装芯片组装的截面图[156]。

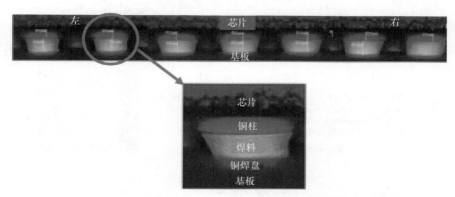

图 2.40　C2 凸点（CUF）的倒装芯片组装小压力热压键合[156]

C4/C2 凸点（CUF）的批量回流和小压力热压键合都属于组装后底部填充的例子，也就是说底部填充是在组装之后完成的。

2.8.7　倒装芯片组装——C2 凸点的大压力热压键合（NCP）

图 2.33c 为 C2 凸点通过采用非导电胶（non-conductive paste，NCP）的大压力热压键合。对于组装前底部填充来说，底部填充料既可以施加在芯片一侧，也可以施加在基板一侧，而且是在倒装芯片组装完成之前进行的。图 2.33c 所示是基板一侧施加非导电胶进行底部填充的 C2 凸点大压力热压键合（TC-NCP），该技术由安靠（Amkor）[157] 首先研究并将其用于高通（Qualcomm）骁龙（SNAPDRAGON）应用处理器上（见图 2.41），该处理器用于三星（Samsung）的 Galaxy 智能手机。NCP 底部填充料可以通过针头点胶或者辅助以真空涂覆的方式在基板上旋涂施加。

2.8.8　倒装芯片组装——C2 凸点的大压力热压键合（NCF）

图 2.33d 为另一种采用组装前底部填充的技术——基于非导电薄膜（non-conductive film，NCF）C2 凸点的大压力热压键合。图 2.42 是在 C2（铜柱＋焊料帽）凸点晶圆上层压 NCF。三星（Samsung）已经实现了单颗带有 NCF 的 C2 芯片大压力热压键合（需先对层压后的晶圆进行切割），并将其用于如图 2.42 所示的基于 TSV 的双倍数据速率 DDR4 DRAM3D IC 集成上。

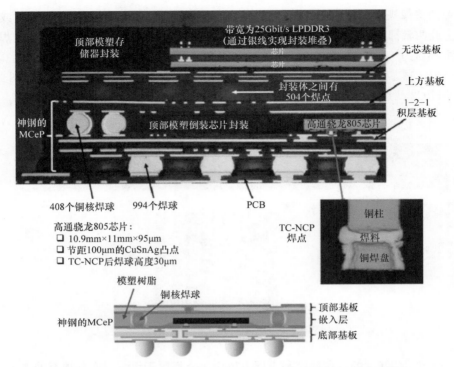

图 2.41 C2 凸点（NCP）的大压力热压键合倒装芯片组装[157]

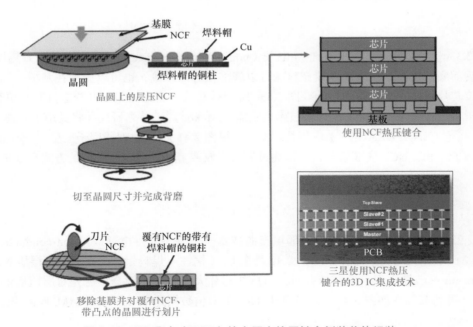

图 2.42 C2 凸点（NCF）的大压力热压键合倒装芯片组装

2.8.9 一种先进的倒装芯片组装——C2 凸点液相接触热压键合

无论是图 2.33b 所示的小压力热压键合，还是图 2.33c 所示的大压力热压键合，它们对于窄节距、高引脚、薄芯片、薄基板倒装芯片封装都十分有效。但是，这两种技术的生产效率并不高。本节介绍一种新的热压键合 - 助焊剂工艺，叫作液相接触（liquid phase contact，LPC）热压键合（thermal compression bonding，TCB）[158]。LPC TCB 与传统 TCB-flux 的显著区别是芯片与焊盘 / 引脚在小压力接触前就将 C2 凸点上的焊料帽加热成了熔融态。LPC TCB 工艺的键合周期时间 [无论是否包括键合头（bond head，BH）冷却时间] 相比传统 TCB- 助焊剂工艺都得到大大缩短，从而可以实现一个高产率的组装过程。LPC TCB 工艺的另一项优势是组装后焊点的厚度（站高）精确可控，这对于焊点的可靠性至关重要。

1. 测试结构

表 2.3 总结了本研究采用的 2 种测试结构，分别是板上芯片（chip-on-substrate，CoS）和芯片堆叠（chip-on-chip，CoC）键合。对于 CoS 键合，芯片的尺寸为 5mm×5mm×0.15mm，共有 31×31 面阵列的 C2（铜柱 +SnAg 焊料帽）凸点（见图 2.43）。铜柱直径为 40μm 或 60μm，铜柱和焊料帽的高度分别为 25μm 和 27μm 或者 25μm 和 17μm。凸点的节距为 160μm。焊料成分为 97.5%Sn2.5%Ag。双马来酰亚胺三嗪（Bismaleimide Triazine，BT）基板的焊盘上凸点（bump-on-pad，BOP）和引脚上凸点（bump-on-lead，BOL）都采用 Cu-OSP 表面处理，焊盘直径为 80μm，引脚宽度为 18μm。

表 2.3　测试结构：CoS 和 CoC

	板上芯片（CoS）	芯片堆叠（CoC）
芯片	焊球直径为 60μm 高度为 25μm 铜柱 + 27μm 焊料帽 焊球节距为 160μm 焊球直径为 40μm 高度为 25μm 铜柱 + 27μm 焊料帽 焊球节距为 160μm	焊球直径为 40μm 高度为 25μm 铜柱 + 27μm 焊料帽 焊球节距为 160μm
基板	基板金属化为 Cu-OSP 铜引脚宽度为 18μm 铜焊盘直径为 80μm 铜引脚 / 焊盘厚度为 15μm	表面处理为 Cu 铜焊盘直径为 60μm 铜焊盘厚度为 10μm 表面处理为 SOP 铜焊盘直径为 60μm 铜焊盘厚度为 10μm

对于 CoC 键合，上方芯片铜柱直径为 40μm，高度为 25μm，焊料高度为 17μm。底部芯片有 2 种表面金属化：一种是裸铜处理，另一种是焊盘上焊料（solder-on-pad，SOP）层。对于裸铜表面处理后的芯片，焊盘直径是 60μm，焊盘厚度是 10μm。对于 SOP 处理的芯片，焊盘直径是 40μm，焊料厚度是 17μm。

2. LPC TCB 工艺

图 2.44 为 LPC TCB 工艺示意图，该工艺遵循以下操作步骤：

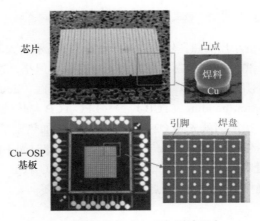

图 2.43 板上芯片（CoS）测试平台

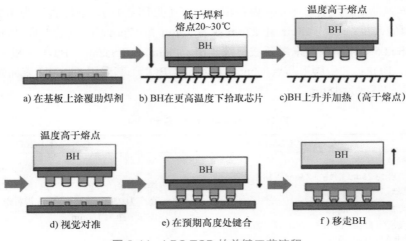

图 2.44 LPC TCB 的关键工艺流程

1）在基板上印刷或喷涂助焊剂。
2）BH 升温后从载具上拾取芯片，该温度低于焊料熔点 20~30℃。
3）BH 加热升温至高于焊料熔点。
4）芯片与基板对准。
5）视觉对准后，芯片在预设好的键合参数水平下与基板接触并润湿。
6）经过一段时间键合后，在键合温度或冷却至焊料熔点以下后移开 BH。

图 2.45 对比了传统 TCB-助焊剂工艺和 LPC TCB 工艺。对于传统 TCB-助焊剂工艺，助焊剂采用浸渍法（dipping method）添加到芯片上。浸渍时温度要低于 100℃，因为高温下助焊剂会因毛细效应或蒸发而损失。在芯片与基板接触后，SnAg 焊料要经历由 100℃到 250℃的大幅温升。所需的加热时间取决于加热器的加热能力/加热速率。另一方面，对于 LPC TCB 工艺，

BH 的待机温度相对较高，只比焊料熔点低 20~30℃。BH 可以在它上升或芯片-基板视觉对准的过程中进行升温。LPC TCB 的整个键合周期可以小于 4s。在一些情况下，如果所需的焊球站高接近焊料平衡水平，甚至可以省略冷却步骤，键合周期也可进一步减小到 3s 以内。LPC TCB 单个 BH 的产率（unit per hour，UPH）可以达到 1200，相比之下，传统 TCB-助焊剂工艺的 UPH 仅约 600。

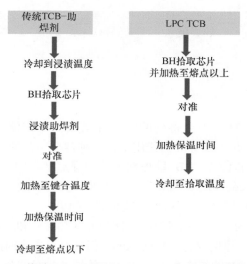

图 2.45 传统 TCB-助焊剂、LPC TCB 键合步骤的差异

3. 键合质量评估

LPC TCB 工艺中，键合温度、键合压力、键合时间、键合高度以及冷却步骤是影响键合质量的关键参数，这些参数需经过改进优化。CoS 和 CoC 键合样品的截面图（分别为图 2.46 和图 2.47）表明焊料厚度变化小于 2μm，键合润湿效果很好。需要注意，CoC 键合时 BH 和工作台（work holder，WH）需要更高的温度。这是因为硅的热导率 [149W/(m·K)] 相比 BT 有机材料 [0.35W/(m·K)] 更高。在 CoC 键合中，从 BH 传过来的热量会通过底部硅基芯片快速耗散进入键合机台。该过程所要施加的额外热量取决于底部芯片的厚度和尺寸。

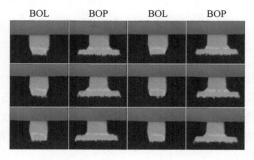

图 2.46 CoS 倒装芯片组装的截面图像：站高 =45.6μm（左）、46.2μm（中）、46.4μm（右）；键合条件：WH=80℃，BH=260℃，BT=1s

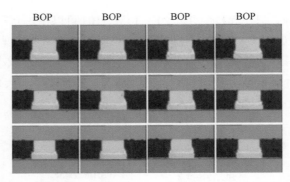

图 2.47　CoC 倒装芯片组装的截面图像：站高 =38.4μm（左）、37.6μm（中）、38.1μm（右）；键合条件：WH = 120℃，BH = 350℃，BT =1s

我们对比了采用 LPC TCB 工艺得到的焊点和采用批量回流（mass reflow，MR）、传统 TCB- 助焊剂工艺得到的焊点。图 2.48 为使用了这三种键合工艺的界面微观组织结构。对于刚键合好的样品，LPC TCB 和传统 TCB- 助焊剂工艺的焊点截面中都可以观察到厚约 1μm 的金属间化合物（intermetallic compound，IMC）。对于 MR 工艺，观察到的 IMC 相对较厚，约为 2μm。这是因为在批量回流过程中，焊料处于液相的时间更长。

为探究 IMC 的生长行为，对焊点进行了一次短时间热时效和三次回流测试。在 150℃下时效为 4h，IMC 的生长有限，厚度仅为 0.1~0.2μm，但在经过 3 次回流后，IMC 生长明显（LPC TCB 和传统 TCB- 助焊剂工艺的焊点生长了 3μm，传统的批量回流工艺焊点生长了 4μm）。三种键合过程的键合截面均观察到扇贝状 Cu_6Sn_5。

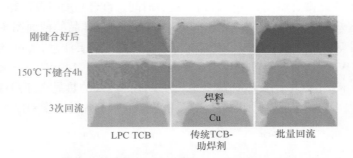

图 2.48　LPC TCB、传统 TCB- 助焊剂和批量回流工艺中的界面微结构

为测试焊点完整性，还对其进行了剪切测试。这三种工艺刚键合好的焊点都有很高的剪切力（13~15kgf⊖）。剪切断裂发生在焊料层内部。经过 150℃时效 4h 的焊点与刚键合好的焊点剪切力也相差不大。但是经过多次回流后的焊点剪切力发生了相对大幅的下降（回流后剪切力为 10~13kgf）。现有的关于界面微观组织结构、剪切力和失效模式的分析都表明 LPC TCB 工艺可以实现一个高可靠的焊点。

⊖　1kgf = 9.80665N。

4. 焊点站高控制

焊点站高控制对于后续底部填充工艺非常重要，也是获得焊点高可靠性的关键。其中最关键的是共面性调整、位置的精确把握和热管理。与传统的 TCB-助焊剂工艺对比，LPC TCB 工艺在芯片与基板接触后不存在键合工具加热的过程，所以键合工具引起的热膨胀可以忽略；同时由于键合时间相对较短，基板的温度也较低，LPC TCB 工艺中基板的热膨胀相比 TCB-助焊剂工艺会更小。

另外，LPC TCB 工艺所需要的键合压力也非常小（<1kg），因为键合过程中焊料已经是液态。小的键合压力还可以减小基板的塑性形变，确保焊点站高相对准确。下面将讨论两个键合参数（冷却步骤和键合温度）对焊点高度的影响。

5. 冷却步骤的影响

正如此前解释的，在键合后 BH 可以不降温直接从芯片上移开，也可以先降温到焊料熔点以下再移开。这两种不同的操作方式会得到不同高度的焊点。图 2.49 左侧为没有冷却步骤的焊点，不同预设键合高度（4～20μm）下的焊点，其最终的焊点高度基本一致；然而经过冷却步骤（BH 移开前降温至 200℃）的焊点，其最终的焊点高度与预设的键合高度基本一致，也与通过 Surface Evolver 模型预测的结果（图 2.49 右侧）表现一致。

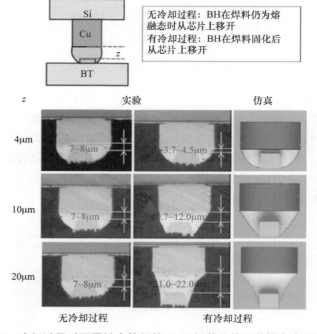

图 2.49 冷却过程对不同键合等级的 CoS 倒装芯片组装焊点高度的影响

为了理解在高于焊料熔化的温度下去除 BH 后，回复力如何驱动熔融焊料到达平衡高度，我们用 Surface Evolver 模型计算了液体焊料的表面张力。基于仿真工作的原理，z 方向的回复力是能量相对于焊料厚度的变化率[159]。在本研究中，采用了中心差分法计算回复力。图 2.50

为回复力与焊料厚度的关系。当焊料被压缩时,力为排斥力;当焊料被拉长时,力为吸引力。过程中只存在一个平衡点,即净力大小为 0。该点为焊料在没有任何外界力施加时的平衡厚度。如果预设的键合高度为 4μm,当 BH 从芯片上方移开后,芯片会在排斥力作用下上移到平衡高度。但是对于预设键合高度为 10μm 和 20μm 的样品,芯片会在吸引力的作用下下移到平衡高度。这就是为什么无论键合高度多大,焊料高度都能保持一致。

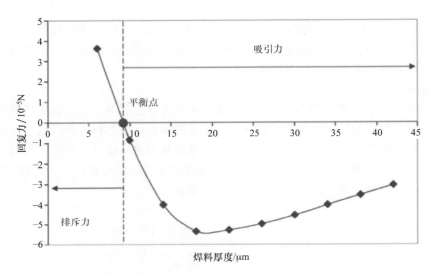

图 2.50　回复力与焊料厚度的关系

6. 键合温度的影响

需要注意在此前讨论的情况下,键合温度都足够高,在键合过程中焊料没有发生固化。但是如果键合温度相对较低,在 BH 与基板接触的过程中,熔融的焊料可能会发生固化,或者部分发生固化。最终在低键合温度下形成的焊点的站高可能高于在高键合温度下焊点的站高。如图 2.51 所示,在 280℃和 250℃的键合温度下,得到的焊点高度分别为 8μm 和 15μm。类似的现象也可在图 2.52 的 CoC 键合中观察到。

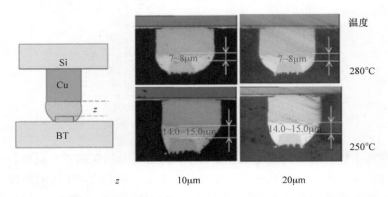

图 2.51　温度对 CoS 倒装芯片组装(无冷却过程)焊点高度的影响

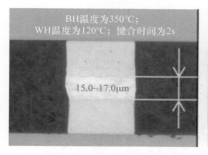
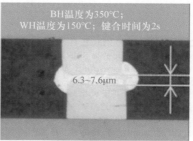

图 2.52 温度对 CoC 倒装芯片组装（无冷却过程）焊点高度的影响

2.8.10 总结和建议

一些重要的结论和建议总结如下。

1）超过 75% 的倒装芯片技术的应用都采用有机基板。

2）大多数倒装芯片工艺都是采用有机基板和 C4 凸点匹配毛细底部填充（capillary underfill，CUF）进行批量回流。引脚数量可以高达 5000，焊盘节距可以小到 60μm（见图 2.53）。

3）随着对薄芯片和薄有机基板的需求提高，基于 CUF 的带有 C2 凸点的倒装芯片小压力热压键合以及基于 NCP/NCF 的带有 C2 凸点的倒装芯片大压力热压键合正在受到越来越多的关注。基于 CUF 的小压力热压键合的 C2 凸点倒装芯片引脚数可以高达 7000，焊盘节距可以小到 50μm；基于 NCP/NCF 的大压力热压键合的 C2 凸点倒装芯片引脚数可以高达 10000，焊盘节距可以小到 30μm（见图 2.53）。

4）有关 SMT 的更多信息，请阅读参考文献 [113]。

5）有关倒装芯片的更多信息，请阅读参考文献 [115-118]。

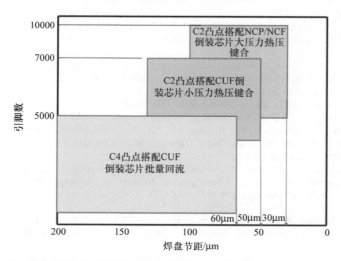

图 2.53 有机基板上 C4 凸点搭配 CUF 倒装芯片批量回流、有机基板上 C2 凸点搭配 CUF 倒装芯片小压力热压键合、有机基板上 C2 凸点搭配 NCP/NCF 倒装芯片大压力热压键合参数对比

参 考 文 献

1. Tsai, M., R. Chiu, D. Huang, F. Kao, E. He, J. Chen, S. Chen, J. Tsai and Y. Wang, "Innovative Packaging Solutions of 3D Double Side Molding with System in Package for IoT and 5G Application", *IEEE/ECTC Proceedings*, May 2019, pp. 700–706.
2. Liu, C., J. Chien, Y. Tseng, K. Liao, A. Chan, D. Chen, M. Shih, and M. Gerber, "Enhanced Reliability of a RF-SiP with Mold Encapsulation and EMI Shielding", *IEEE/ECTC Proceedings*, May 2019, pp. 1902–1908.
3. Milton, B., A. Shah, H. Xu, O. Kwon, G. Schulze, I. Qin, and N. Wong, "Smart Wire Bond Solutions for SiP and Memory Packages", *IEEE/ECTC Proceedings*, May 2019, pp. 55–62.
4. Talebbeydokhti, P., S. Dalmia, T. Thai, S. Tal, and R. Sover, "Ultra Large Area SIPs and Integrated mmW Antenna Array Module for 5G mmWave Outdoor Applications", *IEEE/ECTC Proceedings*, May 2019, pp. 294–299.
5. Dalmia, S., K. Nahalingam, S. Vijayakumar, and P. Talebbeydokhti, "A Zero Height Small Size Low Cost RF Interconnect Substrate Technology for RF Front Ends for M.2 Modules and SiP", *IEEE/ECTC Proceedings*, pp. 1666–1671.
6. Chen, J., X. Yong, D. Trombley, and R. Murugan, "System Co-Design of a 600 V GaN FET Power Stage with Integrated Driver in a QFN System-in-Package (QFN-SiP)", *IEEE/ECTC Proceedings*, pp. 1221–1226.
7. Lee, J., C. Chen, D. Lee, and J. Chen, "Moisture Effect on Physical Failure of Plastic Molded SiP Module", *IEEE/ECTC Proceedings*, May 2020, pp. 2124–2132.
8. Li, J., M. Tsai, R. Chiu, E. He, A. Hsieh, M. Tsai, F. Chu, J. Chen, S. Jian, S. Chen, and Y. Wang, "EMI Shielding Technology in 5G RF System in Package Module", *IEEE/ECTC Proceedings*, May 2020, pp. 931–937.
9. Ouyang, E., Y. Jeong, J. Kim, S. Lin, J. Vang, and A. Yang, "Warpage and Void Simulation of System in Package", *IEEE/ECTC Proceedings*, May 2020, pp. 2066–2071.
10. Chuang, P., M.-L. Lin, S.-T. Hung, Y.-W. Wu, D.-C. Wong, M.-C. Yew, C.-K. Hsu, L.-L Liao, P.-Y. Lai, P.-H. Tsai, S.-M. Chen, S.-K. Cheng, and S.-P. Jeng, "Hybrid Fan-out Package for Vertical Heterogeneous Integration", *IEEE/ECTC Proceedings*, May 2020, pp. 333–338.
11. Fettke, M., T. Kubsch, A. Kolbasow, V. Bejugam, A. Frick, T. Teutsch, "Laser-assisted bonding (LAB) and de-bonding (LAdB) as an advanced process solution for selective repair of 3D and multi-die chip packages", *IEEE/ECTC Proceedings*, May 2020, pp. 1016–1024.
12. Scott, G., J. Bae, K. Yang, W. Ki, N. Whitchurch, M. Kelly, C. Zwenger, J. Jeon, and T. Hwang, "Heterogeneous Integration Using Organic Interposer Technology", *IEEE/ECTC Proceedings*, May 2020, pp. 885–892.
13. Chun, S., T. Kuo, H. Tsai, C. Liu, C. Wang, J. Hsieh, T. Lin, T. Ku, and D. Yu, "InFO_SoW (System-on-Wafer) for High Performance Computing", *IEEE/ECTC Proceedings*, May 2020, pp. 1–6.
14. Peng, C., P. Lin, C. Ko, C. Wang, O. Chuang, and C. Lee, "A Novel Warpage Reinforcement Architecture with RDL Interposer for Heterogeneous Integrated Packages", *IEEE/ECTC Proceedings*, May 2020, pp. 526–531.
15. Liu, P., J. Li, H. van Zeijl, and G. Zhang "Wafer Scale Flexible Interconnect Transfer for Heterogeneous Integration", *IEEE/ECTC Proceedings*, May 2020, pp. 817–823.
16. Ali, M., A. Watanabe, T. Kakutani, P. Raj, R. Tummala, and M. Swaminathan, "Heterogeneous Integration of 5G and Millimeter-Wave Diplexers with 3D Glass Substrates", *IEEE/ECTC Proceedings*, May 2020, pp. 1376–1382.
17. Martins, A., M. Pinheiro, A. Ferreira, R. Almeida, F. Matos, J. Oliveira, H. Santos, M. Monteiro, H. Gamboa, and R. Silva, "Heterogeneous Integration Challenges Within Wafer Level Fan-Out SiP for Wearables and IoT", *IEEE/ECTC Proceedings*, May 2018, pp. 1485–1492.
18. Ko, CT, H. Yang, J. H. Lau, M. Li, M. Li, C. Lin, et al., "Chip-First Fan-Out Panel-Level Packaging for Heterogeneous Integration", *IEEE/ECTC Proceedings*, May 2018, pp. 355–363.

19. Ko, CT, H. Yang, J. H. Lau, M. Li, M. Li, C. Lin, J.W. Lin, T. Chen, I. Xu, C. Chang, J. Pan, H. Wu, Q. Yong, N. Fan, E. Kuah, Z. Li, K. Tan, Y. Cheung, E. Ng, K. Wu, J. Hao, R. Beica, M. Lin, Y. Chen, Z. Cheng, S. Koh, R. Jiang, X. Cao, S. Lim, N. Lee, M. Tao, J. Lo, and R. Lee, "Chip-First Fan-Out Panel-Level Packaging for Heterogeneous Integration", *IEEE Transactions on CPMT*, September 2018, pp. 1561–1572.
20. Hsu, F., J. Lin, S. Chen, P. Lin, J. Fang, J. Wang, and S. Jeng, "3D Heterogeneous Integration with Multiple Stacking Fan-Out Package", *IEEE/ECTC Proceedings*, May 2018, pp. 337–342.
21. Lin, Y., S. Wu, W. Shen, S. Huang, T. Kuo, A. Lin, T. Chang, H. Chang, S. Lee, C. Lee, J. Su, X. Liu, Q. Wu, and K. Chen, "An RDL-First Fan-out Wafer Level Package for Heterogeneous Integration Applications", *IEEE/ECTC Proceedings*, May 2018, pp. 349–354.
22. Lau, J. H., M. Li, M. Li, T. Chen, I. Xu, X. Qing, Z. Cheng, et al., "Fan-Out Wafer-Level Packaging for Heterogeneous Integration", *Proceedings of IEEE/ECTC*, May 2018, pp. 2354–2360.
23. Lau, J. H., M. Li, M. Li, T. Chen, I. Xu, X. Qing, Z. Cheng, N. Fan, E. Kuah, Z. Li, K. Tan, Y. Cheung, E. Ng, P. Lo, K. Wu, J. Hao, S. Koh, R. Jiang, X. Cao, R. Beica, S. Lim, N. Lee, C. Ko, H. Yang, Y. Chen, M. Tao, J. Lo, and R. Lee, "Fan-Out Wafer-Level Packaging for Heterogeneous Integration", *IEEE Transactions on CPMT*, 2018, September 2018, pp. 1544–1560.
24. Knickerbocker, J., R. Budd, B. Dang, Q. Chen, E. Colgan, L.W. Hung, S. Kumar, K.W. Lee, M. Lu, J. W. Nah, R. Narayanan, K. Sakuma, V. Siu, and B. Wen, "Heterogeneous Integration Technology Demonstrations for Future Healthcare, IoT, and AI Computing Solutions", *IEEE/ECTC Proceedings*, May 2018, pp. 1519–1522.
25. Lau, J. H., "Fan-Out Wafer-Level Packaging for 3D IC Heterogeneous Integration", *Proceedings of CSTIC*, March 2018, pp. VII_1–6.
26. Lau, J. H., "Heterogeneous Integration with Fan-Out Wafer-Level Packaging", *Proceedings of IWLPC*, October 2017, pp. 1–25.
27. Panigrahi, A., C. Kumar, S. Bonam, B. Paul, T. Ghosh N. Paul, S. Vanjari, and S. Singh, "Metal-Alloy Cu Surface Passivation Leads to High Quality Fine-Pitch Bump-Less Cu-Cu Bonding for 3D IC and Heterogeneous Integration Applications", *IEEE/ECTC Proceedings*, May 2018, pp. 1555–1560.
28. Faucher-Courchesne, C., D. Danovitch, L. Brault, M. Paquet, and E. Turcotte, "Controlling Underfill Lateral Flow to Improve Component Density in Heterogeneously Integrated Packaging Systems", *IEEE/ECTC Proceedings*, May 2018, pp. 1206–1213.
29. Lau, J. H., "3D IC Heterogeneous Integration by FOWLP", *Chip Scale Review*, Vol. 22, January/February 2018, pp. 16–21.
30. Hu, Y., C. Lin, Y. Hsieh, N. Chang, A. J. Gallegos, T. Souza, W. Chen, M. Sheu, C. Chang, C. Chen, K. Chen, "3D Heterogeneous Integration Structure Based on 40 nm- and 0.18 μm-Technology Nodes", *Proceedings of IEEE/ECTC*, May 2015, pp. 1646–1651.
31. Bajwa, A., S. Jangam, S. Pal, N. Marathe, T. Bai, T. Fukushima, M. Goorsky, and S. S. Iyer, "Heterogeneous Integration at Fine Pitch (≤ 10 μm) using Thermal Compression Bonding", *IEEE/ECTC Proceedings*, May 2017, pp. 1276–1284.
32. Dittrich, M., A. Heinig, F. Hopsch, and R. Trieb, "Heterogeneous Interposer Based Integration of Chips with Copper Pillars and C4 Balls to Achieve High Speed Interfaces for ADC Application", *Proceedings of IEEE/ECTC*, Mat 2017, pp. 643–648.
33. Chuang, Y., C. Yuan, J. Chen, C. Chen, C. Yang, W. Changchien, C. Liu, and F. Lee, "Unified Methodology for Heterogeneous Integration with CoWoS Technology", *IEEE/ECTC Proceedings*, May 2013, pp. 852–859.
34. Ko, C., H. Yang, J. H. Lau, M. Li, M. Li, et al., "Design, Materials, Process, and Fabrication of Fan-Out Panel-Level Heterogeneous Integration", *Proceedings of IMAPS Symposium*, October 2018, pp. TP2_1–7.
35. Lau, J. H., M. Li, Y. Lei, M. Li, I. Xu, T. Chen, Q. Yong, Z. Cheng, et al., "Reliability of Fan-Out Wafer-Level Heterogeneous Integration", *Proceedings of IMAPS Symposium*, October 2018, pp. WA2_1–9.

36. Beal, A., and R. Dean, "Using SPICE to Model Nonlinearities Resulting from Heterogeneous Integration of Complex Systems", *IMAPS Proceedings*, October 2017, pp. 274–279.
37. Lau, J. H., M. Li, Y. Lei, M. Li, I. Xu, T. Chen, Q. Yong, Z. Cheng, et al., "Reliability of Fan-Out Wafer-Level Heterogeneous Integration", *IMAPS Transactions, Journal of Microelectronics and Electronic Packaging*, Vol. 15, Issue 4, October 2018, pp. 148–162.
38. Ko, C. T., H. Yang, and J. H. Lau, "Design, Materials, Process, and Fabrication of Fan-Out Panel-Level Heterogeneous Integration", *IMAPS Transactions, Journal of Microelectronics and Electronic Packaging*, Vol. 15, Issue 4, October 2018, pp. 141–147.
39. Hanna, A, A. Alam, T, Fukushima, S. Moran, W. Whitehead, S. Jangam, S. Pal, G. Ezhilarasu, R. Irwin, A. Bajwa, and S. Iyer, "Extremely Flexible (1 mm Bending Radius) Biocompatible Heterogeneous Fan-Out Wafer-Level Platform with the Lowest Reported Die-Shift (<6 μm) and Reliable Flexible Cu-Based Interconnects", *IEEE/ECTC Proceedings*, May 2018, pp. 1505–1511.
40. Kyozuka, M., T. Kiso, H. Toyazaki, K. Tanaka, and T. Koyama, "Development of Thinner POP base Package by Die Embedded and RDL Structure", *IMAPS Proceedings*, October 2017, pp. 715–720.
41. Yoon, S., J. Caparas, Y. Lin, and P. Marimuthu, "Advanced Low Profile PoP Solution with Embedded Wafer Level PoP (eWLB-PoP) Technology", *IEEE/ECTC Proceedings*, 2012, pp. 1250–1254.
42. Yoon, S., P. Tang, R. Emigh, Y. Lin, P. Marimuthu, and R. Pendse, "Fanout Flipchip eWLB (Embedded Wafer Level Ball Grid Array) Technology as 2.5D Packaging Solutions", *IEEE/ECTC Proceedings*, 2013, pp. 1855–1860.
43. Lin, Y., W. Lai, C. Kao, J. Lou, P. Yang, C. Wang, and C. Hseih, "Wafer Warpage Experiments and Simulation for Fan-Out Chip on Substrate", *IEEE/ECTC Proceedings*, May 2016, pp. 13–18.
44. Lau, J. H., *Fan-Out Wafer-Level Packaging*. Springer Book Company, 2018.
45. Lau, J. H., et al, "Apparatus Having Thermal-Enhanced and Cost-Effective 3D IC Integration Structure with Through Silicon via Interposer". US Patent No: 8,604,603, Date of Patent: December 10, 2013.
46. Chiu, C., Z. Qian, and M. Manusharow, "Bridge Interconnect with AirGap in Package Assembly". US Patent No. 8,872,349, 2014.
47. Mahajan, R., R. Sankman, N. Patel, D. Kim, K. Aygun, Z. Qian, et al., "Embedded Multi-die Interconnect Bridge (EMIB)—A High-Density, High-Bandwidth Packaging Interconnect", *IEEE/ECTC Proceedings*, May 2016, pp. 557–565.
48. Suk, K., S. Lee, J. Kim, S. Lee, H. Kim, S. Lee, P. Kim, D. Kim, D. Oh, and J. Byun, "Low Cost Si-less RDL Interposer Package for High Performance Computing Applications", *IEEE/ECTC Proceedings*, May 2018, pp. 64–69.
49. Podpod, A., J. Slabbekoorn, A. Phommahaxay, F. Duval, A. Salahouedlhadj, M. Gonzalez, K. Rebibis, R. A. Miller, G. Beyer, and E. Beyne, "A Novel Fan-Out Concept for Ultra-High Chip-to-Chip Interconnect Density with 20-μm Pitch", *IEEE/ECTC Proceedings*, May 2018, pp. 370–378.
50. Lau, J. H., C. Lee, C. Zhan, S. Wu, Y. Chao, M. Dai, R. Tain, H. Chien, et al., "Low-Cost Through-Silicon Hole Interposers for 3D IC Integration", *IEEE Transactions on CPMT*, Vol. 4, No. 9, September 2014, pp. 1407–1419.
51. Souriau, J., O. Lignier, M. Charrier, and G. Poupon, "Wafer Level Processing Of 3D System in Package for RF and Data Applications", *IEEE/ECTC Proceedings*, 2005, pp. 356–361.
52. Henry, D., D. Belhachemi, J-C. Souriau, C. Brunet-Manquat, C. Puget, G. Ponthenier, J. Vallejo, C. Lecouvey, and N. Sillon, "Low Electrical Resistance Silicon Through Vias: Technology and Characterization", *IEEE/ECTC Proceedings*, 2006, pp. 1360–1366.
53. Khan, N., V. Rao, S. Lim, H. We, V. Lee, X. Zhang, E. Liao, R. Nagarajan, T. C. Chai, V. Kripesh, and J. H. Lau, "Development of 3-D Silicon Module With TSV for System in Packaging", *IEEE Proceedings of Electronic, Components & Technology Conference*, Orlando, FL, May 27–30, 2008, pp. 550–555. Also, *IEEE Transactions on CPMT*, Vol. 33, No. 1, March 2010, pp. 3–9.

54. Lau, J. H., C.-J. Zhan, P.-J. Tzeng, C.-K. Lee, M.-J. Dai, H.-C. Chien, Y.-L. Chao, et al., "Feasibility Study of a 3D IC Integration System-in-Packaging (SiP) from a 300 mm Multi-Project Wafer (MPW)", *IMAPS International Symposium on Microelectronics*, October 2011, pp. 446–454. Also, *IMAPS Transactions, Journal of Microelectronic Packaging*, Vol. 8, No. 4, Fourth Quarter 2011, pp. 171–178.
55. Zhan, C., P. Tzeng, J. H. Lau, M. Dai, H. Chien1, C. Lee, S. Wu, et al., "Assembly Process and Reliability Assessment of TSV/RDL/IPD Interposer with Multi-Chip-Stacking for 3D IC Integration SiP", *IEEE/ECTC Proceedings*, San Diego, CA, May 2012, pp. 548–554.
56. Che, F., M. Kawano, M. Ding, Y. Han, and S. Bhattacharya, "Co-design for Low Warpage and High Reliability in Advanced Package with TSV-Free Interposer (TFI)", *Proceedings of IEEE/ECTC*, May 2017, pp. 853–861.
57. Hou, S., W. Chen, C. Hu, C. Chiu, K. Ting, T. Lin, W. Wei, W. Chiou, V. Lin, V. Chang, C. Wang, C. Wu, and D. Yu, "Wafer-Level Integration of an Advanced Logic-Memory System Through the Second-Generation CoWoS Technology", *IEEE Transactions on Electron Devices*, October 2017, pp. 4071–4077.
58. Selvanayagam, C., J. H. Lau, X. Zhang, S. Seah, K. Vaidyanathan, and T. Chai, "Nonlinear Thermal Stress/Strain Analysis of Copper Fill TSV (Through Silicon Via) and Their Flip-Chip Microbumps", *IEEE/ECTC Proceedings*, May 27–30, 2008, pp. 1073–1081.
59. Selvanayagam, C., J. H. Lau, X. Zhang, S. Seah, K. Vaidyanathan, and T. Chai, "Nonlinear Thermal Stress/Strain Analyses of Copper Filled TSV (Through Silicon Via) and Their Flip-Chip Microbumps", *IEEE Transactions on Advanced Packaging*, Vol. 32, No. 4, November 2009, pp. 720–728.
60. Lau, J. H., and G. Tang, "Thermal Management of 3D IC Integration with TSV (Through Silicon Via)", *IEEE/ECTC Proceedings*, May 2009, pp. 635–640.
61. Lau, J. H., Y. S. Chan, and R. S. W. Lee, "3D IC Integration with TSV Interposers for High-Performance Applications", *Chip Scale Review*, Vol. 14, No. 5, September/October, 2010, pp. 26–29.
62. Lau, J. H., "TSV Manufacturing Yield and Hidden Costs for 3D IC Integration", *IEEE/ECTC Proceedings,* May 2010, pp. 1031–1041.
63. Zhang, X., T. Chai, J. H. Lau, C. Selvanayagam, K. Biswas, S. Liu, D. Pinjala, et al., "Development of Through Silicon Via (TSV) Interposer Technology for Large Die (21 × 21 mm) Fine-pitch Cu/low-k FCBGA Package", *IEEE Proceedings of ECTC*, May, 2009, pp. 305–312.
64. Chai, T. C., X. Zhang, J. H. Lau, C. S. Selvanayagam, D. Pinjala, et al., "Development of Large Die Fine-Pitch Cu/low-*k* FCBGA Package with Through Silicon Via (TSV) Interposer", *IEEE Transactions on CPMT*, Vol. 1, No. 5, May 2011, pp. 660–672.
65. Chien, H. C., J. H. Lau, Y. Chao, R. Tain, M. Dai, S. T. Wu, W. Lo, and M. J. Kao, "Thermal Performance of 3D IC Integration with Through-Silicon Via (TSV)", *IMAPS Transactions, Journal of Microelectronic Packaging*, Vol. 9, 2012, pp. 97–103.
66. Chaware, R., K. Nagarajan, and S. Ramalingam, "Assembly and Reliability Challenges in 3D Integration of 28 nm FPGA Die on a Large High-Density 65 nm Passive Interposer", *IEEE/ECTC Proceedings*, May 2012, pp. 279–283.
67. Banijamali, B., S. Ramalingam, K. Nagarajan, and R. Chaware, "Advanced Reliability Study of TSV Interposers and Interconnects for the 28 nm Technology FPGA", *IEEE/ECTC Proceedings*, May 2011, pp. 285–290.
68. Banijamali, B., S. Ramalingam, H. Liu, and M. Kim, "Outstanding and Innovative Reliability Study of 3D TSV Interposer and Fine-Pitch Solder Micro-Bumps", *IEEE/ECTC Proceedings*, May 2012, pp. 309–314.
69. Banijamali, B., C. Chiu, C. Hsieh, T. Lin, C. Hu, S. Hou, et al., "Reliability Evaluation of a CoWoS-Enabled 3D IC Package", *IEEE/ECTC Proceedings*, May 2013, pp. 35–40.
70. Xie, J., H. Shi, Y. Li, Z. Li, A. Rahman, K. Chandrasekar, et al., "Enabling the 2.5D Integration", *Proceedings of IMAPS International Symposium on Microelectronics*, October 2012, pp. 254–267.
71. Li, L., P. Su, J. Xue, M. Brillhart, J. H. Lau, P. Tzeng, C. Lee, C. Zhan, et al., "Addressing Bandwidth Challenges in Next Generation High Performance Network Systems with 3D IC Integration", *IEEE/ECTC Proceedings*, May 2012, pp. 1040–1046.

72. Lau, J. H., P. Tzeng, C. Zhan, C. Lee, M. Dai, J. Chen, Y. Hsin, et al., "Large Size Silicon Interposer and 3D IC Integration for System-in-Packaging (SiP)", *Proceedings of the 45th IMAPS International Symposium on Microelectronics*, September 2012, pp. 1209–1214.
73. Wu, S. T., J. H. Lau, H. Chien, Y. Chao, R. Tain, L. Li, P. Su, et al., "Thermal Stress and Creep Strain Analyses of a 3D IC Integration SiP with Passive Interposer for Network System Application", *Proceedings of the 45th IMAPS International Symposium on Microelectronics*, September 2012, pp. 1038–1045.
74. Chien, H., J. H. Lau, T. Chao, M. Dai, and R. Tain, "Thermal Management of Moore's Law Chips on Both sides of an Interposer for 3D IC integration SiP", *IEEE ICEP Proceedings*, Japan, April 2012, pp. 38–44.
75. Chien, H., J. H. Lau, T. Chao, M. Dai, R. Tain, L. Li, P. Su, et al., "Thermal Evaluation and Analyses of 3D IC Integration SiP with TSVs for Network System Applications", *IEEE/ECTC Proceedings*, San Diego, CA, May 2012, pp. 1866–1873.
76. Ji, M., M. Li, J. Cline, D. Seeker, K. Cai, J. H. Lau, P. Tzeng, et al., "3D Si Interposer Design and Electrical Performance Study", *Proceedings of Design Con*, Santa Clara, CA, January 2013, pp. 1–23.
77. Wu, S. T., H. Chien, J. H. Lau, M. Li, J. Cline, and M. Ji, "Thermal and Mechanical Design and Analysis of 3D IC Interposer with Double-Sided Active Chips", *IEEE/ECTC Proceedings*, Las Vegas, NA, May 2013, pp. 1471–1479.
78. Tzeng, P. J., J. H. Lau, C. Zhan, Y. Hsin, P. Chang, Y. Chang, J. Chen, et al., "Process Integration of 3D Si Interposer with Double-Sided Active Chip Attachments", *IEEE/ECTC Proceedings*, Las Vegas, NA, May 2013, pp. 86–93.
79. Stow, D., Y. Xie, T. Siddiqua, and G. H. Loh, "Cost-Effective Design of Scalable High-Performance Systems Using Active and Passive Interposers", *Proceedings of IEEE/ACM International Conference on Computer-Aided Design*, November 2017, pp. 728–735.
80. Hwang, T., D. Oh, E. Song, K. Kim, J. Kim, and S. Lee, "Study of Advanced Fan-Out Packages for Mobile Applications", *IEEE/ECTC Proceedings*, May 2018, pp. 343–348.
81. Hong, J., K. Choi, D. Oh, S Park, S. Shao, H. Wang, Y. Niu, and V. Pham, "Design Guideline of 2.5D Package with Emphasis on Warpage Control and Thermal Management", *IEEE/ECTC Proceedings*, May 2018, pp. 682–692.
82. You, S., S. Jeon, D. Oh, K. Kim, J. Kim, S. Cha, and G. Kim, "Advanced Fan-Out Package SI/PI/Thermal Performance Analysis of Novel RDL Packages", *IEEE/ECTC Proceedings*, May 2018, pp. 1295–1301.
83. Miao, M., L. Wang, T. Chen, X. Duan, J. Zhang, N. Li, L. Sun, R. Fang, X. Sun, H. Liu, and Y. Jin, "Modeling and Design of a 3D Interconnect Based Circuit Cell Formed with 3D SiP Techniques Mimicking Brain Neurons for Neuromorphic Computing Applications", *IEEE/ECTC Proceedings*, May 2018, pp. 490–497.
84. Borel, S., L. Duperrex, E. Deschaseaux, J. Charbonnier, J. Cledière, R. Wacquez, J. Fournier, J.-C. Souriau, G. Simon, and A. Merle, "A Novel Structure for Backside Protection against Physical Attacks on Secure Chips or SiP", *IEEE/ECTC Proceedings*, May 2018, pp. 515–520.
85. Lee, E., M. Amir, S. Sivapurapu, C. Pardue, H. Torun, M. Bellaredj, M. Swaminathan, and S. Mukhopadhyay, "A System-in-Package Based Energy Harvesting for IoT Devices with Integrated Voltage Regulators and Embedded Inductors", *IEEE/ECTC Proceedings*, May 2018, pp. 1720–1725.
86. Li, J., S. Ma, H. Liu, Y. Guan, J. Chen, Y. Jin, W. Wang, L. Hu, and S. He, "Design, Fabrication and Characterization of TSV Interposer Integrated 3D Capacitor for SiP Applications", *IEEE/ECTC Proceedings*, May 2018, pp. 1968–1974.
87. Ki, W., W. Lee, I. Lee, I. Mok, W. Do, M. Kolbehdari, A. Copia, S. Jayaraman, C. Zwenger, and K. Lee, "Chip Stackable, Ultra-thin, High-Flexibility 3D FOWLP (3D SWIFT® Technology) for Hetero-Integrated Advanced 3D WL-SiP", *IEEE/ECTC Proceedings*, May 2018, pp. 580–586.
88. Lee, J., C. Lee, C. Kim, and S. Kalchuri, "Micro Bump System for 2nd Generation Silicon Interposer with GPU and High Bandwidth Memory (HBM) Concurrent Integration", *IEEE/ECTC Proceedings*, May 2018, pp. 607–612.

89. Lim, Y., X. Xiao, R. Vempati, S. Nandar, K. Aditya, S. Gaurav, T. Lim, V. Kripesh, J. Shi, J. H. Lau, and S. Liu, "High Quality and Low Loss Millimeter Wave Passives Demonstrated to 77-GHz for SiP Technologies Using Embedded Wafer-Level Packaging Platform (EMWLP)", *IEEE Transactions on Advanced Packaging*, Vol. 33, 2010, pp. 1061–1071.
90. Manessis, D., L. Boettcher, A. Ostmann, R. Aschenbrenner, and H. Reichl, "Chip Embedding Technology Developments Leading to the Emergence of Miniaturized System-in-Packages", *Proceedings of IEEE/ECTC*, May 2010, pp. 803–810.
91. Lau, J. H., M. S. Zhang, and S. W. R. Lee, "Embedded 3D Hybrid IC Integration System-in-Package (SiP) for Opto-Electronic Interconnects in Organic Substrates", *ASME Transactions, Journal of Electronic Packaging*, Vol. 133, September 2011, pp. 1–7.
92. Lau, J. H., C.-J. Zhan, P.-J. Tzeng, C.-K. Lee, M.-J. Dai, H.-C. Chien, et al., "Feasibility Study of a 3D IC Integration System-in-Packaging (SiP) from a 300 mm Multi-Project Wafer (MPW)", *IMAPS Transactions, Journal of Microelectronic Packaging*, Vol. 8, No. 4, Fourth Quarter 2011, pp. 171–178.
93. Lau, J. H., and G. Y. Tang, "Effects of TSVs (through-silicon vias) on Thermal Performances of 3D IC Integration System-in-Package (SiP)", *Journal of Microelectronics Reliability*, Vol. 52, Issue 11, November 2012, pp. 2660–2669.
94. Ahmad, M., M. Nagar, W. Xie, M. Jimarez, and C. Ryu, "Ultra Large System-in-Package (SiP) Module and Novel Packaging Solution for Networking Applications", *Proceedings of IEEE/ECTC*, May 2013, pp. 694–701.
95. Wu, H., D. S. Gardner, C. Lv, Z. Zou, and H. Yu, "Integration of Magnetic Materials into Package RF and Power Inductors on Organic Substrates for System in Package (SiP) Applications", *Proceedings of IEEE/ECTC*, May 2014, pp. 1290–1295.
96. Qian, R., and Y. Liu, "Modeling for Reliability of Ultra-Thin Chips in a System in Package", *Proceedings of IEEE/ECTC*, May 2014, pp. 2063–2068.
97. Hsieh, C., C. Tsai, H. Lee, T. Lee, H. Chang, "Fan-out Technologies for WiFi SiP Module Packaging and Electrical Performance Simulation", *Proceedings of IEEE/ECTC*, May 2015, pp. 1664–1669.
98. Li, L., P. Chia, P. Ton, M. Nagar, S. Patil, J. Xue, J. DeLaCruz, M. Voicu, J. Hellings, B. Isaacson, M. Coor, and R. Havens, "3D SiP with organic interposer of ASIC and memory integration", *Proceedings of IEEE/ECTC*, May 2016, pp. 1445–1450.
99. Tsai, M., A. Lan, C. Shih, T. Huang, R. Chiu, S. L. Chung, J. Y. Chen, F. Chu, C. Chang, S. Yang, D. Chen, and N. Kao, "Alternative 3D Small Form Factor Methodology of System in Package for IoT and Wearable Devices Application", *Proceedings of IEEE/ECTC*, May 2017, pp. 1541–1546.
100. Das, R., F. Egitto, S. Rosser, E. Kopp, B. Bonitz, and R. Rai, "3D Integration of System-in-Package (SiP) using Organic Interposer: Toward SiP-Interposer-SiP for High-End Electronics", *IMAPS Proceedings*, September 2013, pp. 531–537.
101. Chien, H., C. Chien, M. Dai, R. Tain, W. Lo, Y. Lu, "Thermal Characteristic and Performance of the Glass Interposer with TGVs (Through-Glass Via)", *IMAPS Proceedings*, September 2013, pp. 611–617.
102. Vincent, M., D. Mitchell, J. Wright, Y. Foong, A. Magnus, Z. Gong, S. Hayes, and N. Chhabra, "3D RCP Package Stacking: Side Connect, An Emerging Technology for Systems Integration and Volumetric Efficiency", *IMAPS Proceedings*, September 2013, pp. 447–451.
103. Renaud-Bezot, N., "Size-Matters—Embedding as an Enabler of Next-Generation SiPs", *IMAPS Proceedings*, September 2013, pp. 740–744.
104. Couderc, P., Noiray, J., and C. Val, "Stacking of Known Good Rebuilt Wafers for High Performance Memory and SiP", *IMAPS Proceedings*, September 2013, pp. 804–809.
105. Lim, J., and V. Pandey, "Innovative Integration Solutions for SiP Packages Using Fan-Out Wafer Level eWLB Technology", *IMAPS Proceedings*, October 2017, pp. 263–269.
106. Becker, K., M. Minkus, J. Pauls, V. Bader, S. Voges, T. Braun, G. Jungmann, H. Wieser, M. Schneider-Ramelow, and K.-D., "Non-Destructive Testing for System-in-Package Integrity Analysis", *IMAPS Proceedings*, October 2017, pp. 182–187.

107. Lee, Y., and D. Link, "Practical Application and Analysis of Lead-Free Solder on Chip-On-Flip-Chip SiP for Hearing Aids", *IMAPS Proceedings*, October 2017, pp. 201–207.
108. Milton, B., O. Kwon, C. Huynh, I. Qin, and B. Chylak, "Wire Bonding Looping Solutions for High Density System-in-Package (SiP)", *IMAPS Proceedings*, October 2017, pp. 426–431.
109. Morard, A., J. Riou, and G. Pares, "Flip Chip Reliability and Design Rules for SiP Module", *IMAPS Proceedings*, October 2017, pp. 754–760.
110. Lau, J. H., *3D IC Integration and Packaging*, McGraw-Hill Book Company, New York, 2016.
111. Lau, J. H., *Through-Silicon Via (TSV) for 3D Integration*, McGraw-Hill Book Company, New York, 2013.
112. Lau, J. H., *Handbook of Fine Pitch Surface Mount Technology*, Van Nostrand Reinhold, New York, 1994.
113. Lau, J. H., and N. C. Lee, *Assembly and Reliability of Lead-Free Solder Joints*, Springer, New York, 2020.
114. Prasad, A., L. Pymento, S. Aravamudhan, and C. Periasamy, "Advancement of Solder Paste Inspection (SPI) Tools to Support Industry 4.0 & Package Scaling", *SMTA International Proceedings*, October 2018, pp. 1–5.
115. Lau, J. H., *Flip Chip Technologies*, McGraw-Hill Book Company, New York, 1996.
116. Lau, J. H., and Y. Pao, *Solder Joint Reliability of Flip Chip Assemblies*, McGraw-Hill Book Company, New York, 1997.
117. Lau, J. H., *Low Cost Flip Chip Technologies*, McGraw-Hill Book Company, New York, 2000.
118. Lau, J. H., "Recent Advances and New Trends in Flip Chip Technology", *ASME Transactions, Journal of Electronic Packaging*, September 2016, Vol. 138, Issue 3, pp. 1–23.
119. Davis, E., Harding, W., Schwartz, R., and Corning, J., "Solid Logic Technology: Versatile, High Performance Microelectronics," *IBM J. Res. Dev.*, 8(2), 1964, pp. 102–114.
120. Totta, P., and Sopher, R., "SLT Device Metallurgy and Its Monolithic Extension," *IBM J. Res. Dev.*, 13(3), 1969, pp. 226–238.
121. Lau, J. H., and C. Chang, "Taguchi Design of Experiment for Wafer Bumping by Stencil Printing", *IEEE Transactions on Electronics Packaging Manufacturing*, Vol. 21, No. 3, July 2000, pp. 219–225.
122. Love, D., Moresco, L., Chou, W., Horine, D., Wong, C., and Eilin, S., "Wire Interconnect Structures for Connecting an Integrated Circuit to a Substrate," *U.S. Patent No. 5,334,804*, filed Nov. 17, 1992 and issued Aug. 2, 1994.
123. Tung, F., "Pillar Connections for Semiconductor Chips and Method of Manufacture," *U.S. Patent No. 6,578,754*, filed Apr. 27, 2000 and issued June 17, 2003.
124. Tung, F., "Pillar Connections for Semiconductor Chips and Method of Manufacture," *U.S. Patent No. 6,681,982*, filed June 12, 2002 and issued Jan. 27, 2004.
125. Nakano, F., Soga, T., and Amagi, S., "Resin-Insertion Effect on Thermal Cycle Resistivity of Flip-Chip Mounted LSI Devices," *International Symposium on Microelectronics*, Minneapolis, MN, Sep. 28–30, 1987, pp. 536–541.
126. Tsukada, Y., Tsuchida, S., and Mashimoto, Y., "Surface Laminar Circuit Packaging," *IEEE/ECTC Proceedings*, May 1992, pp. 22–27.
127. Tsukada, Y., and Tsuchida, S., "Surface Laminar Circuit, a Low Cost High Density Printed Circuit Board," *SMTA International Conference*, Aug. 1992, pp. 537–542.
128. Tsukada, Y., "Solder Bumped Flip Chip Attach on SLC Board and Multichip Module," *Chip on Board Technologies for Multichip Modules*, J. H. Lau, ed., Van Nostrand Reinhold, New York, 1994, pp. 410–443.
129. Lau, J. H., and Chang, C., "Characteristics and Reliability of Fast-Flow, Snap-Cure, and Reworkable Underfills for Solder Bumped Flip Chip on Low-Cost Substrates," *IEEE Trans. Electron. Packag. Manuf.*, 25(3), 2002, pp. 231–230.
130. Lau, J. H., and Lee, R., "Fracture Mechanics Analysis of Low Cost Solder Bumped Flip Chip Assemblies With Imperfect Underfills," *ASME J. Electron. Packag.*, 222(4), 2000, pp. 306–310.
131. Lau, J. H., Lee, R., and Chang, C., "Effects of Underfill Material Properties on the Reliability of Solder Bumped Flip Chip on Board With Imperfect Underfill Encapsulants," *IEEE Trans. CPMT*, 23(2), 2000, pp. 323–333.

132. Lau, J. H., and Lee, S. W., "Effects of Underfill Delamination and Chip Size on the Reliability of Solder Bumped Flip Chip on Board," *IMAPS Trans. Int. J. Microcircuit Electron. Packag.*, 23(1), 2000, pp. 33–39.
133. Lau, J. H., Chang, C., and Chen, C., "Characteristics and Reliability of No-Flow Underfills for Solder Bumped Flip Chip Assemblies," *IMAPS Trans. Int. J. Microcircuit Electron. Packag.*, 22(4), 1999, pp. 370–381.
134. Lau, J. H., and Chang, C., "How to Select Underfill Materials for Solder Bumped Flip Chips on Low Cost Substrates?" *IMAPS Trans. Int. J. Microelectron. Electron. Packag.*, 22(1), 1999, pp. 20–28.
135. Lau, J. H., and Chang, C., "Characterization of Underfill Materials for Functional Solder Bumped Flip Chips on Board Applications," *IEEE Trans. CPMT, Part A*, 22(1), 1999, pp. 111–119.
136. Lau, J. H., Chang, C., and Ouyang, C., "SMT Compatible No-Flow Underfill for Solder Bumped Flip Chip on Low-Cost Substrates," *J. Electron. Manuf.*, 8(3–4), 1998, pp. 151–164.
137. Lau, J. H., Chang, C., and Chen, R., "Effects of Underfill Encapsulant on the Mechanical and Electrical Performance of a Functional Flip Chip Device," *J. Electron. Manuf.*, 7(4), 1997, pp. 269–277.
138. Lau, J. H., and Wun, B., "Characterization and Evaluation of the Underfill Encapsulants for Flip Chip Assembly," *J. Inst. Interconnect. Technol.*, 21(1), 1995, pp. 25–27.
139. Lau, J. H., Schneider, E., and Baker, T., "Shock and Vibration of Solder Bumped Flip Chip on Organic Coated Copper Boards," *ASME J. Electron. Packag.*, 118(2), 1996, pp. 101–104.
140. Lau, J. H., 1998, "Flip Chip on PCBs With Anisotropic Conductive Film," Advanced Packaging, July/Aug., pp. 44–48.
141. Lau, J. H., Chang, C., and Lee, R., "Failure Analysis of Solder Bumped Flip Chip on Low-Cost Substrates," *IEEE Trans. Electron. Packag. Manuf.*, 23(1), 2000, pp. 19–27.
142. Lau, J. H., Lee, R., Pan, S., and Chang, C., "Nonlinear Time-Dependent Analysis of Micro Via-In-Pad Substrates for Solder Bumped Flip Chip Applications," *ASME J. Electron. Packag.*, 124(3), 2002, pp. 205–211.
143. Lau, J. H., Q. Zhang, M. Li, K. Yeung, Y. Cheung, N. Fan, Y. Wong, M. Zahn, and M., and M. Koh, "Stencil Printing of Underfill for Flip Chips on Organic-Panel and Si-Wafer Assemblies," *IEEE/ECTC Proceedings*, May 2015, pp. 168–174.
144. Wong, C. P., Baldwin, D., Vincent, M. B., Fennell, P., Wang, L. J., and Shi, S. H., "Characterization of a No-Flow Underfill Encapsulant During the Solder Reflow Process," *IEEE/ECTC Proceedings*, May 25–28, 1998, pp. 1253–1259.
145. Lau, J. H., Krulevitch, T., Schar, W., Heydinger, M., Erasmus, S., and Gleason, J., "Experimental and Analytical Studies of Encapsulated Flip Chip Solder Bumps on Surface Laminar Circuit Boards," *Circuit World*, 19(3), 1993, pp. 18–24.
146. Wun, B., and Lau, J. H., "Characterization and Evaluation of the Underfill Encapsulants for Flip Chip Assembly," *Circuit World*, 21(3), 1995, pp. 25–32.
147. Zhang, Z., and C. P. Wong, "Recent Advances in Flip-Chip Underfill: Materials, Process, and Reliability", September 2004, *IEEE Transactions on Advanced Packaging*, 27(3), pp. 515–524.
148. Cremaldi, J., Gaynes, M., Brofman, P., Pesika, N., and Lewandowski, E., "Time, Temperature, and Mechanical Fatigue Dependence on Underfill Adhesion," *IEEE/ECTC Proceedings*, May 2014, pp. 255–262.
149. Gilleo, K., Cotterman, B., and Chen, I. A., "Molded Underfill for Flip Chip in Package," *Proceedings of High Density Interconnects*, 2000, pp. 28–31.
150. Rector, L. P., Gong, S., Miles, T. R., and Gaffney, K., "Transfer Molding Encapsulation of Flip Chip Array Packages," *Int. J. Microcircuits Electron. Packag.*, 23(4), 2000, pp. 401–406.
151. Lai, Y. M., Chee, C. K., Then, E., Ng, C. H., and Low, M. F., "Capillary Underfill and Mold Encapsulation Method and Apparatus," *U.S. Patent No. 7,262,077*, filed Sep. 30, 2003 and issued Aug. 28, 2007.
152. Lee, J. Y., Oh, K. S., Hwang, C. H., Lee, C. H., and Amand, R. D. S., "Molded Underfill Development for FlipStack CSP," *IEEE/ECTC Proceedings*, May 2009, pp. 954–959.

153. Joshi, M., Pendse, R., Pandey, V., Lee, T. K., Yoon, I. S., Yun, J. S., Kim, Y. C., and Lee, H. R., "Molded Underfill (MUF) Technology for Flip Chip Packages in Mobile Applications," *IEEE/ECTC Proceedings*, June 2010, pp. 1250–1257.
154. Ferrandon, C., Jouve, A., Joblot, S., Lamy, Y., Schreiner, A., Montmeat, P., Pellat, M., Argoud, M., Fournel, F., Simon, G., and Cheramy, S., "Innovative Wafer-Level Encapsulation and Underfill Material for Silicon Interposer Application," *IEEE/ECTC Proceedings*, May 2013, pp. 761–767.
155. Lau, J. H., Zhang, Q., Li, M., Yeung, K., Cheung, Y., Fan, N., Wong, Y., Zahn, M., and Koh, M., "Stencil Printing of Underfill for Flip Chips on Organic-Panel and Si-Wafer Substrates," *IEEE Trans. CPMT*, 5(7), 2015, pp. 1027–1035.
156. Eitan, A., and Jing, K., "Thermo-Compression Bonding for Fine-Pitch Copper-Pillar Flip Chip Interconnect—Tool Features as Enablers of Unique Technology," *IEEE/ECTC Proceedings*, May 2015, pp. 460–464.
157. Lee, M., Yoo, M., Cho, J., Lee, S., Kim, J., Lee, C., Kang, D., Zwenger, C., and Lanzone, R., "Study of Interconnection Process for Fine Pitch Flip Chip," *IEEE/ECTC Proceedings*, May 2009, pp. 720–723.
158. Li, M., D. Tian, Y. Cheung, L. Yang, and J. H. Lau, "A High Throughput and Reliable Thermal Compression Bonding Process for Advanced Interconnections", *IEEE/ECTC Proceedings*, May 2015, pp. 603–608.
159. Brakke, K. A., *Surface Evolver Manual version 2.70,* Susquehanna University, 2013.

第 3 章

扇入型晶圆级/板级芯片尺寸封装

3.1 引言

如图 3.1 所示,扇入型晶圆级/板级芯片尺寸封装技术是一种直接在晶圆或面板上制作封装的技术。该技术的工艺流程与传统的引线键合(wire bonding,WB)、倒装芯片(flip chip,FC)有很大的不同(见图 3.2)。在传统的引线键合或倒装芯片工艺流程中,我们是先将器件晶圆进行划片,再把得到的单个芯片通过引线键合或者倒装到基板上(chip-in-package),或者到 PCB 上形成板上芯片(chip-on-board,COB)或直接芯片粘接(direct chip attach,DCA)。

1998 年 7 月 13 日,Flip Chip Technologies(该公司现在叫 Flip Chip International,是华天科技的子公司)的 Elenius 和 Hollack[1] 提出了一种再布线层(redistribution layer,RDL)的工艺,可以将晶圆上芯片原来周边排布的键合焊盘进行扇入,并用焊球实现到 PCB 或者封装基板的连接(见图 3.1 和图 3.3)。同样的,通过 RDL 还可以对芯片内部的焊盘进行扇出。在诸如存储器产品的某些应用中,因为原始的焊盘是沿着芯片的中线,这样 RDL 也会在晶圆上将焊盘扇出到整个芯片上,如图 3.3 所示。RDL 实现从原始焊盘到焊球(或凸点)之间的电连接,大尺寸焊球可以制作在晶圆上的芯片范围内。引线框架、基板以及底部填充等均可以去除。

需要指出的是,虽然扇入型晶圆级封装(WLP)的概念最先由 Flip Chip International 提出[1, 8, 12],但是该领域的技术早已被日本三菱(Mitsubishi)[2, 6]、Marcous[3]、美国桑迪亚国家实验室(Sandia)[4]、以色列 ShellCase[5]、德国弗劳恩霍夫研究所(Fraunhofer IZM)[7, 9]、DiStefano[10] 以及欧洲光电产业协会(EPIC)[11] 等关注和演示。采用扇入型 WLP 被称为晶圆级芯片尺寸封装(wafer-level chip-scale package,WLCSP),其中最著名的是由 Flip Chip International 提出并获得专利授权的 UltraCSP 技术[1]。

2001 年,在 Amkor 带领下,多家外包封测公司(outsourced semiconductor assembly and test,OSAT)和晶圆厂取得了 UltraCSP 的专利许可,自此开启了晶圆级封装(WLP)的时代。在过去的 20 年里,WLCSP 已经被广泛应用在具有低成本、低端、薄型、低引脚数、小尺寸以及大量使用的产品中,如移动终端(智能手表、智能手机和平板电脑)和便携式设备(数码相机、笔记本电脑)产品,并且进行了大量的研究,尤其是针对其焊点可靠性[14-130]。必须强调,WLP 可以将晶圆上所有的芯片一次性完成封装,是一种高产率的封装技术。本章的第 1 部分将主要介绍晶圆上 RDL 的制作，并给出一个 WLCSP 的案例。

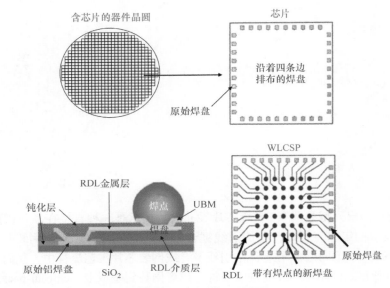

图 3.1 扇入型 WLCSP

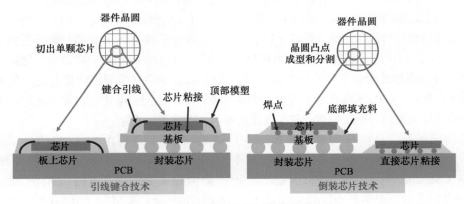

图 3.2 引线键合技术和倒装芯片技术

板级封装是一种非常高产率的封装技术。原因在于，一方面矩形芯片在矩形面板上相对于圆形晶圆要减小 10%～20% 的面积浪费；同时目前临时面板的尺寸（如 508mm×508mm）通常是大于标准尺寸晶圆的（如直径是 300mm），基本是 3.65 倍。本章的第 2 部分将描述一种利用现有的尺寸的 PCB 临时载板和相应的 PCB 设备制作板级芯片尺寸封装（panel level chip scale package，PLCSP）。

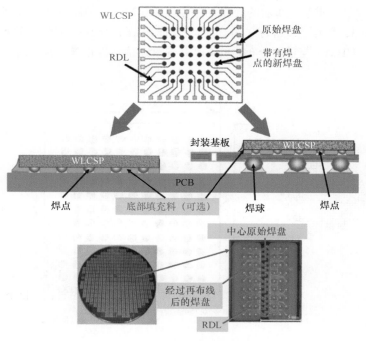

图 3.3 在 PCB 或基板上的 WLCSP

近期,一种高成本的 5 面/6 面模塑 WLCSP 技术得到越来越多的关注,其原因在于:① WLCSP 的正面介质层容易产生分层,尤其是在先进工艺节点下(<14nm)使用脆性的聚酰亚胺材料的情况下;②在机械划分过程中 WLCSP 背面容易出现崩片(chipping)和侧壁开裂(sidewall cracking);③芯片在夹持及 SMT 的拾取和放置过程中容易发生损伤;④引擎罩内经常需要持续的高/低温操作,汽车电子无铅化的趋势对新功能产品 [如先进的辅助驾驶系统(ADAS)] 提出高的焊点可靠性要求。本章第 3 部分会介绍几个 5 面/6 面模塑 WLCSP 案例。

本章第 4 部分会介绍一种 6 面模塑 PLCSP 的设计、材料、加工、组装及其可靠性问题,重点会介绍 PLCSP 技术如何在大的临时面板上同时实现不同器件晶圆上 RDL 的制作。在 RDL 制作完成后,晶圆与 PCB 面板解键合,然后去制作 6 面模塑的 PLCSP。最后,通过跌落测试、热循环测试和相关数值仿真对 PLCSP 进行可靠性评估。

3.2 扇入型晶圆级芯片尺寸封装(WLCSP)

3.2.1 封装结构

图 3.4 所示为一个带有多颗 9mm × 9mm × 0.51mm 芯片的 300mm 晶圆。在芯片边缘分布有 121 个原始焊盘,每个焊盘尺寸为 60μm × 60μm,节距为 100μm。通过在晶圆顶部制作一层额外的金属层,可以将这些原始窄节距的周边阵列的焊盘重新排布到芯片内部,变成比原先节距

大得多的面阵列焊盘。重新分布后的焊盘直径可达到 0.3mm，节距可达到 0.75mm。图 3.5 所示为 RDL 细节示意图，焊点（或焊球）由 Cu 凸台支撑，Cu 凸台经过 Cu/Ti 凸点下金属化层（under bump metallurgy，UBM）与重新分布后的 Cu/Ni 焊盘连接。再布线金属层由 Cu/Ni 制作。

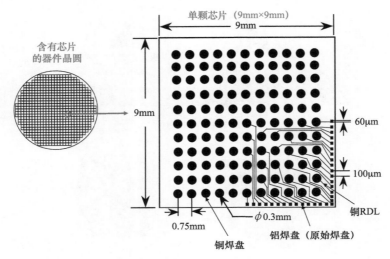

图 3.4　WLCSP 的一个案例

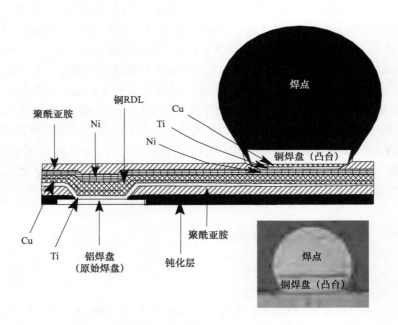

图 3.5　WLCSP 上的 RDL

3.2.2 WLCSP 的关键工艺步骤

制作 WLCSP 上的 RDL（见图 3.4 和图 3.5）的关键工艺步骤（见图 3.6）简单描述如下：

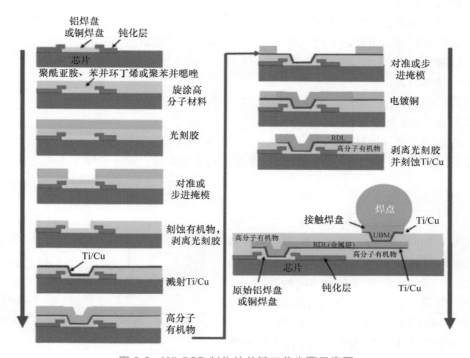

图 3.6 WLCSP 制作的关键工艺步骤示意图

1）超声清洗晶圆。

2）在晶圆上旋涂聚酰亚胺（PI）、苯并环丁烯（BCB）或聚苯并噁唑（PBO）等高分子有机物，并加热固化 1h，最终形成一层 4~7μm 厚的介质层。

3）涂覆光刻胶。

4）光刻机或者步进式光刻机。

5）通过光刻技术（对准、曝光）在光刻胶上进行通孔开窗。

6）刻蚀 PI、BCB 或 PBO。

7）剥离光刻胶。

8）在整片晶圆上溅射 Ti 和 Cu 种子层。

9）涂覆光刻胶，并通过光刻机或步进式光刻机进行光刻，对 RDL 线条进行开窗。

10）在光刻胶开窗内电镀 Cu。

11）电镀 Ni（可选）。

12）剥离光刻胶。

13）刻蚀 Ti/Cu，得到 RDL。

14）重复步骤 2（用于 UBM 制作）。
15）涂覆光刻胶并用模板在凸点目标焊盘位置进行光刻开窗，同时覆盖 RDL 线条。
16）刻蚀 PI、BCB 或 PBO。
17）剥离光刻胶。
18）在整片晶圆上溅射 Ti 和 Cu 种子层。
19）涂覆光刻胶，用模板采用光刻技术在凸点焊盘上开窗露出 UBM。
20）电镀 Cu。
21）电镀焊料。
22）剥离光刻胶。
23）刻蚀 Ti/Cu 种子层。
24）施加助焊剂并回流。

请注意，在以上过程中，若旋涂的有机物是光敏聚酰亚胺，那么步骤 3、5、7 就不是必须的。

3.2.3　WLCSP 在 PCB 上的组装

WLCSP 与 PCB 组装很容易实现。首先，通过两个镜头分别向上和向下的摄像机将带焊料凸点的 WLCSP 与 PCB 对准，然后施加一个很小的力将 WLCSP 面朝下放置在有助焊剂的 PCB 上。放置后的芯片通过回流焊炉完成回流。图 3.7 为一个典型的 WLCSP 与 PCB 组装的截面示意图，如图所示在铝焊盘上通过溅射 Ti/Cu 种子层电镀了 Cu/Ni RDL，Cu/Ni RDL 覆盖了一层聚酰亚胺进行保护。在焊点位置，Ni/Cu 焊盘上的 Cu 触点通过 Cu/Ti UBM 连接，芯片和 Cu/Ti UBM 之间通过一层聚酰亚胺介质层进行隔离。

3.2.4　WLCSP 在 PCB 上组装的热仿真

图 3.8 为带有无铅焊料凸点的 WLCSP 在 PCB 上组装的示意图。图中 PCB 厚度为 1.575mm，由 FR-4 环氧玻璃制作而成。图 3.8 中铜焊盘的厚度为 0.018mm。焊点高度为 0.1524mm，其成分是无铅焊料（96.6Sn3.5Ag 和 100In）和 62Sn36Pb2Ag 焊料合金。图 3.9 为多种焊料合金受到的剪切应力与稳态剪切应变速率关系示意图[29]。

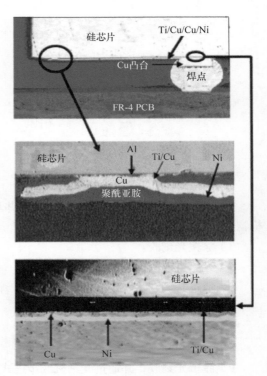

图 3.7　PCB 上不含底部填充料的 WLCSP

第 3 章 扇入型晶圆级/板级芯片尺寸封装 69

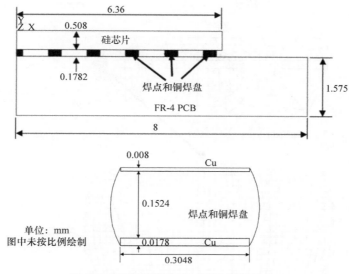

图 3.8 仿真所用的 WLCSP 模型

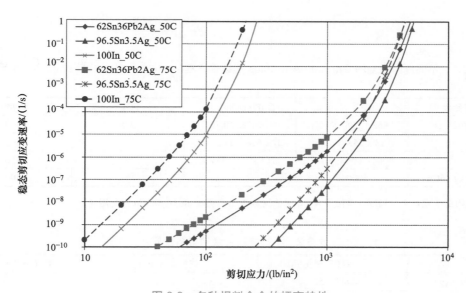

图 3.9 多种焊料合金的蠕变特性

图 3.10 所示为一个无铅焊料凸点 WLCSP 在 PCB 上组装的典型有限元蠕变分析模型。基于结构的对称性,只对一半的结构(沿着对角线)进行了建模,这是一个二维的分析。因为关注的重点是拐角焊点,所以对其采用更细的单元分割进行仿真分析。研究采用 ANSYS 软件进行有限元分析,通过 Garofalo-Arrhenius 连续性方程求解边界值(参数见表 3.1 和图 3.9)。表 3.2 列出了硅芯片、FR-4 PCB、铜和底部填充料的材料特性。

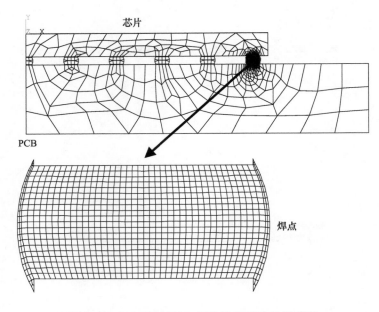

图 3.10 WLCSP PCB 组装的有限元分析模型

表 3.1 不同焊料合金的 Garofalo-Arrhenius 连续性方程 $\dfrac{d\varepsilon}{dt}=C_1\left[\sinh(C_2\sigma)\right]^{C_3}\exp\left(-\dfrac{C_4}{T}\right)$ 系数

焊料合金	C_1/(1/s)	C_2/[1/(lb/in²)]	C_3	C_4/K
62Sn36Pb2Ag	462(508−T)/T	1/(5478−10.79T)	3.3	6360
96.5Sn3.5Ag	18(553−T)/T	1/(6386−11.55T)	5.5	5802
100In	40647(593−T)/T	1/(274−0.47T)	5	8356
96.5Sn3Ag0.5Cu	500000	0.01	5	5800

注：这里的 T 是热力学温度，单位为 K；σ 的单位是 lb/in²，除了 Sn3Ag0.5Cu 在此用的是 MPa。

表 3.2 不同焊料合金的材料性质

材料	杨氏模量/MPa	泊松比（ν）	热膨胀系数（α）/(10^{-6}/℃)
62Sn36Pn2Ag	34441−152T	0.35	24.5
96.5Sn3.5Ag	52708−67.14T−0.0587T^2	0.4	21.85 + 0.02039T
100In	2200	0.4	32.1
底部填充料	9292−35.4T	0.35	31.04 + 0.0923T
硅	131000	0.3	2.8
FR4	22000	0.28	18
铜	76000	0.35	17
积层树脂	20000	0.3	50

注：这里的 T 是摄氏温度，单位为℃；电镀 Cu 假设是弹塑性材料。

对组装在 PCB 上的焊点 WLCSP 施加如图 3.11 所示的温度载荷。每一个循环周期（60min）内，温度的变化区间是 -20 ~ 110℃，温度转变时间为 15min；高温保温时间为 20min，低温保温时间为 10min。整个测试过程共包含 5 次完整的循环。

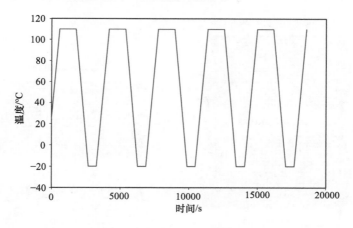

图 3.11　温度曲线（边界条件）

图 3.12a ~ e 是当焊点材料为 62Sn2Ag36Pb 时，组装在 PCB 上的 WLCSP 结构在 588s、4188s、7788s、11388s 和 14988s 时刻发生的几何变形（图中放大了 50 倍）。可以看到在温度循环过程中，由于芯片和 FR-4 PCB 的热膨胀系数不匹配（硅材料热膨胀系数为 2.8×10^{-6}/℃，FR-4 环氧玻璃基板热膨胀系数为 18×10^{-6}/℃），焊点会承受非常大的剪切形变（对于拐角的焊点尤其如此），与此同时在这些时间点（温度在 110℃）整个结构也会发生凹面变形（"笑脸"形）。

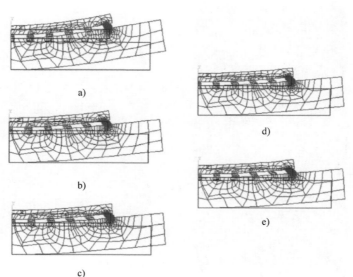

图 3.12　模型形变前后的几何形状

图 3.13a～e 和图 3.14a～e 是拐角焊点中有效蠕变应变分布的等高线图和有效应力分布的等高线图（焊点材料为 62Sn2Ag36Pb），分别代表 588s、4188s、7788s、11388s 和 14988s 时刻焊点内部发生的情况。

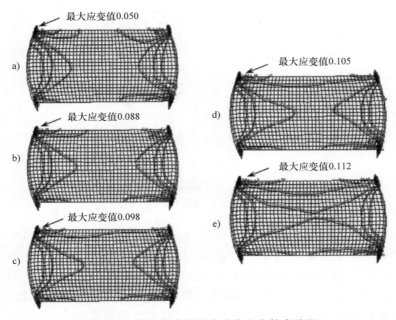

图 3.13　拐角焊点的蠕变应变分布等高线图

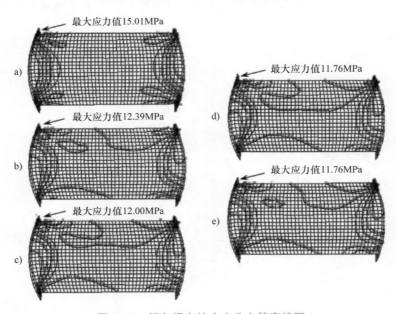

图 3.14　拐角焊点的应力分布等高线图

通过观察我们可以发现一个有趣的现象：在 588s 时，组装的最大偏移量为 0.028mm，此时拐角焊点发生的最大等效蠕变应变和等效应力分别为 0.05 和 15.01MPa。然而在 4188s，组装的最大偏移量减小到 0.022mm（减少了 21%），与此同时拐角焊点的最大等效蠕变应变增大为 0.088（增加了 78%），最大等效应力减小到 12.39MPa（减少了 17.5%）。发生这一变化的主要原因是在热循环过程中，焊点的变形增大，导致芯片和 PCB 的结合减少。

在 7788s，组装的最大偏移量减小为 0.021mm（减少了 5%），而此时拐角焊点的最大等效蠕变应变继续增加至 0.098（增加了 11%），最大等效应力继续减小为 12.00MPa（减少了 3%）。到了 11388s，组装的最大偏移量减小为 0.020mm（减少了 4.7%），此时拐角焊点的最大等效蠕变应变继续增加至 0.105（增加了 7%），最大等效应力继续减少为 11.76MPa（减少了 1.8%）。

到了 14988s，组装的最大偏移量下降为 0.0196mm（减少了 2%），拐角焊点的最大等效蠕变应变继续上升至 0.112（增加了 4.7%），最大等效应力继续下降为 11.76MPa（减少了 0.03%）。基于上述的结果，可以得出，①在经历第一次热循环后，焊点的应力和蠕变应变趋于稳定；②仿真在第 3 次热循环后得到收敛。类似的结果也出现在 96.5Sn3.5Ag 和 100In 无铅焊点的组装仿真中。

图 3.15 为 62Sn36Pb2Ag、96.5Sn3.5Ag 和 100In 三种焊料合金成分的拐角焊点剪切应力变化曲线。从图中可以看出，①在该焊点位置，所有的焊料合金焊点剪切应力变化都与所施加热循环载荷条件的变化一致；② 100In 焊料合金焊点的剪切应力变化范围远小于 62Sn36Pb2Ag 和 96.5Sn3.5Ag 焊料合金；③ 96.5Sn3.5Ag 焊料合金焊点的剪切应力变化范围稍大于 62Sn36Pb2Ag 焊料合金，这是由于 SnAg 材料的刚性（杨氏模量）要高于 SnPb 材料。

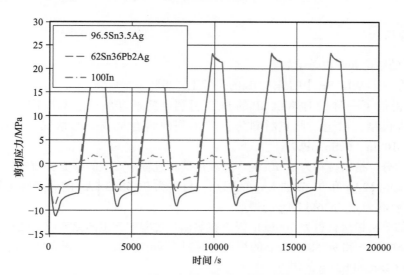

图 3.15　不同焊料合金成分的拐角焊点剪切应力随时间的变化

图 3.16 为 62Sn36Pb2Ag、96.5Sn3.5Ag 和 100In 三种焊料合金成分的拐角焊点的剪切蠕变应变曲线。从图中可以看出，①在该焊点位置，剪切蠕变应变的变化与剪切应力变化一样，都

遵循热循环载荷条件的变化；② 100In 焊料合金焊点的剪切蠕变应变变化范围远大于 62Sn-36Pb2Ag 和 96.5Sn3.5Ag 焊料合金焊点；③ 96.5Sn3.5Ag 焊料合金焊点的剪切蠕变应变变化范围要稍小于 62Sn36Pb2Ag 焊料合金。

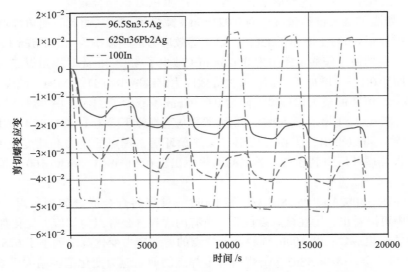

图 3.16　不同焊料合金成分的拐角焊点剪切蠕变应变随时间的变化

3.2.5　总结和建议

一些重要的结论和建议总结如下：

1）本节介绍了 WLCSP 中制备 RDL 的关键工艺步骤。

2）本节提供了一种针对 WLCSP 的 PCB 组装结构的非线性、与时间和温度相关的三维有限元分析模型，讨论了凸点材料为三种不同焊料合金成分的情况，三种合金分别为 96.5Sn3.5Ag、100In、62Sn36Pb2Ag。

3）本节的分析指出材料为 100In 的焊点在热循环过程中产生的蠕变剪切应变变化范围要大于材料为 62Sn36Pb2Ag 和 96.5Sn3.5Ag 的焊点。材料为 96.5Sn3.5Ag 的焊点蠕变剪切应变变化范围要稍小于材料为 62Sn36Pb2Ag 的焊点。

4）本节的分析指出材料为 100In 的焊点在热循环过程中产生的蠕变应力变化范围要远小于材料为 62Sn36Pb2Ag 和 96.5Sn3.5Ag 的焊点。材料为 96.5Sn3.5Ag 的焊点蠕变应力变化范围要稍大于材料为 62Sn36Pb2Ag 的焊点。

5）100In 的焊料合金适合于连接脆性材料（如 GaAs、Si 和玻璃）和低温应用场景。另一方面，96.5Sn3.5Ag 的焊料合金适合于连接如陶瓷等的硬质材料和高温应用场景（如应用于汽车引擎盖内）。

3.3 扇入型板级芯片尺寸封装（PLCSP）

板级组装与制造是一项效率非常高的封装技术。这是由于矩形的面板相比于圆形的晶圆能更好地提高表面的空间利用率（10%～20%），同时一般临时面板的面积也比标准12in晶圆（直径300mm）要大[131-140]。

多个晶圆的板级封装是更为高效的封装技术（具体取决于晶圆数量）。这主要是因为除了有大面积的面板之外，还可以省去将每个晶圆切割成单片以及将单个芯片拾取并向临时的大面板放置的步骤。多晶圆的板级封装技术目前所面临的一个瓶颈是缺少能在大面积的临时面板上完成如制备RDL和微通孔等关键工艺的设备和方法。

本节将介绍一种利用现有尺寸的临时PCB面板和相关的设备进行多晶圆板级封装的方法[85, 86]。由于大多数的面板是矩形的，所以在进行多晶圆加工时我们会先将部分或者所有的晶圆（带或不带定向平边）切成两片或者多片。晶圆切割的比例（每一片的尺寸）取决于晶圆的尺寸、可使用的面板的尺寸和相应的PCB设备。这种方法的可行性通过相应的PCB设备，在面积为20in×20in（508mm×508mm）的临时面板上制作RDL来演示。300mm直径的晶圆被切割成2片或者4片。

在本研究中，首先在20in×20in的PCB临时面板上层压双面热剥离胶膜（thermal release film）或胶带。然后将另一个制作好空腔后的PCB面板粘接到该PCB上，通过空腔将已切割或者未切割的晶圆粘到带胶带的PCB上（面朝上）。之后再依次层压上一层Ajinomoto积层膜（Ajinomoto build-up film，ABF）、激光钻孔、去胶渣、化学镀铜、干膜层压、激光直写图形（laser direct imaging，LDI）、显影、PCB电镀铜、蚀刻，然后涂覆阻焊层并完成焊盘表面处理。最后，将晶圆从临时面板解键合，进行植球，将晶圆切成单颗的常规板级芯片尺寸封装（panel-level chip-scale package，PLCSP）。因为使用了大的PCB面板且所有的设备都是PCB设备（而非半导体设备），所以该工艺不仅产率高而且成本低。对PLCSP的PCB组装进行了可靠性测试（如跌落测试、热循环测试），相应的测试过程和结果讨论（包括失效分析）将在下面进行描述和讨论。

3.3.1 测试芯片

图3.17为测试所用的芯片，芯片尺寸为5mm×5mm（在直径300mm的晶圆上共有2324颗）。芯片厚度为798μm。芯片有2排周边排布的焊盘：外侧焊盘节距为180μm，共88个；内侧焊盘节距为100μm，共160个。从图3.17还可看到，①对于外侧周边排布的焊盘，铜焊盘尺寸为130μm×130μm，Ti/Cu（0.1/0.2μm）UBM焊盘直径为110μm，钝化层（PI2）开窗直径为40μm，铜触点焊盘直径为110μm，厚度为8μm；②对于内侧周边排布的焊盘，铜焊盘尺寸为70μm×70μm，UBM焊盘直径为50μm，钝化层（PI2）开窗直径为35μm，铜触点焊盘直径为50μm，厚度为8μm。两种焊盘的铜焊盘都通过菊花链连接，用于连接的铜层厚度为3μm。

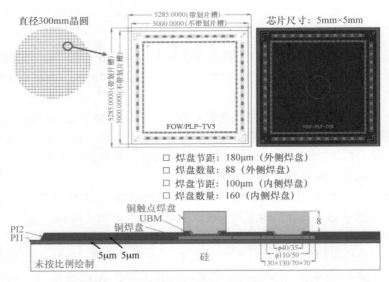

图 3.17 测试芯片的俯视图和截面图

3.3.2 测试封装体

图 3.18 所示为测试封装体,从图中可以看到仅外侧的边缘焊盘被重新排布(扇入)到了芯片内部。最终新的焊盘直径为 300μm,阻焊层开窗直径为 220μm。总共有 99 个焊盘和焊球(其中 11 个为非功能焊盘)。焊球的直径为 200μm,节距为 0.4mm。图 3.19 所示为测试封装体的截面图,可以看到一层 RDL,该 RDL 由绝缘介质层和金属导体层组成。介质层和导体层的厚度都是 20μm。RDL 金属布线的线宽/线距分别为 20/20μm,介质层通孔开窗直径为 50μm。

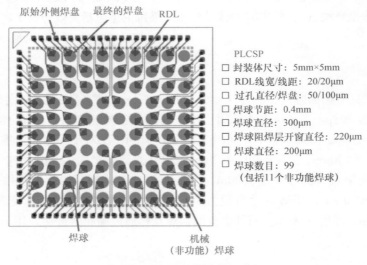

图 3.18 测试封装体的示意图

第 3 章 扇入型晶圆级／板级芯片尺寸封装

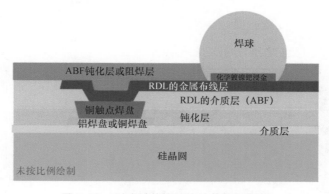

图 3.19 测试封装体结构的截面示意图

3.3.3 PLCSP 工艺流程

1. 面板和晶圆准备

如图 3.20 和图 3.21 所示,在制作 PLCSP 的 RDL 之前,需要先在两块临时面板上放置切割／未切割的晶圆。首先,在室温下将一块 PCB 面板与双面热剥离胶带进行层压。然后,在第二块 PCB 面板上开出空腔(比如直径 300mm 或直径 15mm 的圆),并将两块面板通过胶粘在一起。之后将切割或者未切割的晶圆(面朝上)嵌入面板上的空腔并通过胶粘住。这是图 3.22a 所示流程的前三步。图 3.23 所示为一块 20in × 20in 的面板,面板上放置了一片 12in (300mm) 晶圆、两片 1/2 片 12in 晶圆和一片 1/4 片 12in 晶圆。这样就可以开始制作常规 PLCSP 的 RDL 了。

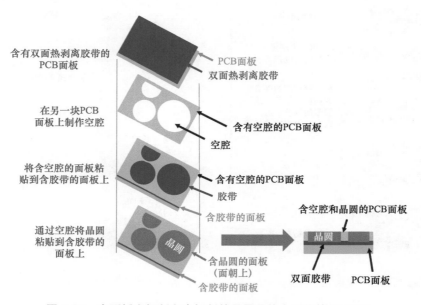

图 3.20 在面板上切割和未切割的晶圆芯片上制作的 PLCSP

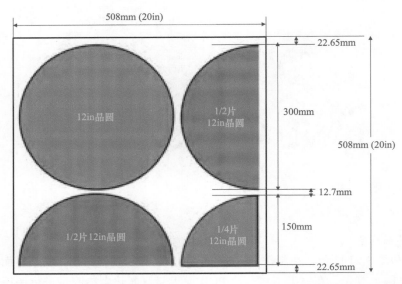

图 3.21　20in×20in 面板上切割和未切割晶圆的布局。封装是通过面板上多个晶圆的芯片制作获得

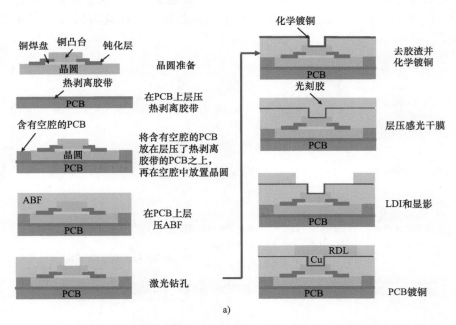

图 3.22　a）采用切割/未切割晶圆的 PLCSP 制备工艺步骤；b）PLCSP 工艺步骤

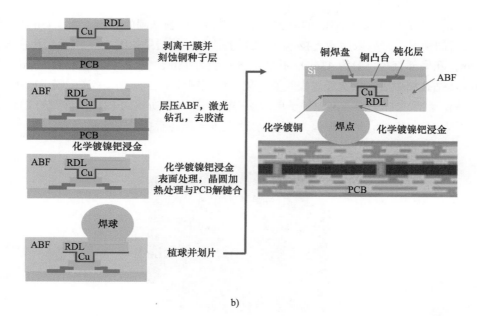

图 3.22　a）采用切割 / 未切割晶圆的 PLCSP 制备工艺步骤；b）PLCSP 工艺步骤（续）

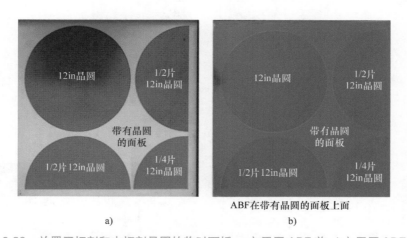

图 3.23　放置了切割和未切割晶圆的临时面板：a）层压 ABF 前；b）层压 ABF 后

2. RDL 制作

第一步，在放置有晶圆的面板上层压厚度为 20μm 的 ABF（见图 3.23a）。ABF 是一种用于 PCB/封装基板上复杂电路的电绝缘材料（介质材料）[85, 86]，它起到如同 WLCSP 中用于 RDL 介质层的聚酰亚胺的作用，其成本低于聚酰亚胺。采用 ABF 制作得到的导体布线层厚度、布线宽度和间距要远大于采用聚酰亚胺。ABF 层压包括 2 个阶段：第一阶段，在真空条件下，保持 120℃的温度 30s，然后采用 0.68MPa 的压力压合 30s；第二阶段，在 100℃的温度下，采用 0.58MPa 的压力压合 60s。

图 3.23a 所示为尚未层压 ABF 的放置好了切割和未切割晶圆的面板,图 3.23b 所示为层压了 ABF 之后的情况。之后采用激光技术在 ABF 上钻孔,其结果如图 3.24 所示。孔顶部直径一般为 63μm,孔底部直径一般为 43μm,孔的深度一般为 18μm。孔的目标直径为 50μm,目标厚度为 20μm。钻孔结束后,依次经过去胶渣、化学沉积铜种子层、层压感光干膜、LDI、显影以及 PCB 镀铜。然后去除干膜、刻蚀种子层后就得到了图 3.25 所示的 RDL。得到的 RDL 的金属线宽典型值为 16~25μm,但目标值为 20μm,因此该工艺还有提升的空间。

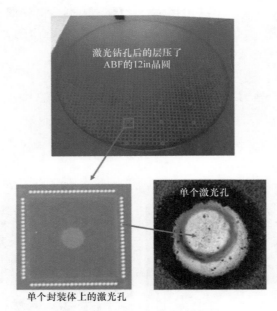

图 3.24 ABF 上的激光钻孔

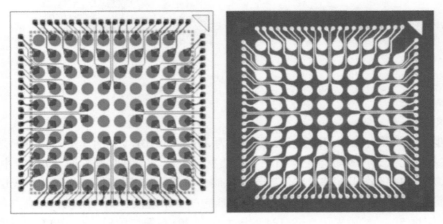

图 3.25 PLCSP 上的 RDL

3. 阻焊层开窗和植球

接下来在整个面板上再压一层厚为 20μm 的 ABF（用于制作阻焊层），依次经过激光钻孔、去胶渣和化学镀镍钯浸金（electroless nickel electroless palladium immersion gold，ENEPIG）操作，如图 3.26 和图 3.27 所示。阻焊层开窗（solder resist opening，SRO）直径目标为 0.22mm。我们一开始尝试了紫外（ultraviolet，UV）激光，但是由于 UV 激光的能量过大，激光穿透了铜焊盘（见图 3.26a）。后来我们尝试了 CO_2 激光，得到了图 3.26b 所示的可接受的结果：顶部开窗直径为 199.3μm、底部开窗直径为 157.1μm、开孔厚度为 20μm。去胶渣的工艺是在 60℃条件下用溶胀剂浸泡 2min，然后在 80℃的 $KMnO_4$ 溶液中清洗 8min，去胶渣的结果如图 3.26c～e 所示，可以看出，①在阻焊层开窗去胶渣后，大约有 3μm 的 ABF 被蚀刻掉；②绝大部分的孔底部均清洗干净；③孔顶部直径为 202.6μm，孔底部直径为 165.2μm，孔的厚度为 17μm。

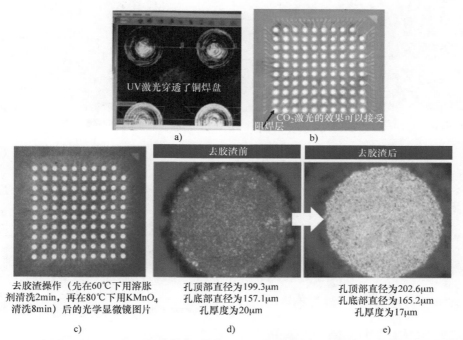

图 3.26 利用 UV 激光和 CO_2 激光分别进行阻焊层开窗以及去胶渣前后的对比。a）UV 激光穿透了铜焊盘；b）CO_2 激光钻孔的结果可以接受；c）去胶渣后的阻焊层开窗图片；d）放大图：去胶渣前；e）放大图：去胶渣后

图 3.27 所示为 ENEPIG 表面处理工艺所要用到的化学品以及对应操作条件（如温度、时间）。化学镀镍层的目标厚度为 5μm，化学镀钯层的目标厚度为 0.05μm，浸金厚度为 0.05μm。ENEPIG 工艺处理完毕之后，在 230℃下加热 10min 完成晶圆从面板上的解键合，然后在晶圆上进行植球。图 3.28 为在 1/2 片 12in 晶圆上进行植球操作所用的夹具，包括 2 块模板（1 块用于印刷助焊剂，1 块用于植球）、真空系统以及晶圆工装夹具。植球结束后就可将晶圆进行划片，

以得到单个的 PLCSP。图 3.29 所示为单个的常规 PLCSP。

工艺	清洗剂	微蚀刻剂	酸浸剂	预浸剂	活化剂	化学镀镍液	化学镀钯液	浸金液
化学品	ACL-738	SPS	H_2SO_4	H_2SO_4	MFD-5	NPR-4	TPD-21	TWX-40
温度/℃	50	30	室温	室温	30	80	50	80
目标厚度/μm	NA	0.3~0.5	NA	NA	NA	5	0.05	0.05
浸润时间/min	1	0.5	1	1	1	25	10	15

ENEPIG 后封装单元的光学显微镜图像

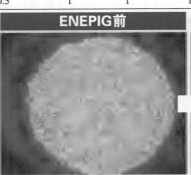
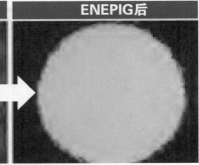
ENEPIG 前后过孔的光学显微镜图像

图 3.27　ENEPIG 表面处理

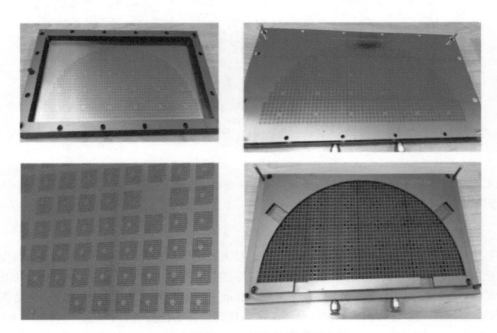

图 3.28　用于 1/2 片晶圆植球的夹具

图 3.29 单个的常规 PLCSP

3.3.4　PLCSP 的 PCB 组装

1. PCB

图 3.30 所示为扇入型板级封装所用的 PCB，是采用 FR-4 制作的。PCB 上有 42 颗（6×7）PLCSP 的组装位置，PCB 的几何尺寸为 132mm×77mm×0.65mm，共有 6 层。每个 PLCSP 对应有 99 个焊盘（焊盘节距为 0.4mm）。每个焊盘直径为 0.3mm，无阻焊层定义的焊盘表面用有机可焊性保护层（organic solderability preservative，OSP）进行处理。

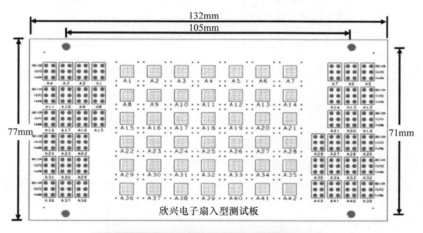

图 3.30　PLCSP 的测试板

2. PLCSP 在 PCB 上的表面安装技术

采用标准的无铅化表面安装技术（surface-mount technology，SMT）工艺实现 PLCSP 在 PCB 上的组装[99]。图 3.31 所示为回流温度曲线、组装后的 PCB、单个封装体在 PCB 上组装的 X 射线图像、截面图像以及焊点图像。可以看到 PCB 组装效果良好，通过了电学一致性测试。

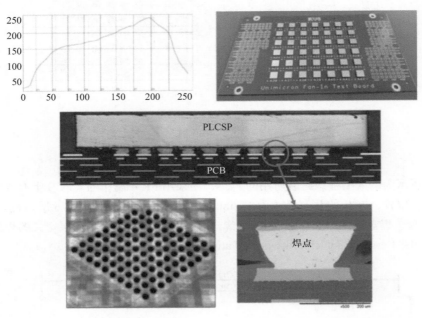

图 3.31 PLCSP 的 PCB 回流温度曲线、组装、X 射线图像和截面图像

3.3.5 PLCSP PCB 组装的跌落试验

1. 跌落测试装置和跌落谱

无底部填充的 PLCSP PCB 组装的可靠性评估采取跌落试验。试验按照 JEDEC 标准 JESD22-B111 进行设计，如图 3.32 所示。试验包括一个跌落塔和一个安装 PCB 样品的跌落平台。装置一共有 25 个独立通道，外加 1 个公共通道和 1 个数据采集系统（data acquisition system，DAS）。跌落平台的高度为 69.85mm，可产生相应 1500g/ms 的跌落谱（1500g，0.5ms 半正弦冲击脉冲），如图 3.32 所示。整个跌落试验共进行 30 次，失效判据定义为测得的样品电阻值达到 1000Ω，如图 3.33b 所示。阻值低于 1000Ω 就认为没有发生失效，如图 3.33a 所示。

2. 跌落测试结果和失效分析

样品在经过 30 次跌落测试后没有检测到任何失效迹象，其阻值始终如图 3.33a 所示保持在小于 1000Ω 的状态，因此 PLCSP PCB 组装通过了跌落试验。但是，部分未失效样品的截面图中也观察到微小的裂纹，如图 3.34 所示。当我们将测试次数增加到 50 次后，有一个样品在第 31 次跌落后发生了失效，测得的电阻（>1000Ω）如图 3.34b 所示。图 3.35 为该样品截面图，可以看到焊点已经完全断开。

第 3 章 扇入型晶圆级 / 板级芯片尺寸封装 85

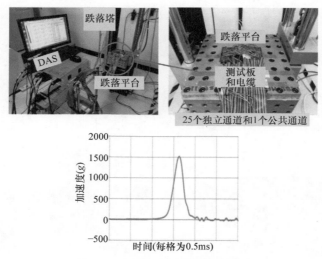

图 3.32 跌落测试装置和跌落谱

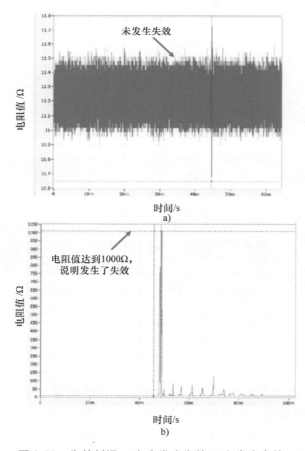

图 3.33 失效判据：a）未发生失效；b）发生失效

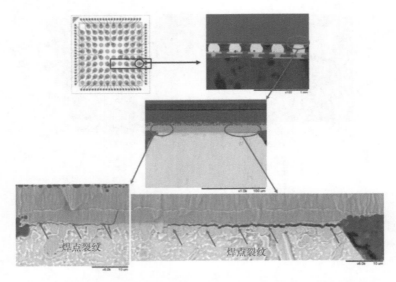

图 3.34　DAS 结果表明没有发生失效，但是实际上焊点已经发生微小开裂

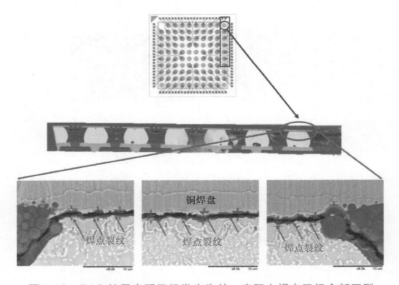

图 3.35　DAS 结果表明已经发生失效，实际上焊点已经全部开裂

3.3.6　PLCSP PCB 组装的热循环试验

如图 3.36 和图 3.37[123] 所示，将 PLCSP PCB 组装放入热循环试验箱中。试验过程中，试验箱中的温度通过热电偶实时监控并在图 3.38 中展示。从图中可以看到温度曲线的变化，每 50min 完成一个 $-55℃ \leftrightarrows 125℃$ 的循环，试验的样品数为 48。

第 3 章 扇入型晶圆级／板级芯片尺寸封装

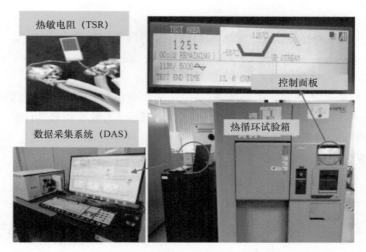

图 3.36 热循环试验箱、热电偶和数据采集系统

图 3.37 在热循环试验箱内部的 PLCSP 测试板

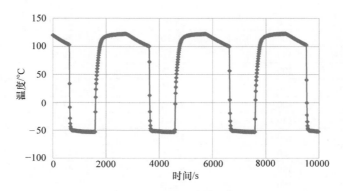

图 3.38 热循环温度曲线

1. 失效判据

热循环试验的失效判据定义为 PLCSP PCB 组装的菊花链阻值升高 50%。PLCSP 的失效周期（cycle-to-failure）定义为 PLCSP 发生第一个焊点失效的周期。

2. PLCSP PCB 组装的试验结果

PLCSP PCB 组装热循环试验结果见表 3.3。从表中可以看到 48 个样品都发生了失效。中位秩[99]可由 $F(x)=(j-0.3)/(n+0.4)$ 计算得出，其中 j 为发生失效的序数，n 代表样品数目（48）。这些数据可拟合成图 3.39 所示的 Weibull 分布，其中特征寿命（尺度参数）θ（63.2% 的样品失效）为 368 个周期，Weibull 斜率（形状参数）β 为 3.34。

表 3.3　中位秩、5% 秩、95% 秩条件下 PLCSP 热循环试验结果

失效序数	失效周期	F(x)		
		中位秩（50%）	90% 置信度	
			5% 秩	95% 秩
1	155	1.44	0.11	6.05
2	177	3.51	0.75	9.51
3	180	5.57	1.73	12.54
4	186	7.64	2.90	15.37
5	193	9.70	4.20	18.06
6	195	11.77	5.59	20.66
7	199	13.84	7.05	23.19
8	200	15.90	8.57	25.65
9	203	17.97	10.15	28.07
10	205	20.04	11.76	30.44
11	213	22.10	13.42	32.77
12	213	24.17	15.10	35.07
13	250	26.23	16.83	37.34
14	250	28.30	18.57	39.57
15	250	30.37	20.35	41.78
16	250	32.43	22.15	43.97
17	250	34.50	23.98	46.13
18	250	36.56	25.83	48.27
19	300	38.63	27.70	50.38
20	300	40.70	29.59	52.48
21	300	42.76	31.50	54.55
22	300	44.83	33.43	56.60
23	300	46.90	35.39	58.63
24	300	48.96	37.36	60.65
25	350	51.03	39.35	62.64
26	350	53.09	41.37	64.61
27	350	55.16	43.40	66.57
28	350	57.23	45.45	68.50
29	350	59.29	47.52	70.41
30	350	61.36	49.62	72.30

（续）

失效序数	失效周期	F(x)		
		中位秩（50%）	90% 置信度	
			5% 秩	95% 秩
31	400	63.43	51.73	74.17
32	400	65.49	53.87	76.02
33	400	67.56	56.03	77.85
34	400	69.62	58.22	79.65
35	400	71.69	60.43	81.43
36	400	73.76	62.66	83.17
37	450	75.82	64.93	84.90
38	450	77.89	67.23	86.58
39	450	79.16	69.56	88.24
40	450	82.02	71.93	89.85
41	450	84.09	74.35	91.43
42	450	86.15	76.81	92.95
43	500	88.22	79.40	94.41
44	500	90.29	81.94	95.80
45	500	92.35	84.63	97.10
46	500	94.42	87.46	98.27
47	500	96.48	90.49	99.25
48	500	98.55	93.95	99.89

注：样本数量为 48。

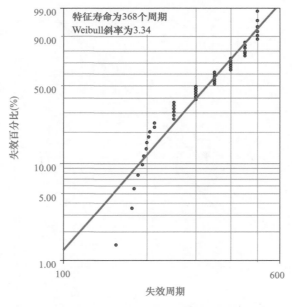

图 3.39　中位秩下 PLCSP 焊点的 Weibull 分布

PLCSP 的低可靠性是有一些原因的。第一，这是第一次使用面板作为多个切割/未切割晶圆的临时载板。第二，这是第一次采用 PCB 制造设备实现扇入型封装。第三，PLCSP 的厚度太厚（0.79mm），这样硅芯片（2.5×10^{-6}/℃）和 FR-4 PCB（18.5×10^{-6}/℃）之间有非常大的热失配。

对于一个目标秩（G），相应的失效百分比（z）可以由下式决定（z 是 n 个样品中第 j 个值的失效百分比）[99]：

$$1-(1-z)^n - nz(1-z)^{n-1} - \frac{n(n-1)}{2!}z^2(1-z)^{n-2} - \cdots$$

$$- \frac{n(n-1)\cdots(n-j+1)}{(j-1)!}z^{j-1}(1-z)^{n-j+1} = G$$

对于 90% 的置信度，需要 5% 的秩（$G = 5\%$）和 95% 的秩（$G = 95\%$），我们也将其列在了表 3.3 中。举个例子，当 $n = 48$ 且 $G = 5\%$ 时，$j = 1$ 样品的失效百分比（z）为 0.11，$j = 2$ 样品的失效百分比（z）为 0.75；相应地，当 $n = 48$ 且 $G = 95\%$ 时，$j = 1$ 样品的失效百分比（z）为 6.05，$j = 2$ 样品的失效百分比（z）为 9.51。

置信度的定义为根据测试数据给定的区间得到总体的参数的可能性（以 Weibull 分布为例，相应参数为特征寿命和 Weibull 斜率）。可以这么说，置信度的概念适用于试验本身，而可靠性的概念适用于产品！

图 3.40 为 90% 置信度下 PLCSP PCB 组装焊点的 Weibull 分布。从图中可以看出，90% 置信度下实际的特征寿命（θ_t）为 326～403 个周期。

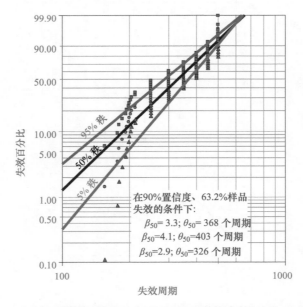

图 3.40　90% 置信度下 PLCSP PCB 组装焊点的 Weibull 分布

为预测总体的 Weibull 斜率（β_t），需要估计与样本数目、置信度有关的 Weibull 斜率误差。该误差（E）取决于失效数目（N）和选取的置信度（C），具体如图 3.41[99] 所示。对于 PLCSP，置信度为 90%、失效数目为 48 时，E 为 0.22。这样真实的 Weibull 斜率为 $2.6 \leqslant \beta_t \leqslant 4.1$。

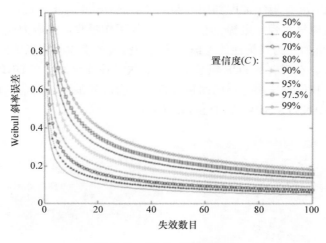

图 3.41　Weibull 斜率误差

3. PLCSP PCB 组装的失效位置和失效模式

PLCSP 焊点的失效位置和失效模式如图 3.42 所示，可以发现 PLCSP 组装的失效位置是外圈排布的邻近拐角焊盘的焊点。PLCSP PCB 组装的失效模式是邻近芯片和焊点之间界面的焊料开裂。

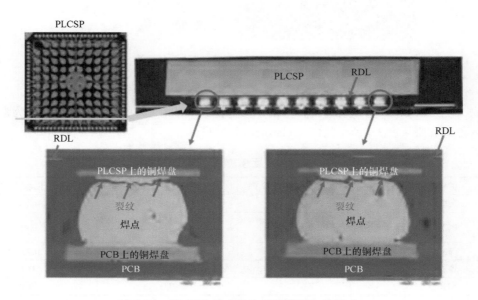

图 3.42　PLCSP 焊点的失效位置和失效模式

3.3.7 PLCSP PCB 组装的热循环仿真

1. 结构和动力学边界条件

图 3.43 为 PLCSP 的 PCB 组装示意图。由于结构在 x 轴、y 轴和对角线上存在对称性，所以在仿真时只考虑按 1/8 建模。考虑到中心距离（distance to neutral point，DNP）的影响[99]，拐角焊点不带有任何电气属性（电源/地/信号），仅用作吸收最大应力应变的非功能（机械）焊点。除了采用 1/8 模型作为经济性的考虑，还对预测失效发生的焊点的网格单元进行集中的细化。如图 3.43 所示，在现有的 PCB 组装工艺中，失效往往会发生在 DNP 最大且靠近芯片拐角的焊点处。因此，我们对这些焊点的网格进行了高密度的细化，其余焊点的网格划分较粗。仿真所用具体单元为 ANSYS 2019 的 3D Solid 186 单元。

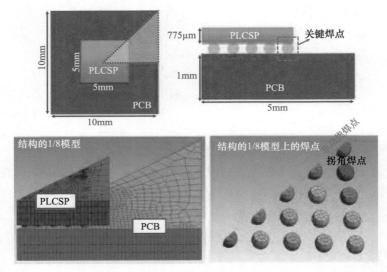

图 3.43　PLCSP 的 PCB 组装及其建模

2. 材料特性

表 3.4 为 PLCSP PCB 组装的材料特性。我们假定除焊料之外的所有材料特性都为常数，假定 Sn3Ag0.5Cu 服从表 3.1 所示的广义 Garofalo 蠕变模型。

表 3.4　普通和 6 面模塑 PLCSP 的材料特性

材料	热膨胀系数/(10⁶/℃)	杨氏模量/GPa	泊松比
铜	16.3	121	0.34
PCB	$\alpha_x = \alpha_y = 18$ $\alpha_z = 70$	$E_x = E_y = 22$ $E_z = 10$	0.28
硅	2.8	131	0.278
焊料	$21 + 0.017T$	$49 - 0.07T$	0.3
ABF	7.0	7.0	0.3

注：温度 T 的单位为℃。

3. 动力学边界条件

图3.44所示为施加在PLCSP PCB组装模型上的温度曲线。整个温度曲线共包含5次循环，温度范围是-40~85℃，每个循环是60min，其中包括升温、降温、高温保持、低温保持4个阶段，每个阶段时长都是15min。

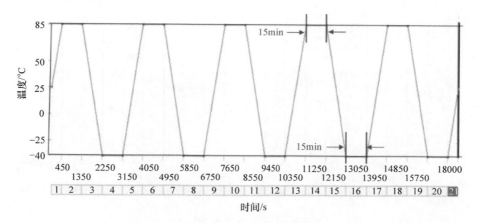

图3.44 仿真所用的温度曲线

4. 热循环仿真结果

图3.45a、b分别是450s（85℃）和2250s（-40℃）时的仿真结果，图中彩色云图代表模型发生形变后的形状，黑色线条代表模型本来的形状。从图中还可以看到，①在85℃时，PCB的膨胀量大于硅芯片，因此模型发生向外弯曲；焊点（尤其是最大DNP的拐角焊点）承受了很大的剪切形变，整个结构呈现出凹面变形（"笑脸"形）；②在-40℃时，PCB的收缩量大于硅芯片，因此模型发生向内弯曲，整个结构呈现出凸面变形（"哭脸"形）。这些结果表明，建模时的运动学和动力学边界条件假设没有错误。

最大累积蠕变应变（失效位置）发生在非功能焊点和拐角焊点（见图3.43）处。图3.46为在450s（85℃）和2250s（-40℃）时的最大累积蠕变应变云图。从图中可以观察到以下现象：

1）无论是非功能焊点还是拐角焊点，最大累积蠕变应变都发生在焊点和芯片接触的界面。该位置的蠕变应变是导致焊点最终开裂的主要驱动力，因此焊点任何的失效都应该发生在这个位置。

2）每个循环中非功能焊点的最大累积蠕变应变量（0.063）要大于拐角焊点的最大累积蠕变应变量（0.056）。因此，最好保留非功能焊点用于吸收最大应力应变以延长拐角焊点的寿命（见图3.46）。

3）图3.47为非功能焊点和拐角焊点的蠕变应变能密度随时间变化的曲线。可以看到每个周期中非功能焊点的蠕变应变能密度变化（6.9MPa）要大于拐角焊点的蠕变应变能密度变化（5.9MPa）。

事实上，图3.47对比的是有无非功能焊点情况下的蠕变应变能密度随时间的变化，可以看到，在没有非功能焊点的情况下，拐角焊点所经受的蠕变应变能更大。

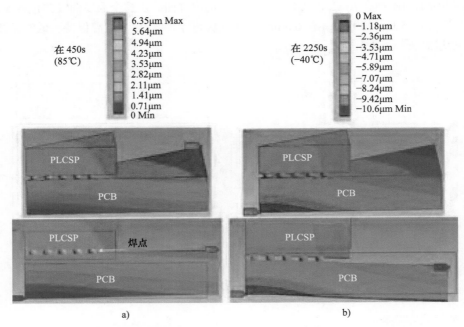

图 3.45 形变前后的模型：a）450s（85℃）；b）2250s（-40℃）

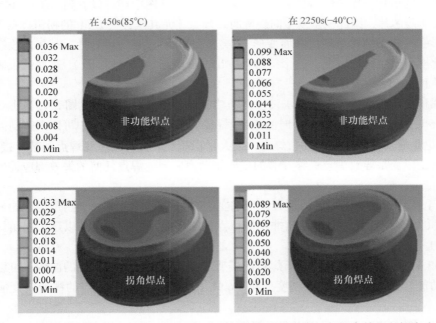

图 3.46 在 450s（85℃）和 2250s（-40℃）时的非功能焊点和拐角焊点的累积蠕变应变云图

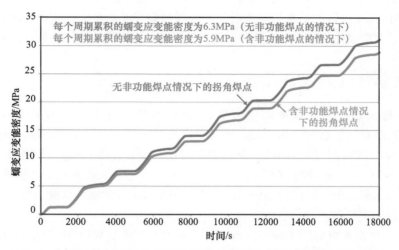

图 3.47 非功能焊点和拐角焊点中蠕变应变能密度随时间变化的曲线

3.3.8 总结和建议

一些重要的结论和建议总结如下。

1) 开发了一种非常高产率、低成本制作扇入型芯片尺寸封装的方法。该方法利用了现有的 PCB 面板和相应的 PCB 设备。由于现有的 PCB 面板是矩形的,部分器件晶圆被切割成 2 片或更多片以尽可能充分利用面板面积。

2) 设计了一种可用于测试的芯片,并制备了相应的测试晶圆。

3) 设计和演示了一种 PLCSP 的封装。

4) PLCSP 的面板和晶圆准备工作非常简单。首先,在一块面板上层压一片双面热剥离胶带;然后,在另一块面板上开出空腔并将其粘在有胶带的面板上;最后,将切割和未切割的晶圆以面朝上的形式透过空腔放置在有胶带的面板上。

5) PLCSP 的 RDL 可以完全依靠 PCB 工艺和设备制作完成。其中,RDL 的介质层材料为 ABF,通过激光钻孔和去胶渣工艺可加工得到过孔;RDL 的金属层材料为铜,布线依次通过化学镀铜种子层、涂覆感光胶膜、激光直写图形和 PCB 电镀铜得到。制得 PLCSP 的 RDL 典型线宽、线距在 16~25μm 之间,而目标参数是 20μm,因此该工艺仍有提升空间。

6) PLCSP 的阻焊层采用了 ABF 材料。对于阻焊层开窗,建议采用 CO_2 激光,这是因为 UV 激光的功率太大,容易穿透铜焊盘。焊盘表面处理工艺采用 ENEPIG。

7) 针对 PLCSP 的植球工艺,设计并制造了相应的固定装置。该装置包含真空系统、夹具和两块模板;其中一块模板用于印刷助焊剂,另一块模板用于植球。

8) 设计并制备了含有菊花链结构的 PCB 测试板,用于对 PLCSP 进行可靠性测试。

9) 实现了 PLCSP 在 PCB 测试板上的组装并成功通过了电学一致性测试。

10) 热循环试验失效分析结果表明,PLCSP 的失效位置都沿着焊点外侧发生(靠近拐角处）。失效的机制是焊料和芯片界面位置处的焊料发生机械开裂。

11）PLCSP PCB 组装的仿真结果表明，最大累积蠕变应变发生在芯片和焊料的界面位置。因此，任何失效都应发生（萌生）在这个位置。该仿真结果得到了实际热循环试验结果的印证。

3.4　6 面模塑晶圆级芯片尺寸封装

近期，一种高成本的 5 面 /6 面模塑 WLCSP 技术得到越来越多的关注，其原因在于：① WLCSP 的正面介质层容易产生分层，尤其是在先进工艺节点下（<14nm）使用脆性的聚酰亚胺材料；②在机械切割过程中 WLCSP 背面容易出现崩片（chipping）和侧壁开裂（sidewall cracking）；③ WLCSP 在夹持与 SMT 的拾取和放置过程中容易发生损伤；④引擎罩内经常需要高 / 低温操作，汽车电子无铅化的趋势对新功能产品 [如先进的辅助驾驶系统（ADAS）] 提出高的焊点可靠性要求 [87-98, 123]。

2011 年，一种由星科金朋（Statschippac）开发的 5 面模塑 WLCSP 的专利首次获得授权并公开（专利号 US8456002[87]）[56, 88-90]，具体如图 3.48 所示，该项技术被称为 eWLCSP（encapsulated WLCSP，包封式 WLCSP）。芯片的背面和侧壁都通过模塑保护，这与 eWLB（embedded wafer-level ball grid array，嵌入式晶圆级球栅阵列）或先上晶（面朝下）的扇出型晶圆级封装方式基本上是类似的 [141]。

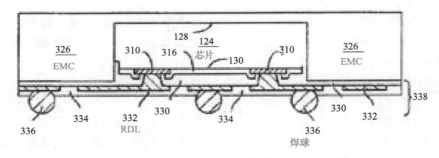

图 3.48　星科金朋的 5 面模塑 WLCSP 专利 [87]

2015 年，联合科技（UTAC）发表了 6 面保护的 WLCSP 的论文 [91]。WLCSP 的正面和侧壁都通过模塑保护，而在背面用了如层压胶带（薄膜）的传统方法进行保护。2016 年，矽品科技（SPIL）就 5 面保护（四周侧壁和芯片正面）发表了一篇类似的论文，他们将该技术称为 mWLCSP（molded WLCSP，模塑 WLCSP）[92]。他们同时提出了在芯片背面层压环氧胶膜进行保护。

2018 年开始，华天科技（Huatian Technology）发表了一系列有关 5 面或 6 面模塑 WLCSP 的论文 [93-95]，并推出了实际 6 面模塑的 WLCSP。他们的工艺与联合科技和矽品科技非常类似，只是将芯片背面也换成了用模塑进行保护。

2020 年，联发科（MediaTek）和矽品科技联合发表了一篇有关 5 面和 6 面模塑 WLCSP 的论文。他们与星科金朋使用临时晶圆的工艺类似，但在具体细节上有所差异。

本节我们简要介绍星科金朋、联合科技、矽品科技、华天科技和联发科这几家公司的 5 面或者 6 面模塑 WLCSP。

3.4.1 星科金朋的 eWLCSP

图 3.49 所示为星科金朋的 eWLCSP[88-90]，可以看出其基本上是一种面朝下先上晶的扇出型封装。如图 3.49 所示的流程，首先对晶圆进行测试并切割挑选出合格芯片（known good dies，KGD），然后将它们放在一块临时晶圆上进行重新排布，再用环氧模塑料（epoxy molding compound，EMC）对重构晶圆进行压模。接下来，将临时晶圆解键合，制作再布线层（redistribution layer，RDL）并植球，最后将重构的晶圆进行切割得到单颗 eWLCSP（见图 3.49）[88-90]。可以看到在这项工艺中芯片的四边和背面都被 EMC 所保护。

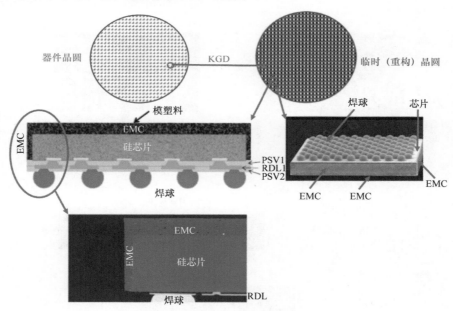

图 3.49　星科金朋的 eWLCSP[88]

3.4.2 联合科技的 WLCSP

联合科技制备 5 面模塑 WLCSP 的方式[91]与星科金朋的区别很大。它们不经过划片、KGD 拾取和放置的步骤，直接在原始的器件晶圆上制作 RDL。RDL 制作完成后再依次进行植球、划片、模塑以及研磨，最后再切割晶圆得到单颗的 WLCSP 产品（见图 3.50）。

3.4.3 矽品科技的 mWLCSP

矽品科技的 5 面模塑 WLCSP 技术（4 面侧壁 + 芯片正面）与联合科技的十分类似，他们将该技术称为 mWLCSP[92]（见图 3.51）。除此之外，矽品科技还提出过一种通过在背面层压环氧树脂薄膜进行芯片 6 面保护的封装方法。

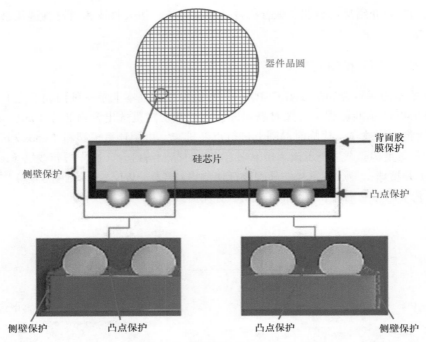

图 3.50 联合科技的 WLCSP

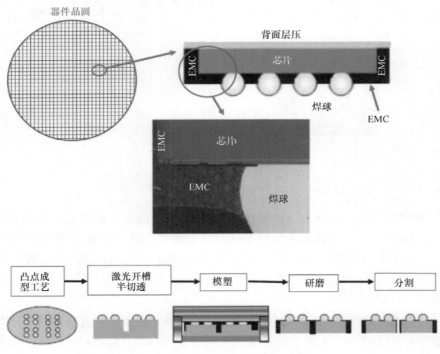

图 3.51 矽品科技的 mWLCSP[92]

3.4.4 华天科技的 WLCSP

华天科技的 6 面模塑 WLCSP 技术（4 面侧壁 + 芯片正反面）[93-95] 与联合科技 [91] 和矽品科技 [92] 的也十分类似。图 3.52 是华天科技的封装结构，图 3.53 是相应的工艺流程。采用如 ASM 的激光切割技术 [101-103]，划片道可以小至 60μm。

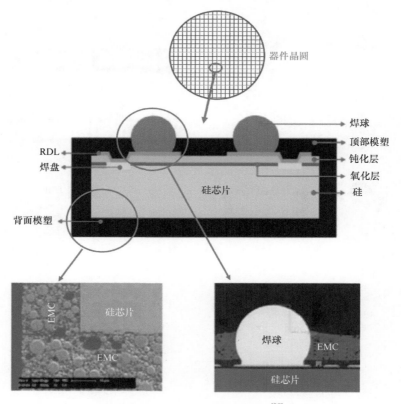

图 3.52　华天科技的 WLCSP[95]

3.4.5 矽品科技和联发科的 mWLCSP

图 3.54 为矽品科技和联发科的 5 面和 6 面模塑 mWLCSP[96]。图 3.55 和图 3.56 分别对应 5 面和 6 面模塑 WLCSP 的工艺流程。与星科金朋 [88-90] 的技术类似，都使用了临时晶圆，但是在工艺细节上有所不同 [96]。

如图 3.55 所示，对于 5 面模塑 WLCSP，首先在器件晶圆上制作 RDL 层（金属层和介质层）和 UBM 层，然后再将晶圆切割成单颗芯片。这些芯片（面朝下）拾取到临时晶圆上进行放置，并用 EMC 进行模塑。模塑完成后取下临时晶圆，植球。最后对重构后的晶圆进行划片得到单颗 5 面模塑 mWLCSP[96]。

如图 3.56 所示，对于 6 面模塑 mWLCSP 工艺，首先在器件晶圆上制作 RDL，然后将晶圆

100 半导体先进封装技术

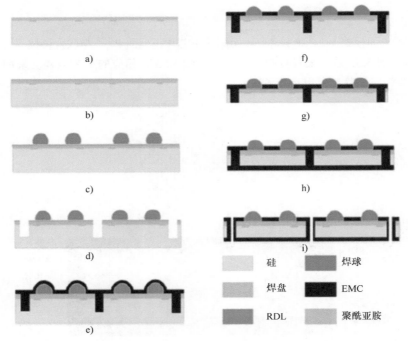

图 3.53 华天科技的 WLCSP 关键工艺步骤[95]

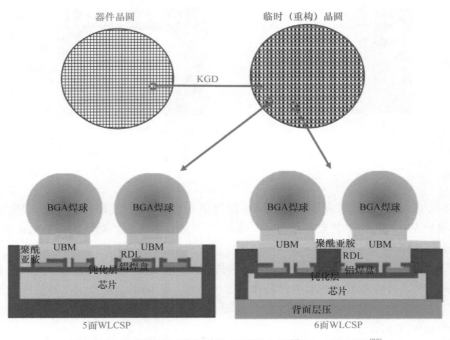

图 3.54 矽品科技/联发科的 5 面和 6 面模塑 mWLCSP[96]

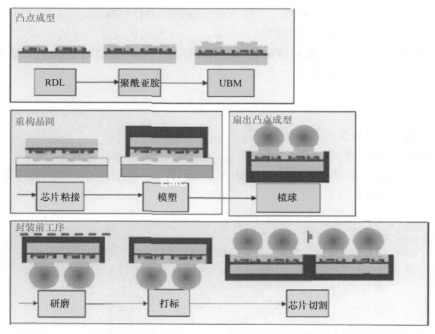

图 3.55　矽品科技/联发科的 5 面模塑 mWLCSP 的关键工艺步骤[96]

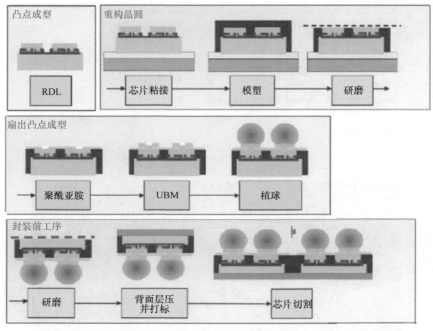

图 3.56　矽品科技/联发科的 6 面模塑 mWLCSP 的关键工艺步骤[96]

切割成单个芯片，之后将芯片（面朝上）拾取到临时晶圆上进行放置，然后用 EMC 进行模塑。然后对 EMC 进行背面磨削露出芯片的接触焊盘。之后再依次进行晶圆解键合、制作钝化层、UBM 以及进行植球操作。再之后对芯片进行背面减薄并层压背面胶膜，最后再将重构的晶圆切割就得到了最终单颗的 6 面模塑 mWLCSP[96]。

3.4.6 总结和建议

本节的一些重要结论和建议总结如下：

1）本节简要介绍了不同公司的 5 面或者 6 面模塑 WLCSP。

2）这些模塑 WLCSP 的制作方法基本可分为 2 大类：一类是通过临时晶圆排布 KGD 后再制作；另一类是直接在器件晶圆上制作。

3.5　6 面模塑板级芯片尺寸封装

在本节中我们将 3.3 节中所制备的普通 PLCSP 进一步加工成 6 面模塑 PLCSP。

3.5.1　6 面模塑 PLCSP 的结构

图 3.57 所示为所测试封装体的截面示意图[97, 98]。从图中可以看到，普通 PLCSP 的 6 个表面（4 个侧壁 + 正反面）现在已经被保护（模塑）起来。封装体中有一层 RDL，包括介质层和金属层，它们的厚度为 20μm，RDL 的金属线宽/线距为 20/20μm，RDL 介质层的开窗直径为 50μm。

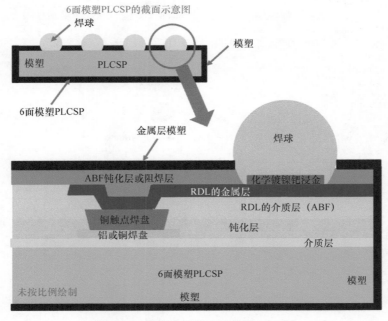

图 3.57　欣兴电子的 6 面模塑 PLCSP

第 3 章 扇入型晶圆级/板级芯片尺寸封装

图 3.58a、b 为制作 6 面模塑 PLCSP 的关键工艺步骤。在 3.3 节中已经完成了其中的大部分。在本节中，首先要做的是通过加热（230℃加热 10min）使得晶圆从面板上解键合并在晶圆上植球（见图 3.59a、b），图 3.59c 为一个普通 PLCSP 的 X 射线图像，其中焊球平均高度为 150μm。

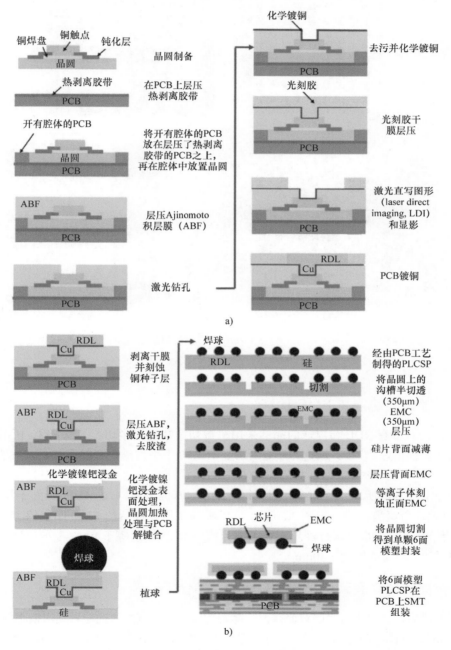

图 3.58 a) 欣兴电子的 6 面模塑 PLCSP 关键工艺步骤；b) 续图

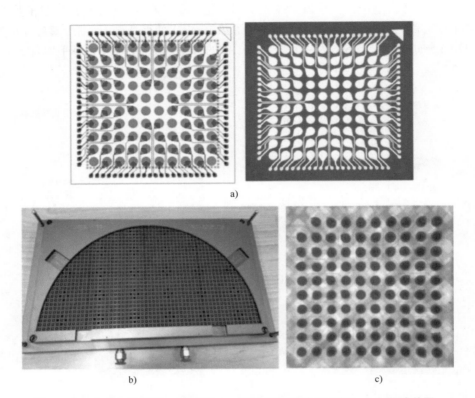

图 3.59 a）由一块 508mm×508mm 面板制作好的 PLCSP；b）植球装置；
c）带有焊球的 PLCSP 的 X 射线图像

3.5.2 晶圆正面切割和 EMC 模塑

接下来，对带有 RDL 和焊球的晶圆采用机械切割从正面半切，如图 3.60a 所示。采用激光开槽的效果会更好[101-103]，但这并非必要的。图 3.60 中平均沟槽宽度为 170μm、平均深度为 394μm。

然后如图 3.60b 所示，在晶圆正面层压 200μm 厚的 EMC（或 ABF）。由于层压时的压力作用，EMC 会填满沟槽（见图 3.60b）。可以看到层压完成后的 EMC 厚度约 150μm（不是 200μm，因为部分 EMC 材料已经填充在了划片槽中），该厚度等于 EMC 层的厚度（202.4μm）减去阻焊层和介质层的厚度（50μm）。

3.5.3 背面减薄和晶圆背面模塑

层压完成后，就要进行晶圆背面减薄。晶圆背面需要一直减薄到能看到沟槽中的 EMC（约减至 390μm），如图 3.61 所示。这样，6 面模塑 PLCSP 的芯片厚度变成 390μm，之后在晶圆背面层压一厚度为 25μm 的 EMC（ABF）。

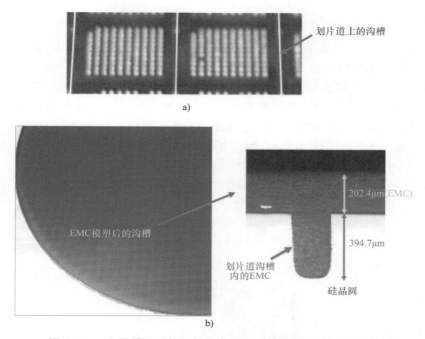

图 3.60　a）晶圆正面的切割照片；b）晶圆正面模塑后的照片

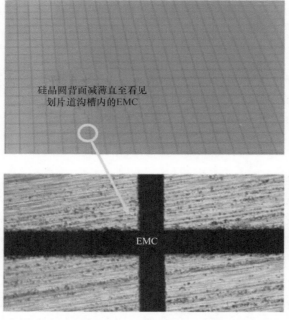

图 3.61　硅晶圆背面减薄直至看见划片道沟槽内的 EMC

3.5.4 等离子体刻蚀和划片

如图 3.58b 所示,下一步是采用等离子体刻蚀的方法将焊球上的 EMC 去除。等离子体的工艺条件为:加热温度 132℃、等离子体功率 2700W、真空度 0.33Torr[⊖]、O_2/CF_4 进气速率为 2000mL/min。该条件下刻蚀的速率约为 1.76μm/min。图 3.62a 所示为普通 PLCSP,图 3.62b 所示为带有正面 EMC 模塑的 6 面模塑 PLCSP,图 3.62c 所示为等离子体刻蚀后的 6 面模塑 PLCSP 的焊点。普通 PLCSP 的焊球的平均高度为 148μm,如图 3.62d 所示;5 面或者 6 面模塑的 PLCSP 的焊球的平均高度为 103μm,如图 3.62e 所示。

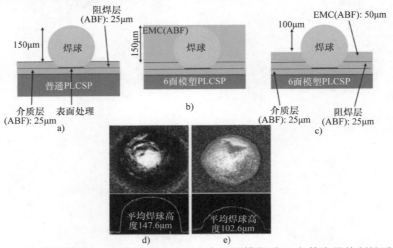

图 3.62 焊球高度:a)普通 PLCSP;b)正面模塑后;c)等离子体刻蚀后;
d)普通 PLCSP 焊球实物图;e)5 面或 6 面模塑 PLCSP 焊球实物图

最后,对晶圆进行划片得到单颗的 6 面模塑 PLCSP。图 3.63a、b 分别为 6 面模塑 PLCSP 的 3D 视图和截面视图。从图中可以看到侧壁平均模塑厚度约为 78μm,正面平均模塑厚度约为 53μm,焊球的平均站高约为 100μm。芯片厚度为 390μm。

3.5.5 测试的 PCB

如图 3.64 所示,用于 6 面模塑 PLCSP 测试的 PCB 与之前普通 PLCSP 测试的 PCB 一样。PCB 几何尺寸为 132mm × 77mm × 0.65m,是 6 层板。每个 PLCSP 共有 99 个焊盘(焊盘节距为 0.4mm)。无阻焊层定义的焊盘直径为 0.3mm,焊盘使用有机可焊性保护层(organic solderability preservative,OSP)进行了表面处理。

3.5.6 6 面模塑 PLCSP 在 PCB 上的 SMT 组装

采用无铅表面安装技术(surface-mount technology,SMT)将 6 面模塑 PLCSP 组装到测试的 PCB 上。图 3.64a 是工艺的回流温度曲线。图 3.64b 是普通 PLCSP 的 PCB 组装实物图,

---

⊖ 1Torr = 133.322Pa。

图 3.64c 是 6 面模塑 PLCSP 的 PCB 组装实物图。

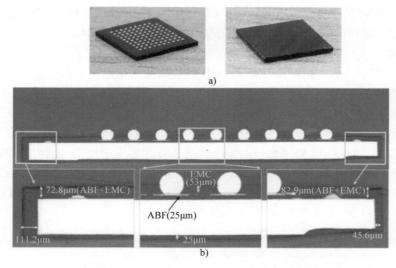

图 3.63　a）3D 视图；b）6 面模塑 PLCSP 的截面示意图

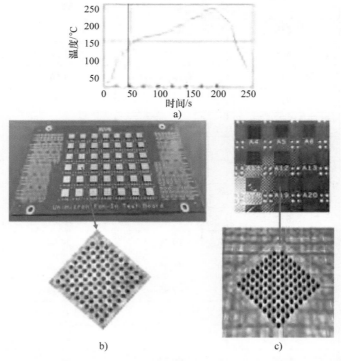

图 3.64　6 面模塑 PLCSP 的 SMT 组装：a）回流温度曲线；b）普通 PLCSP 的 PCB 组装；
c）6 面模塑 PLCSP 的 PCB 组装

3.5.7 6面模塑 PLCSP 的热循环试验

6面模塑 PLCSP 的 PCB 组装被放入图 3.36 和图 3.37 所示的热循环试验箱中[123]。图 3.38 是试验所用的温度循环曲线。试验的 6面模塑 PLCSP 样品数目为 24。

1. 失效判据

在本节的研究中，6面模塑 PLCSP 的 PCB 组装失效判据定义为其上的菊花链电阻阻值增加 50%。6面模塑 PLCSP 的失效周期定义为第一个焊点发生失效的周期。

2. 6面模塑 PLCSP 的试验结果

表 3.5 列出了 6面模塑 PLCSP 的热循环试验结果，共有 15 个样品发生了失效。表 3.5 同样也列出了中位秩、5% 秩和 95% 秩下的情况。图 3.65 是根据中位秩绘制出的 Weibull 分布图。从图中可以看到样品的特征寿命（θ）（63.2% 样品失效）为 1037 个循环周期，Weibull 斜率（β）为 5.53。图 3.66 是 90% 置信度下 6面模塑 PLCSP 的 Weibull 分布图。从图中可以看到 90% 置信度下的总体特征寿命（θ_t）范围在 958~1069 个周期之间。

表 3.5 中位秩、5% 秩和 95% 秩计算条件下的 6面模塑 PLCSP 热循环试验结果

失效序数	失效周期	F(x)		
		中位秩	90% 置信度	
			5% 秩	95% 秩
1	638	2.88	0.21	11.73
2	638	6.98	1.50	18.29
3	712	11.08	3.50	23.98
4	712	15.17	5.90	29.23
5	788	19.27	8.59	34.18
6	872	23.37	11.49	38.91
7	872	27.46	14.57	43.47
8	872	31.56	17.80	47.87
9	872	35.66	21.26	52.14
10	872	39.75	24.64	56.29
11	872	43.85	28.24	60.32
12	872	47.95	31.94	64.24
13	872	52.04	35.76	68.06
14	1096	56.14	39.68	71.76
15	1096	60.24	43.71	75.36

注：样品数目为 24。

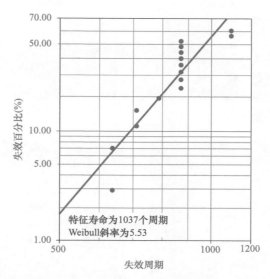

图 3.65　中位秩计算条件下 6 面模塑 PLCSP 的 Weibull 分布图

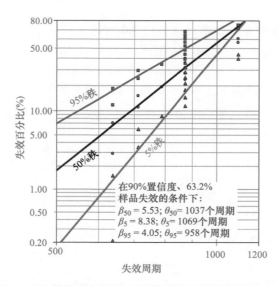

图 3.66　90% 置信度条件下 6 面模塑 PLCSP 的 Weibull 分布图

通过图 3.41，确定 $C = 90\%$、$N = 15$，得到 $E = 0.29$。因此，全体的 Weibull 斜率是 $3.93 \leqslant \beta_t \leqslant 7.13$。

3. 6 面模塑 PLCSP 的 PCB 组装失效位置以及失效模式

如图 3.67 所示，6 面模塑 PLCSP 的失效位置一般都发生在外侧（靠近拐角）的焊点。这些焊点中大部分的失效模式是在焊料与 PCB 界面发生裂纹，这与普通 PLCSP 的失效模式有很大不同（见图 3.42）。更多的细节会在热循环仿真部分进行讨论。

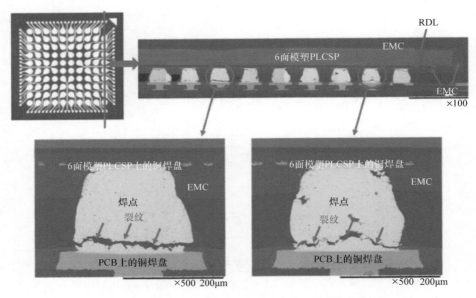

图 3.67　6 面模塑 PLCSP 焊点的失效位置和失效模式

4. 6 面模塑 PLCSP 与普通 PLCSP 焊点平均寿命对比

图 3.68 为 6 面模塑 PLCSP 与普通 PLCSP 焊点寿命的 Weibull 分布图（数据以中位秩形式给出）。从图中可以明显看出，6 面模塑 PLCSP 的特征寿命（1037 个周期）长于普通 PLCSP（368 个周期）；同时，6 面模塑 PLCSP 的平均寿命（958 个周期）也长于普通 PLCSP（330 个周期）。[平均寿命计算公式：平均寿命 = $\theta \Gamma(1+1/\beta)$]。

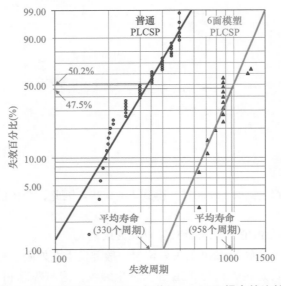

图 3.68　6 面模塑 PLCSP 与普通 PLCSP 焊点的比较

如果一个产品在试验中的表现优于另一个产品，那么如何去比较两个样品的总体情况呢？两种样品总体表现与样本表现一致的置信度（P）又有多少呢？以下是一个确定两种样品平均寿命孰优孰劣的简单方法，该方法无需掌握真实的差异[99]：

$$P = \frac{1}{1 + \dfrac{\log 1/q}{\log 1/(1-q)}}$$

式中

$$q = 1 - \frac{1}{\left[1 + \left(\dfrac{t + 4.05}{6.12}\right)^5\right]^{40/7}}$$

$$t = \frac{\sqrt{1 + \sqrt{T}}(\rho - 1)}{\rho \Omega_2 + \Omega_1}$$

$$\rho = \frac{M_2}{M_1}$$

$$T = (r_1 - 1)(r_2 - 1)$$

$$\Omega_1 = \sqrt{\frac{\Gamma\left(1 + \dfrac{2}{\beta_1}\right)}{\Gamma^2\left(1 + \dfrac{1}{\beta_1}\right)} - 1}$$

$$\Omega_2 = \sqrt{\frac{\Gamma\left(1 + \dfrac{2}{\beta_2}\right)}{\Gamma^2\left(1 + \dfrac{1}{\beta_2}\right)} - 1}$$

M_1 是样品 1（S1）的平均寿命，M_2 是样品 2（S2）的平均寿命，r_1 是 S1 中失效样品的数量，r_2 是 S2 中失效样品的数量，β_1 是 S1 的 Weibull 斜率，β_2 是 S2 的 Weibull 斜率。在我们的实验中，这些参数的值分别为 $M_1 = 958$、$M_2 = 330$、S1 = 24、S2 = 48、$r_1 = 15$、$r_2 = 48$、$\beta_1 = 5.53$、$\beta_2 = 3.34$。通过计算不难发现 $M_1/M_2 = 2.9$、$P = 0.999$，也就是说在 1000 组样品的对比中，大约有 999 组的结果都是 6 面模塑 PLCSP 焊点平均寿命长于普通 PLCSP（约 2.9 倍）。

3.5.8　6 面模塑 PLCSP 的 PCB 组装热循环仿真

1. 结构和动力学边界条件

图 3.69 为 PLCSP 的 PCB 组装示意图。由于结构在 x 轴、y 轴和对角线上存在对称性，所以在模拟时可以只考虑按 1/8 建模。考虑到中心距离（distance to neutral point，DNP）的影响[99]，拐角焊点不带有任何电气属性（电源 / 地 / 信号），仅用作吸收最大应力应变的非功能（机械）焊点。

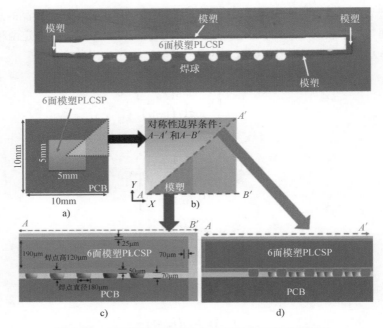

图 3.69 6 面模塑 PLCSP 的几何结构以及边界条件

除了采用 1/8 模型作为经济性的考虑，还对预测失效发生的焊点的网格单元进行集中的细化。在现有的 PCB 组装工艺中，失效往往会发生在 DNP 最大且靠近芯片拐角的焊点处，如图 3.70 所示。因此，我们对这些焊点的网格进行了高密度的细化，其余焊点的网格划分较粗，仿真所用具体单元为 ANSYS 2020 的 3-D Solid 186 单元。

2. 材料特性

表 3.1 和表 3.4 给出了 6 面模塑 PLCSP PCB 组装的材料特性。我们假设除了焊料之外所有材料的特性都是常数，假定 Sn3Ag0.5Cu 服从广义 Garofalo 蠕变模型[99]。焊料的热膨胀系数（coefficient of thermal expansion，CTE）和杨氏模量分别是 $21+0.017T$ 和 $49-0.07T$，其中温度 T 采用摄氏温度。

3. 动力学边界条件

图 3.44 是施加在 PLCSP PCB 组装模型上的温度载荷曲线。整个温度载荷曲线共包含五次循环，每次循环温度范围是 -40~85℃，每个循环所经历的时间是 60min，其中包括升温、降温、高温保持、低温保持 4 个阶段，每个阶段时长都是 15min。

4. 热循环模拟结果

图 3.71a、b 分别是 450s（85℃）和 2250s（-40℃）时的仿真结果，图中彩色云图代表模型发生形变后的形状，黑色线条代表模型本来的形状。从图中可以看到，①在 85℃时，PCB 的膨胀量大于硅芯片，因此模型发生向外弯曲；焊点（尤其是最大 DNP 的拐角焊点）承受了很大的剪切形变，整个结构呈现出凹面变形（"笑脸"形）;②在 -40℃时，PCB 的收缩量大于硅芯片，因此模型发生向内弯曲，整个结构呈现出凸面变形（"哭脸"形）。

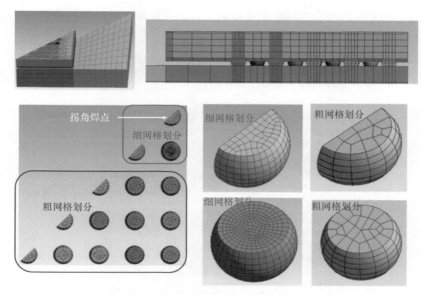

图 3.70 仿真所用的有限单元模型

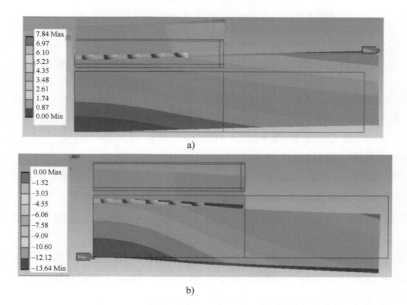

图 3.71 6 面模塑 PLCSP 形变前后的模型：a）450s（85℃）；b）2250s（-40℃）

6 面模塑 PLCSP 组装的最大累积蠕变应变（失效位置）发生在拐角焊点处（见图 3.72）。图 3.72a、b 分别为 450s（85℃）和 2250s（-40℃）时的累积蠕变应变分布图。从图中可以看出最大累积蠕变应变发生在 PCB 和拐角焊点界面位置，它也是焊点发生开裂的最初驱动力，所以焊点的任何失效都应该率先发生在这个位置。但是需要注意的是，最大蠕变应变发生的区域相对整个拐角焊点而言还是非常小的（图中红色区域）。事实上，拐角焊点剩下绝大部分区域

(图中蓝色区域)的累积蠕变应变都是非常小的(包括芯片和焊料的界面位置)。这是由于焊点的上半部分(大约 50μm)都嵌入在 EMC 中。

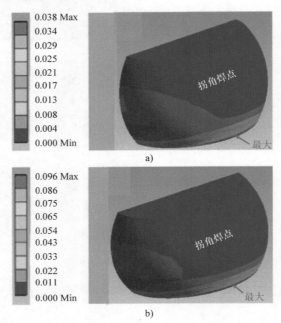

图 3.72　6 面模塑 PLCSP 的 PCB 组装拐角焊点的累积蠕变应变：a) 450s (85℃); b) 2250s (-40℃)

3.3.7 节分析了在相同边界条件下，普通 PLCSP 的 PCB 组装情况。结果表明在普通 PLCSP 样品中，最大累积蠕变应变同样发生在拐角焊点位置，这与 6 面模塑 PLCSP 样品的情况一样；但是，普通 PLCSP 样品与 6 面模塑 PLCSP 样品在拐角焊点的累积蠕变应变分布(失效模式)上有着很大不同(见图 3.46 和图 3.72)。

第一点，拐角焊点最大累积蠕变应变发生的位置已经从普通 PLCSP 中芯片和焊料的界面(见图 3.46)转移到了 6 面模塑 PLCSP 中 PCB 与焊料的界面(见图 3.72)。这一方面是因为普通 PLCSP 中芯片和焊料的热失配更加明显，另一方面也是由于 6 面模塑 PLCSP 的焊点上半部分(约 50μm)是嵌入在 EMC 中的。

第二点，普通 PLCSP 与 6 面模塑 PLCSP 拐角焊点的失效模式不同。在普通 PLCSP 中，焊点裂纹萌生于芯片和焊料的界面位置(见图 3.42 和图 3.46)，而在 6 面模塑 PLCSP 中，焊点裂纹萌生于 PCB 和焊料的界面位置(见图 3.67 和图 3.72)。

参考文献 [142, 143] 给出了一个预测热疲劳寿命的简单模型：

$$N_f = \sum_j \alpha_j \left[\frac{\sum_i \Delta W_i \times V_i}{\sum_i V_i} \right]^{\beta_j}$$

式中，N_f 是热疲劳寿命，α_j 和 β_j 针对某一特定器件/封装、焊点为常数，由实验测得(多数由等温疲劳试验测得)。ΔW_i 是第 i 个单元在每个周期中的蠕变应变能密度(或累积等效蠕变应变)

变化，V_i 是第 i 个单元的体积。

根据目前的仿真来看，拐角焊点划分成了 320 个单元。其中部分 6 面模塑 PLCSP 焊点单元的累积蠕变应变（ΔW_i）最大值（85℃下为 0.038，-45℃下为 0.96）与普通 PLCSP 拐角焊点单元（85℃下为 0.036，-45℃下为 0.99）的最大值接近，但是这些单元数量也只占很小部分。6 面模塑 PLCSP 拐角焊点的大部分单元（体积）的 ΔW_i 还是比普通 PLCSP 焊点更小。因此，可合理地给出结论：6 面模塑 PLCSP 的拐角焊点比普通 PLCSP 的拐角焊点寿命更长。比较可惜的是，上式中 α_j 和 β_j 的具体数值尚未获得，因此还不能准确预测两种焊点的热疲劳寿命。

3.5.9 总结和建议

一些重要结论和建议总结如下：

1）演示了 6 面模塑 PLCSP 在设计、材料、工艺、制备和可靠性上的可行性。

2）本节 PLCSP 的 RDL 制备过程是在一块放置了切割和未切割晶圆的 508mm×508mm 面板上进行的，面板的面积是一片 12in 晶圆的 2.25 倍，是一种十分高效的工艺方法。

3）PLCSP 的 RDL 制备全部采用 PCB 制备的工艺和设备，制备成本低。

4）在制作完 RDL 后，晶圆从面板上解键合。然后进行植球、从正面半切晶圆得到切割沟槽、晶圆正面层压 EMC、晶圆背面磨削露出切割沟槽、背面层压 EMC、等离子体刻蚀 EMC 露出焊球，之后将晶圆切割成单颗的 6 面模塑 PLCSP。

5）6 面模塑 PLCSP 的失效位置发生在外侧（靠近拐角处）的焊点。失效模式是在 PCB 和焊料界面位置产生裂纹。

6）通过比较 6 面模塑 PLCSP 焊点和普通 PLCSP 焊点的寿命（前者 958 个循环，后者 330 个循环），得出结论：在 1000 组试验中，会约有 999 组试验 6 面模塑 PLCSP 的平均寿命是普通 PLCSP 的 2.9 倍。

7）构建了一种与温度、时间有关的非线性 3D 有限模型，研究得出在 6 面模塑 PLCSP 的 PCB 组装中，最大累积蠕变应变发生在 PCB 和焊料的界面位置。因此所有的失效都应萌生于该位置，这得到了实际热循环试验结果的失效模式验证。

8）6 面模塑 PLCSP 中焊点的失效模式从原先芯片和焊料界面转变到了 PCB 和焊料界面，原因在于 6 面模塑 PLCSP 焊点上半部分受到了 EMC 模塑的保护。

9）6 面模塑 PLCSP 和普通 PLCSP 拐角焊点的累积蠕变应变最大值基本一致。但是，在 6 面模塑 PLCSP 中，这些最大值单元体积仅占焊点整体的很小一部分；且 6 面模塑 PLCSP 焊点中大部分区域的累积蠕变应变值都小于普通 PLCSP 焊点。因此可以推断，6 面模塑 PLCSP 的热疲劳寿命要长于普通 PLCSP。

参考文献

1. Elenius, P., and H. Hollack, "Method for forming chip scale package," *US patent 6,287,893*, filed on July 13, 1998; patented on September 11, 2001.
2. Yasunaga, M., "Chip-scale package: a lightly dressed LSI chip," *Proc. of IEEE/CPMT IEMTS*, 1994, pp. 169–176.

3. Marcoux, P., "A minimal packaging solution for known good die and direct chip attachment," *Proc. Of SMTC*, 1994, pp. 19–26.
4. Chanchani, R., "A new mini ball grid array (m-BGA) multichip module technology," *Proc. Of NEPCON West*, 1995, pp. 938–945.
5. Badihi, A., "Shellcase—a true miniature integrated circuit package," *Proc. of International FC, BGA, Adv. Packaging Symp.*, 1995, pp. 244–252.
6. Baba, S., et al., "Molded chip-scale package for high pin count," *Proc. of IEEE/ECTC*, 1996, 1251–1257.
7. Topper, M., "Redistribution technology for chip scale package using photosensitive BCB," *Future Fab International*, 1996, pp. 363–368.
8. Elenius, P., "FC2SP-(Flip Chip-Chip Size Package),"*Proc. of NEPCON West*, 1997, pp. 1524–1527.
9. Auersperg, J., "Reliability evaluation of chip-scale packages by FEA and microDAC," *Proc. of Symp. On Design and Reliability of Solder and Solder Interconnections*, TMS Annual Meeting, 1997, pp. 439–445.
10. DiStefano, T., "Wafer-level fabrication of IC packages," *Chip Scale Review*, 1997, pp. 20–27.
11. Kohl, J. E., "Low-cost chip scale packaging and interconnect technology," *Proc. of the CSP Symp.*, 1997, pp. 37–43.
12. Elenius, P., "Flip-chip bumping for IC packaging contractors," *Proc. of NEPCON West*, 1998, pp. 1403–1407.
13. Lau, J. H., and S. W. R. Lee, *Chip Scale Package*, McGraw-Hill Book Company, New York, 1999.
14. Lau, J. H., T. Chung, R. Lee, C. Chang, and C. Chen, "A Novel and Reliable Wafer-Level Chip Scale Package (WLCSP)", *Proceedings of the Chip Scale International Conference*, SEMI, September 1999, pp. H1–8.
15. Lau, J. H., S. W. R. Lee, and C. Chang, "Solder Joint Reliability of Wafer Level Chip Scale Packages (WLCSP): A Time-Temperature-Dependent Creep Analysis", *ASME Transactions, Journal of Electronic Packaging*, Vol. 122, No. 4, May 2000, pp. 311–316.
16. Lau, J. H., "Critical Issues of Wafer Level Chip Scale Package (WLCSP) with Emphasis on Cost Analysis and Solder Joint Reliability", *IEEE Transactions on Electronics Packaging Manufacturing*, Vol. 25, No. 1, 2002, pp. 42–50.
17. Lau, J. H., and R. Lee, "Effects of Build-Up Printed Circuit Board Thickness on the Solder Joint Reliability of a Wafer Level Chip Scale Package (WLCSP)," *IEEE Transactions on Components & Packaging Technologies*, Vol. 25, No. 1, March 2002, pp. 3–14.
18. Lau, J. H., S. Pan, and C. Chang, "A New Thermal-Fatigue Life Prediction Model for Wafer Level Chip Scale Package (WLCSP) Solder Joints", *ASME Transactions, Journal of Electronic Packaging*, Vol. 124, September 2002, pp. 212–220.
19. Lau, J. H., and R. Lee, "Modeling and Analysis of 96.5Sn-3.5Ag Lead-Free Solder Joints of Wafer Level Chip Scale Package (WLCSP) on Build-Up Microvia Printed Circuit Board," *IEEE Transactions on Electronics Packaging Manufacturing*, Vol. 25, No. 1, 2002, pp. 51–58.
20. Lau, J. H., R. Lee, S. Pan, and C. Chang, "Nonlinear Time-Dependent Analysis of Micro Via-in-Pad Substrates for Solder Bumped Flip Chip Applications," *ASME Transactions, Journal of Electronic Packaging*, Vol. 124, September 2002, pp. 205–211.
21. Lau, J. H., C. Chang, and R. Lee, "Solder Joint Crack Propagation Analysis of Wafer-Level Chip Scale Package on Printed Circuit Board Assemblies," *IEEE Transactions on Components & Packaging Technologies*, Vol. 24, No. 2, 2001, pp. 285–292.
22. Lau, J. H., and R. Lee, "Computational Analysis on the Effects of Double-Layer Build-Up Printed Circuit Board on the Wafer Level Chip Scale Package (WLCSP) Assembly with Pb-Free Solder Joints," *International Journal of Microcircuits & Electronic Packaging, IMAPS Transactions*, Vol. 24, No. 2, 2001, pp. 89–104.
23. Lau, J. H., and R. Lee, "Effects of Microvia Build-Up Layers on the Solder Joint Reliability of a Wafer Level Chip Scale Package (WLCSP)," *IEEE Proceedings of Electronic Components & Technology Conference*, May 29-June 1, Orlando, Florida, U.S.A., 2001, pp. 1207–1215.

24. Lau, J. H., and R. Lee, "Reliability of 96.5Sn-3.5Ag Lead-Free Solder-Bumped Wafer Level Chip Scale Package (WLCSP) on Build-Up Microvia Printed Circuit Board," *Proceedings of the 2nd International Conference on High Density Interconnect and System Packaging*, April 17–20, Santa Clara, California, U.S.A., 2001, pp. 314-322.
25. Lau, J. H., and R. Lee, "Effects of Build-Up Printed Circuit Board Thickness on the Solder Joint Reliability of a Wafer Level Chip Scale Package (WLCSP)," *Proceeding of the International Symposium on Electronic Materials & Packaging*, November 30-December 2, Kowloon, Hong Kong, China, 2000, pp. 115–126.
26. Lau, J. H., S. Pan, and C. Chang, "Nonlinear Fracture Mechanics Analysis of Wafer-Level Chip Scale Package Solder Joints with Creaks", *Proceedings of IMAPS Microelectronics Conference*, Boston, MA, September 2000, pp. 857–865.
27. Lau, J. H., and R. Lee, "Reliability of Wafer Level Chip Scale Package (WLCSP) with 96.5Sn-3.5Ag Lead-Free Solder Joints on Build-Up Microvia Printed Circuit Board," *Proceeding of the International Symposium on Electronic Materials & Packaging*, November 30-December 2, Kowloon, Hong Kong, China, 2000, pp. 55–63.
28. Lau, J. H., S. Pan, and C. Chang, "A New Thermal-Fatigue Life Prediction Model for Wafer Level Chip Scale Package (WLCSP) Solder Joints", *Proceeding of the 12th Symposium on Mechanics of SMT & Photonic Structures*, ASME International Mechanical Engineering Congress & Exposition, November 5–10, Orlando, Florida, USA, 2000, pp. 91-101.
29. Lau, J. H., S. Pan, and C. Chang, "Creep Analysis of Wafer Level Chip Scale Packages (WLCSP) with 96.5Sn-3.5Ag and 100In Lead-Free Solder Joints and Microvia Build-Up Printed Circuit Board", *Proceeding of the 12th Symposium on Mechanics of SMT & Photonic Structures*, ASME International Mechanical Engineering Congress & Exposition, November 5–10, Orlando, Florida, USA, 2000, pp. 79-89.
30. Lau, J. H., C. Chang, and R. Lee, "Solder Joint Crack Propagation Analysis of Wafer-Level Chip Scale Package on Printed Circuit Board Assemblies," *IEEE Proceeding of the 50th Electronic Components & Technology Conference*, Las Vegas, NA, 2000, pp. 1360–1368.
31. Lau, J. H. and R. Lee, "Fracture Mechanics Analysis of Low Cost Solder Bumped Flip Chip Assemblies with Imperfect Underfills," *Proceedings of NEPCON West*, Anaheim, CA, 2000, pp. 653–660.
32. Lau, J. H., T. Chung, T., R. Lee, and C. Chang, "A Low Cost and Reliable Wafer Level Chip Scale Package," *Proceedings of. NEPCON West*, Anaheim, CA, 2000, pp. 920–927.
33. Lau, J. H., Lee, S.W.R., Ouyang, C., Chang, C. and Chen, C.C., "Solder Joint Reliability of Wafer Level Chip Scale Packages (WLCSP): A Time-Temperature-Dependent Creep Analysis," ASME Winter Annual Meeting, *ASME Paper No. 99-IMECE/EEP-5,* Nashville, TN, 1999.
34. Lau, J. H., C. Ouyang, and R. Lee, "A Novel and Reliable Wafer-Level Chip Scale Package (WLCSP)", *Proceedings of Chip Scale International Conference*, San Jose, CA, September 1999, pp. H1-H9.
35. Chen, C., K. H. Chen, Y. S. Wu, P. H. Tsao and S. T. Leu, "WLCSP Solder Ball Interconnection Enhancement for High Temperature Stress Reliability", *IEEE/ECTC Proceedings*, May 2020, pp. 1212–1217.
36. Zhang, H., Z. Wu, J. Malinowski, M. Carino, K. Young-Fisher, J. Trewhella, and P. Justison, "45RFSOI WLCSP Board Level Package Risk Assessment and Solder Joint Reliability Performance Improvement", *IEEE/ECTC Proceedings*, May 2020, pp. 2151–2156.
37. Ma, S., Y. Liu, F. Zheng, F. Li, D. Yu, A. Xiao, and X. Yang, "Development and Reliability study of 3D WLCSP for automotive CMOS image sensor using TSV technology", *IEEE/ECTC Proceedings*, May 2020, pp. 461–466.
38. Machani, K., F. Kuechenmeister, D. Breuer, C. Klewer, J. Cho, and K. Fisher, "Chip Package Interaction (CPI) risk assessment of 22FDX® Wafer Level Chip Scale Package (WLCSP) using 2D Finite Element Analysis modeling", *IEEE/ECTC Proceedings*, May 2020, pp. 1100–1105.
39. Chiu, J., K.C. Chang, S. Hsu, P. Tsao and M. J. Lii, "WLCSP Package and PCB Design for Board Level Reliability", *IEEE/ECTC Proceedings*, May 2019, pp. 763–767.

40. Yu, D., Y. Zou, X. Xu, A. Shi, X. Yang, and Z. Xiao, "Development of 3D WLCSP with Black Shielding for Optical Finger Print Sensor for the Application of Full Screen Smart Phone", *IEEE/ECTC Proceedings*, May 2019, pp. 884–889.
41. Zhou, Y., L. Chen, Y. Liu, and S. Sitaraman, "Thermal Cycling Simulation and Sensitivity Analysis of Wafer Level Chip Scale Package with Integration of Metal-Insulator-Metal Capacitors", *IEEE/ECTC Proceedings*, May 2019, pp. 1521–1528.
42. Chou, P., H. Hsiao, and K. Chiang, "Failure Life Prediction of Wafer Level Packaging using DoS with AI Technology", *IEEE/ECTC Proceedings*, May 2019, pp. 1515–1520.
43. Chen, Z., B. Lau, Z. Ding, E. Leong, C. Wai, B. Han, L. Bu, H. Chang, and T. Chai, "Development of WLCSP for Accelerometer Packaging with Vertical CuPd Wire as Through Mold Interconnection (TMI)", *IEEE/ECTC Proceedings*, May 2018, pp. 1188–1193.
44. Tsao, P. H., T. H. Lu, T. M. Chen, K. C. Chang, C. M. Kuo, M. J. Lii and L. H. Chu, "Board Level Reliability Enhancement of WLCSP with Large Chip Size", *IEEE/ECTC Proceedings*, May 2018, pp. 120–1205.
45. Ramachandran, V., K. C. Wu, C. C. Lee, and K. N. Chiang, "Reliability Life Assessment of WLCSP Using Different Creep Models", *IEEE/ECTC Proceedings*, May 2018, pp. 1017–1022.
46. Sheikh, M., A. Hsiao, W. Xie, S. Perng, E. Ibe, K. Loh, and T. Lee, "Multi-axis loading impact on thermo-mechanical stress-induced damage on WLCSP and components with via-in pad plated over (VIPPO) board design configuration", *IEEE/ECTC Proceedings*, May 2018, pp. 911–915.
47. Tsao, P. H., T. M. Chen, Y. L. Kuo, C. M. Kuo, S. Hsu, M. J. Lii, and L. H. Chu, "Investigation of Production Quality and Reliability Risk of ELK Wafer WLCSP Package", *IEEE/ECTC Proceedings*, May 2017, pp. 371–375.
48. Lin, W., Q. Pham, B. Baloglu, and M. Johnson, "SACQ Solder Board Level Reliability Evaluation and Life Prediction Model for Wafer Level Packages", *IEEE/ECTC Proceedings*, May 2017, pp. 1058–1064.
49. Yang, S., C. Chen, W. Huang, T. Yang, G. Huang, T. Chou, C. Hsu, C. Chang, H. Huang, C. Chou, C.. Ku, C. Chen, C. Chen, K. Liu, A. Kalnitsky, and M. Liao, "Implementation of Thick Copper Inductor Integrated into Chip Scaled Package", *IEEE/ECTC Proceedings*, May 2017, pp. 306–311.
50. Lee, T., Y. Chang, C. Hsu, S. Hsieh, P. Lee, Y. Hsieh, L. Wang, and L. Zhang, "Glass Based 3D-IPD Integrated RF ASIC in WLCSP", *IEEE/ECTC Proceedings*, May 2017, pp. 631–636.
51. Hsu, M., K. Chiang, C. Lee, "A Modified Acceleration Factor Empirical Equation for BGA Type Package", *IEEE/ECTC Proceedings*, May 2017, pp. 1020–1026.
52. Jalink, J., R. Roucou, J. Zaa, J. Lesventes, R. Rongen, "Effect of PCB and Package Type on Board Level Vibration using Vibrational Spectrum Analysis", *IEEE/ECTC Proceedings*, May 2017, pp. 470–475.
53. Xu, J., Z. Ding, V. Chidambaram, H. Ji, and Y. Gu, "High Vacuum and High Robustness Al-Ge Bonding for Wafer Level Chip Scale Packaging of MEMS Sensors", *IEEE/ECTC Proceedings*, May 2017, pp. 956–960.
54. Max K., C. Wu, C. Liu, and D. Yu, "UFI (UBM-Free Integration) Fan-In WLCSP Technology Enables Large Die Fine Pitch Packages", *IEEE/ECTC Proceedings*, May 2017, pp. 1154–1159.
55. Takyu, S., Y. Fumita, D. Yamamoto, S. Yamashita, K. Furuta, Y. Yamashita, K. Tanaka, N. Uchiyama, T. Ogiwara, and Y. Kondo, "A Novel Dicing Technologies for WLCSP Using Stealth Dicing through Dicing Tape and Back Side Protection-Film", *IEEE/ECTC Proceedings*, May 2016, pp. 1241–1246.
56. Lin, Y., E. Chong, M. Chan, K. Lim, and S. Yoon, "WLCSP + and eWLCSP in Flex-Line: Innovative Wafer Level Package Manufacturing", *IEEE/ECTC Proceedings*, May 2015, pp. 865–870.
57. Chen, J. H., Y. L. Kuo, P. H. Tsao, J. Tseng, M. Chen, T. M. Chen, Y. T. Lin, and A. Xu, "Investigation of WLCSP Corrosion Induced Reliability Failure on Halogens Environment for Wearable Electronics", *IEEE/ECTC*, May 2015, pp. 1599–1603.

58. Chatinho, V., A. Cardoso, J. Campos, and J. Geraldes, "Development of Very Large Fan-In WLP/ WLCSP for Volume Production", *IEEE/ECTC*, May 2015, pp. 1096–1101.
59. Nomura, H., K. Tachibana, S. Yoshikawa, D. Daily, and A. Kawa, "WLCSP CTE Failure Mitigation via Solder Sphere Alloy", *IEEE/ECTC*, May 2015, pp. 1257–1261.
60. Yang, S., C. Wu, Y. Hsiao, C. Tung, D. Yu, "A Flexible Interconnect Technology Demonstrated on a Wafer-Level Chip Scale Package", *IEEE/ECTC*, May 2015, pp. 859–864.
61. Yang, S., B. Tsai, C. Lin, E. Yen, J. Lee, W. Hsieh, and V. Wu, "Advanced Multi-sites Testing Methodology after Wafer Singulation for WLPs Process", *IEEE/ECTC*, May 2015, pp. 871–876.
62. Keser, B., R. Alvarado, M. Schwarz, and S. Bezuk, "0.35 mm Pitch Wafer Level Package Board Level Reliability: Studying Effect of Ball De-population with Varying Ball Size", *IEEE/ECTC*, May 2015, pp. 1090–1095.
63. Arumugam, N., G. Hill, G. Clark, C. Arft, C. Grosjean, R. Palwai, J. Pedicord, P. Hagelin, A. Partridge, V. Menon, and P. Gupta, "2-die Wafer-level Chip Scale Packaging enables the smallest TCXO for Mobile and Wearable Applications", *IEEE/ECTC*, May 2015, pp. 1338–1342.
64. Liu, Y., Y. Liu, and S. Qu, "Bump Geometric Deviation on the Reliability of BOR WLCSP", *IEEE/ECTC Proceedings*, May 2014, pp. 808–814.
65. Anzai, N., M. Fujita, and A. Fujii, "Drop Test and TCT Reliability of Buffer Coating Material for WLCSP", *IEEE/ECTC Proceedings*, May 2014, pp. 829–835.
66. Cui, T., A. Syed, B. Keser, R. Alvarado, S. Xu, and M. Schwarz, "Interconnect Reliability Prediction for Wafer Level Packages (WLP) for Temperature Cycle and Drop Load Conditions", *IEEE/ECTC Proceedings*, May 2014, pp. 100–107.
67. Keser, B., R. Alvarado, A. Choi, M. Schwarz, and S. Bezuk, "Board Level Reliability and Surface Mount Assembly of 0.35 mm and 0.3 mm Pitch Wafer Level Packages", *IEEE/ECTC Proceedings*, May 2014, pp. 925–930.
68. Xiao, Z., J. Fan, Y. Ren, Y. Li, X. Huang, D. Yu, W. Zhang, "Development of 3D Thin WLCSP Using Vertical Via Last TSV Technology with Various Temporary Bonding Materials and Low Temperature PECVD Process", *IEEE/ECTC Proceedings*, May 2017, pp. 302–309.
69. Zoschke, K., M. Klein, R. Gruenwald, C. Schoenbein, And K. Lang, "LiTaO3 Capping Technology for Wafer Level Chip Size Packaging of SAW Filters", *IEEE/ECTC Proceedings*, May 2017, pp. 889–896.
70. Kuo, F., J. Chiang, K. Chang, J. Shu, F. Chien, K. Wang and R. Lee, "Studying The Effect Of Stackup Structure Of Large Die Size Fan-in Wafer Level Package At 0.35 mm Pitch With Varying Ball Alloy To Enhance Board Level Reliability Performance", *IEEE/ECTC Proceedings*, May 2017, pp. 140–146.
71. Tsou, C., T. Chang, K. Wu, P. Wu and K. Chiang, "Reliability Assessment using Modified energy based model for WLCSP Solder Joints", *IEEE/ICEP2017*, Yamagata, Japan, April 2017.
72. Rogers, B., and C. Scanlan, "Improving WLCSP Reliability Through Solder Joint Geometry Optimization" *International Symposium on Microelectronics*, October 2013, pp. 546–550,
73. Hsieh, M. C., "Modeling correlation for solder joint fatigue life estimation in wafer-level chip scale packages", *International Microsystems, Packaging, Assembly and Circuits Technology Conference (IMPACT)*, Oct. 2015, pp. 65–68.
74. Hsieh, M. C., and S. L. Tzeng, "Solder joint fatigue life prediction in large size and low cost wafer-level chip scale packages," *IEEE Electronic Packaging Technology (ICEPT)*, November 2015, pp. 496–501.
75. Liu, Y. M., and Y. Liu, "Prediction of board-level performance of WLCSP," *IEEE/ECTC Proceedings*, June 2013, pp. 840–845.
76. Liu, Y., Q. Qian, M. Ring, J. Kim, and D. Kinzer, "Modeling for Critical Design of Wafer Level Chip Scale Package," *IEEE/ECTC Proceedings*, June 2012, pp. 959–964.
77. Chan, Y.., S. Lee, F. Song, J. Lo, and T. Jiang, "Effect of UBM and BCB layers on the thermomechanical reliability of wafer level chip scale package (WLCSP)," *Proc. Microsystems, Packaging, Assembly and Circuits Technology Conf. (IMPACT)*, 2009, pp. 407–412.

78. Tee, T., L. Tan, R. Anderson, H. Ng, J. Low, C. Khoo, R. Moody, and B. Rogers "Advanced Analysis of WLCSP Copper Interconnect Reliability under Board Level Drop Test," *IEEE/ECTC Proceedings*, May 2008, pp. 1086–1095.
79. Fan, X., and Q. Han, "Design and Reliability in Wafer Level Packaging," *IEEE/ECTC Proceedings*, May 2008, pp. 834–841.
80. Jung, B. Y., et al., "MEMS WLCSP development using vertical interconnection," *Electronics Packaging Technology Conference (EPTC)*, IEEE 18th, December 2016, pp. 455–458.
81. Ding, M., B. Lau, and Z. Chen, "Molding process development for low-cost MEMS-WLCSP with silicon pillars and Cu wires as vertical interconnections," *Electronics Packaging Technology Conference (EPTC)*, IEEE 19th, 2017.
82. Zeng, K., and A. Nangia, "Thermal cycling reliability of SnAgCu solder joints in WLCSP," *Proc. 2014 IEEE 16th Electronics Packaging Technology Conference*, December 2014, pp. 503–511.
83. Peng Sun, "Package & board level reliability study of 0.35 mm fine pitch wafer level package," *Proc. 2017 18th International Conference on Electronic Packaging Technology*, pp. 322–326.
84. Yeung, T., "Material characterization of a novel lead-free solder material – SACQ," *IEEE/ECTC Proceedings, May 2014*, pp. 518–522.
85. Lau, J. H., C. Ko, T. Tseng, K. Yang, C. Peng, T. Xia, P. Lin, E. Lin, L. Chang, H. Liu, and D. Cheng, "Fan-In Panel-Level with Multiple Diced Wafers Packaging", *IEEE/ECTC Proceedings*, May 2020, pp. 1146–1153.
86. Lau, J. H., C. Ko, T. Tseng, K. Yang, C. Peng, T. Xia, P. Lin, E. Lin, L. Chang, H. Liu, and D. Cheng, "Panel-Level Chip-Scale Package with Multiple Diced Wafers", *IEEE Transactions on CPMT*, Vol. 10, No. 7, July 2020, pp. 1110–1124.
87. Lin, Y., P. Marimuthu, K. Chen, H. Goh, Y. Gu, I. Shim, R. Huang, S. Chow, J. Fang, and X. Feng, "Semiconductor Device and Method of Forming Insulating layer Disposed Over the Semiconductor Die for Stress Relief", *US Patent 8,456,002B2*, filling date: December 21, 2011.
88. Strothmann, T., S. Yoon, and Y. Lin, "Encapsulated Wafer Level Package Technology (eWLCSP)", *Proceedings of IEEE/ECTC*, May 2014, pp. 931–934.
89. Lin, Y., K. Chen, K. Heng, L. Chua, and S. Yoon, "Encapsulated Wafer Level Chip Scale Package (eWLCSP™) for Cost Effective and Robust Solutions in FlexLine™", *Proceeding of IEEE/IMPACT*, September 2014, pp. 316–319.
90. Lin, Y., K. Chen, K. Heng, L. Chua and S. Yoon, "Challenges and Improvement of Reliability in Advanced Wafer Level Packaging Technology", *Proceedings of IEEE 23rd International Symposium on the Physical and Failure Analysis (IPFA)*, Singapore, July 2016, pp. 47–50.
91. Smith, L., and J. Dimaano Jr., "Development Approach & Process Optimization for Sidewall WLCSP Protection", *Proceedings of IWLPC*, October 2015, pp. 1–4.
92. Tang, T., A. Lan, J. Wu, J. Huang, J. Tsai, J. Li, A. Ho, J. Chang, W. Lin, "Challenges of Ultra-thin 5 Sides Molded WLCSP", *Proceedings of IEEE/ECTC*, May 2016, pp. 1167–1771.
93. Ma, S., T. Wang, Z. Xiao, D. Yu, "Process development of five-and six-side molded WLCSP", *Proceedings of China Semiconductor Technology International Conference (CSTIC)*, March 2018, pp. 1–3.
94. Zhao, S., F. Qin, M. Yang, M. Xiang, and D. Yu, "Study on warpage evolution for six-side molded WLCSP based on finite element analysis", *Proceeding of the International Conference on Electronic Packaging Technology (ICEPT)*, August 2019, pp. 1–4.
95. Qin, F., S. Zhao, Y. Dai, M. Yang, M. Xiang, and D. Yu, "Study of Warpage Evolution and Control for Six-Side Molded WLCSP in Different Packaging Processes", *IEEE Transactions on CPMT*, Vol. 10, No. 4, April 2020, pp. 730–738.
96. Chi, Y., C. Lai, C. Kuo, J. Huang, C. Chung, Y. Jiang, H. Chang, N. Liu, and B. Lin, Board Level Reliability Study of WLCSP with 5-Sided and 6-Sided Protection", *Proceedings of IEEE/ECTC, May 2020*, pp. 807–810.
97. Lau, J. H., C. Ko, T. Tseng, T. Peng, K. Yang, C. Xia, P. Lin, E. Lin, L.N. Liu, C. Lin, D. Cheng, and W. Lu, "Six-Side Molded Panel-Level Chip-Scale Package with Multiple Diced Wafers", *IMAPS Proceedings*, October 2020, pp. 1–10.

98. Lau, J. H., C. Ko, T. Tseng, T. Peng, K. Yang, C. Xia, P. Lin, E. Lin, L.N. Liu, C. Lin, D. Cheng, and W. Lu, "Six-Side Molded Panel-Level Chip-Scale Package with Multiple Diced Wafers", *IMAPS Transactions, Journal of Microelectronics and Electronic Packaging*, Vol. 17, December 2020, pp. 111–120.
99. Lau, J. H., and N. C. Lee, *Assembly and Reliability of Lead-Free Solder Joints*, Springer, New York, 2020.
100. Lau, J. H., "Recent Advances and Trends in Fan-Out Wafer/Panel-Level Packaging", *ASME Transactions, Journal of Electronic Packaging*, V. 141, December 2019, pp. 1–27.
101. Borkulo, J., E. Tan, and R. Stam, "Laser Multi Beam Full Cut Dicing of Dicing of Wafer Level Chip-Scale Packages", *Proceedings of IEEE/ECTC*, May 2017, pp. 338–342.
102. Borkulo, J, and R. Stam, "Laser-Based Full Cut Dicing Evaluations for Thin Si wafers", *Proceedings of IEEE/ECTC*, May 2018, pp. 1945–1949.
103. Borkulo, J., R. Evertsen, R. Stam, "A More than Moore Enabling Wafer Dicing Technology", *IEEE/ECTC Proceedings*, May 2019, pp. 423–427.
104. Qu, S., J. Kim, G. Marcus, M. Ring, "3D Power Module with Embedded WLCSP", *IEEE/ECTC Proceedings,* May 2013, pp. 1230–1234.
105. Syed, A., K. Dhandapani, C. Berry, R. Moody, and R. Whiting, "Electromigration Reliability and Current Carrying Capacity of various WLCSP Interconnect Structures", *IEEE/ECTC Proceedings*, May 2013, pp. 714–724.
106. Arfaei1, B., S. Mahin-Shirazi, S. Joshi, M. Anselm1, P. Borgesen, E. Cotts, J. Wilcox, and R. Coyle, " Reliability and Failure Mechanism of Solder Joints in Thermal Cycling Tests", *IEEE/ECTC Proceedings*, May 2013, pp. 976–985.
107. Yang, S., C. Wu, D. Shih, C. Tung, C. Wei, Y. Hsiao, Y. Huang, and D. Yu, "Optimization of Solder Height and Shape to Improve the Thermo-mechanical Reliability of Wafer-Level Chip Scale Packages", *IEEE/ECTC Proceedings*, May 2013, pp. 1210–1218.
108. Hau-Riege, C., B. Keser, Y. Yau, S. Bezuk, "Electromigration of Solder Balls for Wafer-Level Packaging with Different Under Bump Metallurgy and Redistribution Layer Thickness", *IEEE/ECTC Proceedings*, May 2013, pp. 707–713.
109. Lai, Y., C. Kao, Y. Chiu, and B. Appelt, "Electromigration Reliability of Redistribution Lines in Wafer-level Chip-Scale Packages", *IEEE/ECTC Proceedings*, May 2011, pp. 326–331.
110. Darveaux, R., S. Enayet, C. Reichman, C. Berry, and N. Zafar, "Crack Initiation and Growth in WLCSP Solder Joints", *IEEE/ECTC Proceedings*, May 2011, pp. 940–953.
111. Yadav, P., S. Kalchuri, B. Keser, R. Zang, M. Schwarz, and B. Stone, "Reliability Evaluation on Low k Wafer Level Packages", *IEEE/ECTC Proceedings*, May 2011, pp. 71–77.
112. Franke, J., R. Dohle, F. Schüßler, T. Oppert, T. Friedrich, and S. Härter, "Processing and Reliability Analysis of Flip-Chips with Solder Bumps Down to 30 µm Diameter", *IEEE/ECTC Proceedings*, May 2011, pp. 893–900.
113. Bao, Z., J. Burrell, B. Keser, P. Yadav, S. Kalchuri, and R. Zang, "Exploration of the Design Space of Wafer Level Packaging Through Numerical Simulation", *IEEE/ECTC Proceedings*, May 2011, pp. 761–766.
114. England, L., "Solder Joint Reliability Performance of Electroplated SnAg Mini-Bumps for WLCSP Applications", *IEEE/ECTC Proceedings*, May 2010, pp. 599–604.
115. Walls, J., S. Kuo, E. Gelvin, and A. Rogers, "High-Sensitivity Electromigration Testing of Lead-Free WLCSP Solder Bumps", *IEEE/ECTC Proceedings*, May 2010, pp. 293–296.
116. Zhang, Y., and Y. Xu, "The Experimental and Numerical Investigation on Shear Behaviour of Solder Ball in a Wafer Level Chip Scale Package", *IEEE/ECTC Proceedings*, May 2010, pp. 1746–1751.
117. Liu, Y., Q. Qian, J. Kim, and S. Martin, "Board Level Drop Impact Simulation and Test for Development of Wafer Level Chip Scale Package", *IEEE/ECTC Proceedings*, May 2010, pp. 1186–1194.
118. Chen, L., Y. Hsu, P. Fang, and R. Chen, "Packaging Effect Investigation for MEMS-based Sensors WL-CSP with a Central Opening", *IEEE/ECTC Proceedings*, May 2010, pp. 1689–1695.

119. Okayama, Y., M. Nakasato, K. Saitou, Y. Yanase, H. Kobayashi, T. Yamamoto, R. Usui, and Y. Inoue, "Fine Pitch Connection and Thermal Stress Analysis of a Novel Wafer Level Packaging Technology Using Laminating Process", *IEEE/ECTC Proceedings*, May 2010, pp. 287–292.
120. Chen, L., C. Chen, T. Wilburn, and G. Sheng, "The Use of Implicit Mode Functions to Drop Impact Dynamics of Stacked Chip Scale Packaging", *IEEE/ECTC Proceedings*, May 2011, pp. 2152–2157.
121. Chang, S., C. Cheng, L. Shen, and K. Chen, "A Novel Design Structure for WLCSP With High Reliability, Low Cost, and Ease of Fabrication", *IEEE Transactions on Advanced Packaging*, September 2007, 30(3), pp. 377 – 383.
122. Zhou, T., S. Ma, D. Yu, M. Li, and T. Hang, "Development of Reliable, High Performance WLCSP for BSI CMOS Image Sensor for Automotive Application", *Sensors* 2020, 20(15), July 2020, pp. 4077–4083.
123. Lau, J. H., C. Ko, T, Peng, T. Tseng, K. Yang, T. Xia, B. Lin, E. Lin, L. Chang, H. Liu, C. Lin, Y. Fan, D. Cheng, and W. Lu, "Reliability of 6-side Molded Panel-Level Chip-Scale Packages (PLCSPs), *IEEE/ECTC Proceeding,* May 2021.
124. Garrou, P., "Wafer level chip scale packaging (WL-CSP): an overview", *IEEE Transactions on Advanced Packaging*, Vol. 23, Issue: 2, May 2000, pp. 198–205.
125. Rogers, B., M. Melgo, M. Almonte, S. Jayaraman, C. Scanlan, and T. Olson, "Enhancing WLCSP Reliability Through Build-up Substrate Improvements and New Solder Alloys", IWLPC Proceedings, October 2014, pp. 1–7.
126. Wu, Z., H. Zhang, and J. Malinowski, "Understanding and Improving Reliability for Wafer Level Chip Scale Package: A Study Based on 45 nm RFSOI Technology for 5G Applications", *IEEE Journal of the Electron Devices Society*, September 2020, pp. 1–10.
127. Liu, T., C. Chen, S. Liu, M. Chang, and J. Lin, "Innovative methodologies of circuit edit by focused ion beam (FIB) on wafer-level chip-scale-package (WLCSP) devices", *Journal of Materials Science: Materials in Electronics*, Vol. 22, No. 10, pp. 1536–1541.
128. Rahangdale, U., B. Conjeevaram, A. Doiphode, and S. Kummerl, " "Solder ball reliability assessment of WLCSP — Power cycling versus thermal cycling", *IEEE/ITHERM Proceedings*, June 2017, pp. 1361–1368.
129. Hsiao, A., M. Sheikh, K. Loh, E. Ibe, and T. Lee, "Impact of Conformal Coating Induced Stress on Wafer Level Chip Scale Package Thermal Performance", *SMTA Journal,* Volume 33 Issue 2, 2020, pp. 7–13.
130. Hsiao, A., G. Baty, E. Ibe, K. Loh, S. Perng, W. Xie, and T. Lee, "Edgebond and Edgefill Induced Loading Effect on Large WLCSP Thermal Cycling Performance", *SMTA Journal,* Volume 33 Issue 2, 2020, pp. 22–27.
131. Braun, T., K.-F. Becker, O. Hoelck, R. Kahle, M. Wohrmann, L. Boettcher, M. Topper, L. Stobbe, H. Zedel, R. Aschenbrenner, S. Voges, M. Schneider-Ramelow, and K.-D. Lang, "Panel Level Packaging – A View along the Process Chain", *Proceedings of IEEE/ECTC*, May 2019, pp. 70–78.
132. Braun, T., K. Becker, O. Hoelck, S. Voges, R. Kahle, M. Dreissigacker, and M. Ramelow, "Fan-Out Wafer and Panel Level Packaging as Packaging Platform for Heterogeneous Integration", *Micromachines*, May 2019, pp. 1–9.
133. Ueno, K., K. Dohi, Y. Suzuki, and M. Hirose, "Development of Sheet Typemolding Compounds for Panel Level Package", *Proceedings of IEEE/ECTC*, May 2019, pp. 2162–2167.
134. Fujinaga, T., "High rate and low damage etching method as pretreatment of seed layer sputtering for fan out panel level packaging", *Proceedings of IEEE/ECTC*, May 2019, pp. 358–362.
135. Weichart, J., J. Weichart, A. Erhart, and K. Viehweger, "Preconditioning Technologies for Sputtered Seed Layers in FOPLP", *Proceedings of IEEE/ECTC*, May 2019, pp. 1833–1841.
136. Selhofer, H., A. Mayr, and H. Pristauz, "Large Panel Size Bonder with High Performance and High Accuracy", *Proceedings of IEEE/ECTC*, May 2019, pp. 1492–1497.
137. Bu, L., F. X. Che, V. Rao, and X. Zhang, "Mechanism of Moldable Underfill (MUF) Process for RDL-1st Fan-Out Panel Level Packaging (FOPLP)", *Proceedings of IEEE/ECTC*, May 2019, pp. 1152–1158.

138. Che, F. X., K. Yamamoto, V. Rao, and V. Sekha, "Study on Warpage of Fan-Out Panel Level Packaging (FO-PLP) using Gen-3 Panel", *Proceedings of IEEE/ECTC*, May 2019, pp. 842–849.
139. Ko, C. T., H. Yang, J. H. Lau, et al., "Chip-First Fan-Out Panel Level Packaging for Heterogeneous Integration", *IEEE Transactions on CPMT*, 2018, Vol. 8, Issue 9, September 2018, pp. 1561–1572.
140. Ko, C. T., H. Yang, J. H. Lau, et al., "Design, Materials, Process, and Fabrication of Fan-Out Panel-Level Heterogeneous Integration", *IMAPS Transactions, Journal of Microelectronics and Electronic Packaging*, Vol. 15, Issue: 4, October 2018, pp. 141–147.
141. Lau, J. H., *Fan-Out Wafer-Level Packaging*, Springer, New York, 2018.
142. Lau, J. H., *Heterogeneous Integrations*, Springer, New York, 2019.
143. Lau, J. H., "State of the Art of Lead-Free Solder Joint Reliability", *ASME Transactions, Journal of Electronic Package,* Vol. 143, June 2021, pp. 803–1 – 802-36.

第 4 章

扇出型晶圆级 / 板级封装

4.1 引言

首先明确，无论是扇入型晶圆级 / 板级芯片尺寸封装（W/PLCSP 或者统称为 WLCSP）（见图 4.1a），还是扇出型晶圆级 / 板级封装（FOW/PLP 或者统称为 FOWLP）（见图 4.1b），都属于晶圆级 / 板级封装或者统称为 WLP（wafer-level packaging）。WLCSP 和 FOWLP 最大的区别在于 FOWLP 需要临时支撑片而 WLCSP 不用。由于器件晶圆上芯片的划片道非常窄（约 50μm），晶圆上没有足够的空间来对再布线层（redistribution-layer，RDL）进行扇出。因此，对于扇出型晶圆级封装，需要先对器件晶圆进行合格芯片（known good die，KGD）测试，然后将晶圆切割得到单颗的 KGD，再把 KGD 拾放在临时支撑片（圆形或矩形）上相隔一定距离重排，如图 4.1c 所示。

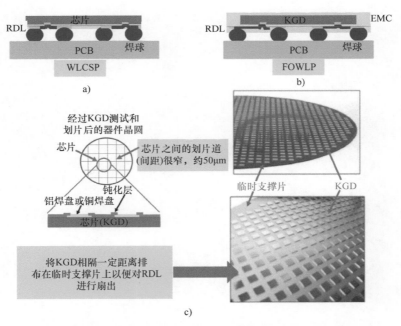

图 4.1　WLCSP 和 FOWLP 对比：FOWLP 需要一块临时支撑片

目前有很多扇出型封装方法[1-150]。它们基本上可分为三大类，分别是先上晶（芯片面朝下）[1-72]，先上晶（芯片面朝上）[73-94]以及后上晶或者先RDL[95-150]。本章会介绍这三种方法，每种方法包含使用圆形支撑片（晶圆）和矩形支撑片（面板）两类情况。本章最后还会介绍一种基于扇出型先上晶且面朝下技术的mini-LED RGB显示器板级封装的设计、材料、工艺、制备以及可靠性。

4.2 扇出型（先上晶且面朝下）晶圆级封装（FOWLP）

本节主要介绍扇出型（先上晶芯片面朝下）晶圆级封装。第一个关于扇出型晶圆级封装（fan-out wafer-level packaging，FOWLP）的美国专利是英飞凌（Infineon）在2001年10月31日提交的，最早的技术论文也是英飞凌与其行业合作伙伴日本长濑（Nagase）、日东电工（NittoDenko）和日本山田（Yamada）等公司在ECTC2006和EPTC2006上共同发表的。当时他们将这项技术命名为嵌入式晶圆级焊球（embedded wafer-level ball，eWLB）阵列。该方法消除了引线键合或者晶圆凸点成型以及引线框或者封装基板的使用，并有望实现一种较低成本、更优性能、更小型化的封装。该方法需要采用临时支撑片来进行合格芯片（known-good die，KGD）的晶圆重构、环氧模塑料（epoxy molding compound，EMC）、模塑工艺以及再布线层（redistribution layer，RDL）的制作。

在ECTC2007会议上，飞思卡尔（Freescale）发表了一项类似的技术并称其为重构芯片封装（redistributed chip package，RCP）[5]。在ECTC2008会议上，新加坡微电子所（IME）开发出了支持多芯片和多芯片3D堆叠的FOWLP技术[6]；在ECTC2009会议上，IME发表了4篇FOWLP相关的论文，它们分别是：①一种预测模压成型过程中芯片移位的新方法[7]；②超薄芯片的横向放置和垂直堆叠[8]；③3D FOWLP的可靠性[9]；④高质量、低损耗毫米波无源器件FOWLP的演示[10]。

参考文献[12-15]演示了用于异质集成的扇出型（先上晶面朝下）晶圆级封装。液态环氧模塑料（epoxy molding compound，EMC）是模压成型的，所制得的RDL金属线宽/线距为10/15μm。本节采用一种全新的干膜EMC层压方法代替参考文献[12，13]的液态EMC模塑的方式。本节制得的RDL金属线宽/线距可以缩小到10~5μm。与此同时，为了节省昂贵的EMC材料和实现小型化（薄型）封装，提出了一种特殊的组装工艺。

4.2.1 测试芯片

图4.2为所考虑的两颗测试芯片。图4.2a为大测试芯片及其焊盘排布，从图中可以看到，大芯片的尺寸为5mm×5mm×150μm，共包含160个焊盘，内侧焊盘节距为100μm。铝焊盘上SiO_2钝化层开窗尺寸为50μm×50μm，铝焊盘的尺寸为70μm×70μm。小芯片尺寸为3mm×3mm×150μm，图4.2b为制作的芯片，可以看到共包含80个焊盘，内侧焊盘节距为100μm。小测试芯片的焊盘截面和尺寸与大测试芯片的相同。

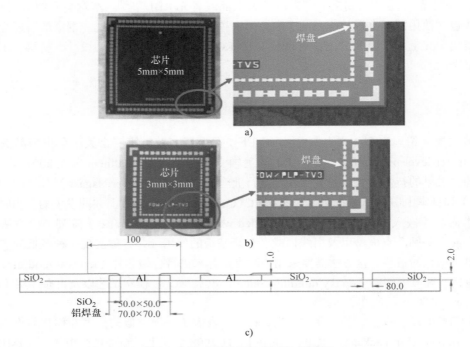

图 4.2 测试芯片的俯视图和截面图

4.2.2 测试封装体

图 4.3a 为测试封装体的示意图。封装体的尺寸为 10mm×10mm,它包括一颗大芯片(5mm×5mm)和三颗小芯片(3mm×3mm)。大芯片和小芯片之间的间距为 100μm。封装体有 2 层再布线层(RDL),图 4.3b 为芯片到 RDL1 间的再布线层结构,图 4.3c 为 RDL1 到 RDL2 间的再布线层结构,图 4.3d 为 RDL2 到 PCB 间的结构,图 4.3e 为封装体的轮廓。封装体全部在一块 12in 的临时玻璃重构晶圆上制作完成,晶圆上封装体的节距为 10.2mm。在实际应用中,大芯片可能是应用处理器芯片,小芯片可能是存储芯片。这种制作的效率很高,一次性可以实现 629 个封装体(10mm×10mm,每个含四个芯片)的异质集成。

图 4.4 为测试封装体的截面示意图。从图中可以看到 2 层 RDL,其中 RDL1 的金属层厚度为 3μm,RDL2 的金属层厚度为 7.5μm。RDL1 的金属线宽和线距为 5μm,RDL2 的金属线宽和线距为 10μm。DL1 和 DL2 的介质层厚度为 5μm,DL3 的为 10μm。

连接测试芯片铜焊盘和 RDL1 的穿过介质层(DL1)的孔(VC1)的直径为 20~30μm,RDL1 焊盘直径为 55μm,通过直径为 30~40μm 的过孔(V12)与 RDL2 连接,RDL2 焊盘直径为 65μm。最终在 RDL2 上形成直径为 220μm 的焊球铜焊盘,钝化层(DL3)开窗直径为 180μm。焊球直径为 200μm,节距为 0.4mm。

第 4 章 扇出型晶圆级 / 板级封装

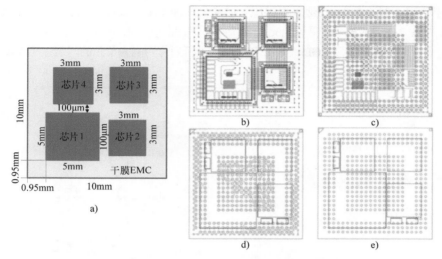

图 4.3 测试封装体的焊盘布局示意图（俯视图）

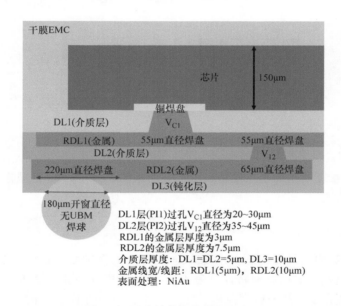

图 4.4 测试封装体的截面示意图

4.2.3 传统的先上晶（面朝下）晶圆级工艺

图 4.5 是传统的先上晶面朝下 FOWLP 工艺流程图。首先，对器件晶圆进行合格芯片（known-good die，KGD）测试，并切割得到单颗的芯片。拾取 KGD，将其面朝下放置到一块圆形（晶圆）或矩形（面板）的临时支撑片上（该支撑片可以是金属、硅或玻璃），如图 4.5b 所示。临时载具上事先粘有双面热剥离胶带，如图 4.5a 所示。放置好 KGD 后，对重构支撑片

进行 EMC 模塑（见图 4.5c），模塑采用模压 + 后固化处理（post mold cure，PMC）。模塑完成后将临时支撑片去除，并剥离胶带（见图 4.5d）。然后在重构晶圆上制作 RDL（见图 4.5e）。最后，进行植球并对整个重构晶圆（包含 KGD、RDL 和焊球）进行划片切割为单独的封装体（见图 4.5f）。

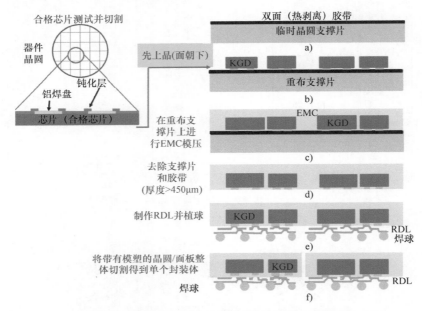

图 4.5　传统先上晶（芯片面朝下）关键工艺步骤

4.2.4　异质集成封装的新工艺

本节不采用图 4.5c 所示的模压模塑和图 4.5e 所示的 RDL 制作等常规方法。本节采用的是图 4.6 所示的新工艺流程。

4.2.5　干膜 EMC 层压

在参考文献 [22] 中，我们用了一种液态 EMC 和模压的方法。在本研究中，我们将使用一种新的干膜 EMC 材料，它的材料特性见表 4.1，通过层压方法进行模塑。如图 4.6a、b 所示，将 KGD 拾取和放置在支撑片上后，在其上层压一厚 200μm 的干膜 EMC（见图 4.6c 和图 4.7），层压的温度为 100℃，层压的时间为 30min。

4.2.6　临时键合另一块玻璃支撑片

对于传统 FOWLP，在完成 EMC 模塑后，就要解键合支撑片并剥离胶带（见图 4.5d），整个重构晶圆的厚度通常 ≥ 450μm。之后就是制作 RDL 并植球（见图 4.5e）。在本研究中，为了节约昂贵的 EMC 材料并实现一个小型化（薄型）封装体，在没有支撑片时重构晶圆的总厚度仅有 300μm，这么薄的重构晶圆很易碎，使得 RDL 制作和植球都变得非常困难。

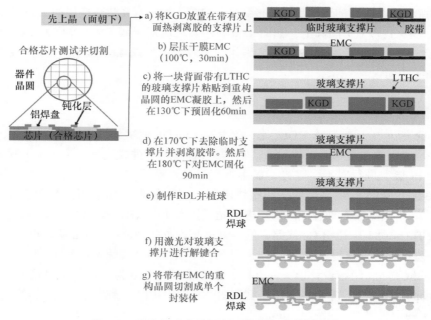

图 4.6 先上晶（芯片面朝下）新工艺关键步骤

表 4.1 干膜 EMC 的材料特性

EMC 材料	过去的 液态 EMC（R4507）	现在的 干膜 新的 EMC
最大填料直径 /μm	25	5
填料占比（%）	85	80
玻璃化转变温度 T_g（DMA 法测试）/℃	150	163
热膨胀系数 1/(10^{-6}/K)	10	15
热膨胀系数 2/(10^{-6}/K)	41	23
杨氏模量 /GPa	19	10

图 4.7 图 4.6c 完成后实物照片:（左）玻璃面;（右）干膜 EMC 面

一种解决方法是将薄的重构晶圆粘接到另一块事先涂覆有光热转换层的 1mm 厚的玻璃（热膨胀系数为 $6.4\times10^{-6}/℃$）晶圆上，如图 4.6d 所示。之后对干膜 EMC 进行 130℃、60min 的预固化处理。然后，去除第一块临时支撑片并剥离胶带，如图 4.6e 所示。再对干膜 EMC 进行 180℃、90min 的固化。

4.2.7 再布线层

如图 4.6e 所示，开始制作 RDL。图 4.8 为制作 RDL 的关键工艺步骤。首先，在重构晶圆上旋涂一层光敏聚酰亚胺（polyimide，PI）。然后应用步进式光刻机（每 4 个封装体作为一个单元）在 PI 层上进行对准、曝光、显影得到过孔。然后对 PI 材料在 200℃固化 1h，从而获得一层 4~5μm 厚的 PI 层。之后，在整块晶圆上利用物理气相沉积法溅射上 Ti 层和 Cu 层（175~200℃）。再次涂覆光刻胶，采用步进式光刻技术获得 RDL 布线的图形，通过电化学方式在开窗后的 Ti/Cu 层上电镀出 Cu 线路，然后剥离光刻胶并刻蚀掉 Ti/Cu 层就得到了 RDL1。重复以上的工艺步骤可以获得 RDL2。图 4.9 所示为 RDL1 的金属线宽/线距，从图中可以看到，RDL1 金属的线宽/线距与设计值（5μm）非常接近。图 4.10 为通过扫描电镜（scanning electron microscope，SEM）观察到的 RDL1 截面图，从图中可以看到芯片、光敏聚酰亚胺和 RDL1 结构，RDL1 的金属线宽/线距为 5μm。

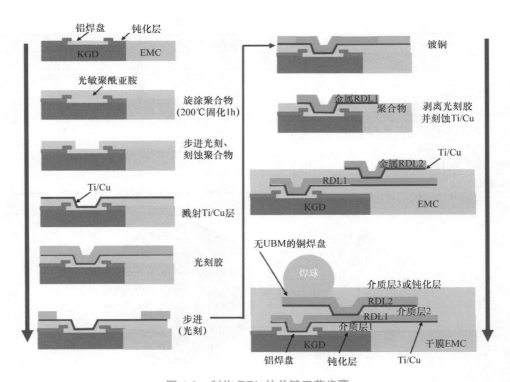

图 4.8　制作 RDL 的关键工艺步骤

图 4.9 5μm 线宽和线距 RDL 的顶视图

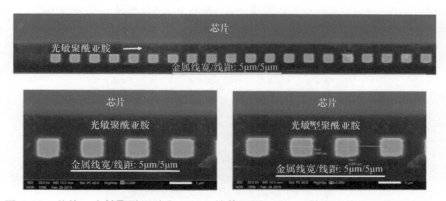

图 4.10 芯片、光敏聚酰亚胺和 RDL1 的截面图。RDL1 的金属线宽和线距为 5μm

4.2.8 焊球植球

在进行图 4.6e 所示的植球工艺时，需要用到 2 套不同的模板，一套是用于助焊剂模板印刷，另一套用于焊球模板植球。本研究采用 SAC305（Sn3wt%Ag0.5%Cu）焊球（直径 200μm），焊球节距 0.4mm。焊料回流的峰值温度为 245℃。

4.2.9 最终解键合

图 4.6f 所示的玻璃支撑片解键合是从玻璃支撑片一面用激光（355nm DPSS Nd∶YAG UV）扫描完成的。激光束斑为 240μm，扫描速度为 500mm/s，扫描步进距离为 100μm。LTHC 材料

在激光照射下会转化成粉末状,这样玻璃支撑片很容易移除,解键合完成后需要进行化学清洗。图 4.11 为没有任何支撑片的重构晶圆以及某个封装体的特写照片。从图中可以看到,一个封装体中包含 4 颗芯片,这些芯片的排布很准确。我们将重构晶圆进行图 4.6g 所示的划片,得到单个芯片单元,芯片单元的 X 射线图像如图 4.12 所示,图中还可以看到 RDL1 和 RDL2。

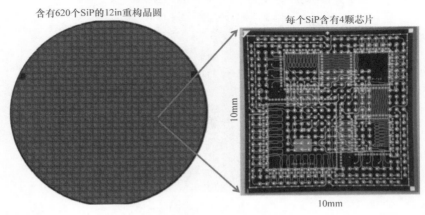

图 4.11　含有 620 个合格 SiP 的 12in 重构晶圆。每个 SiP(10mm×10mm)含有一颗大芯片(5mm×5mm)和三颗小芯片(3mm×3mm)

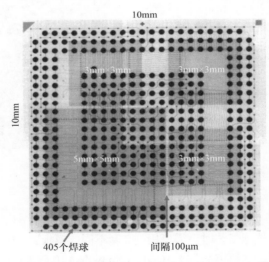

图 4.12　一颗大芯片(5mm×5mm)和三颗小芯片(3mm×3mm)异质集成的 X 射线图像。SiP 的尺寸为 10mm×10mm,其上共有 405 个直径 200μm、节距 0.4mm 的焊球

图 4.13 是一个典型的异质集成封装体截面图。从图中可以看到有 2 层 RDL,RDL2 的金属层厚度(7.5μm)大于 RDL1(3μm)。RDL2 较厚的原因是为了在无 UBM 时得到较厚的铜焊盘,从而防止在焊球回流和工作过程中铜被耗尽。

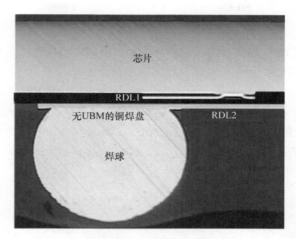

图 4.13 单个封装体的截面图

采用 C 模式扫描超声波显微镜（C-mode scanning acoustic microscopy，C-SAM）检测层压干膜 EMC 中的模塑空洞。为平衡良好的分辨率和信号穿透深度，采用了 75MHz 的超声换能器进行空洞检测。经过一系列扫描参数的研究，均没有在晶圆中检测到任何空洞（包括 EMC 和间隔空间），如图 4.14 所示。

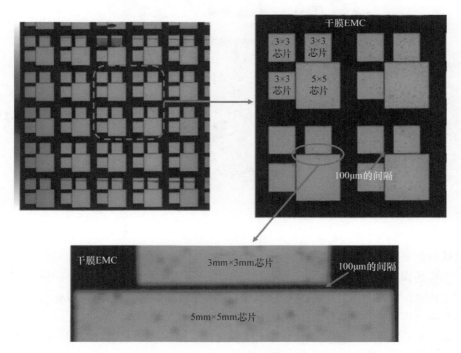

图 4.14　C-SAM 图像表明在干膜 EMC 中没有明显的空洞，甚至在大芯片和小芯片之间宽 100μm 的间隔空间中也没有空洞

4.2.10 PCB 组装

本节首先简要介绍 PCB 和模板的设计，然后介绍模板印刷焊膏、芯片拾取和放置、焊料回流等工艺步骤，并给出 PCB 组装后的 X 射线图像和截面图片。

1. PCB

用于扇出型晶圆级封装的 PCB 采用 FR-4 制作，如图 4.15 所示。从图中可以看到，板子上有四个放置封装体的位置，PCB 的尺寸为 103mm×52mm×0.65mm，为 6 层板。每个封装体上有 405 个焊盘（焊盘节距 0.4mm）。无阻焊层定义的焊盘直径为 0.2mm，焊盘经过有机可焊性保护层（OSP）的表面处理。阻焊层开窗直径为 0.28mm。

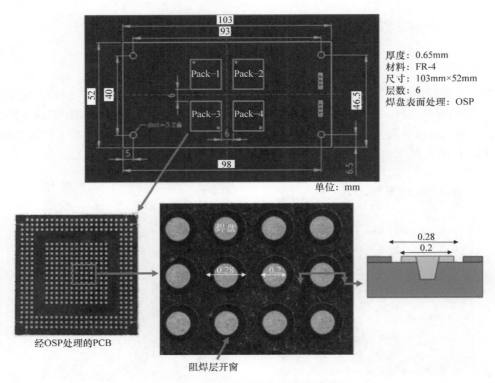

图 4.15　PCB 的布局和制作完成的 PCB

2. 模板和印刷

模板印刷的模板由不锈钢制成，晶粒尺寸为 2μm（有利于焊膏转移），模板厚 0.08mm。钢板通过激光开窗，并经过电化学抛光处理。模板印刷设备型号为 DEK Horizon 9。

3. 芯片拾放和回流

芯片拾取和放置操作通过 SiPlace x4s 设备完成。回流操作通过 10 温区 BYU Pyram nitrogen 150N 设备完成。图 4.16 为回流的温度曲线。从图中可以看到，最高回流温度为 245℃，温度在 217℃ 以上的时间为 85s。

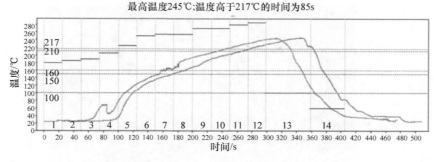

图 4.16 SMT 回流曲线。异质集成封装体 PCB 组装的截面图

4.2.11 异质集成的可靠性（跌落试验）

图 4.16 所示的 4 芯片异质集成 FOWLP PCB 组装的可靠性评估通过跌落试验进行。

1. 跌落试验设置

跌落试验按照 JEDEC 标准 JESD22-B111 设置，如图 4.17 所示。在经过 20 多次尝试后，确定了能产生图 4.18 所示的 1500g/ms 跌落谱的高度。

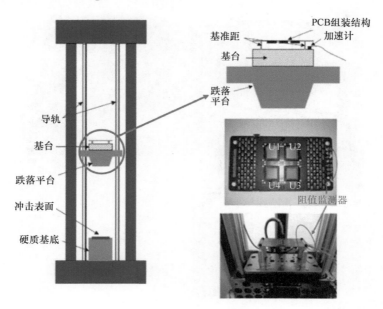

图 4.17 跌落试验设置

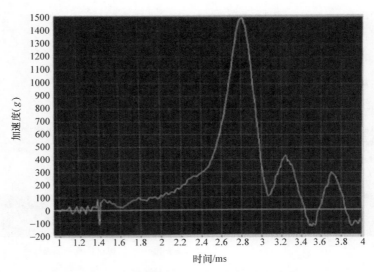

图 4.18 跌落谱

2. 跌落试验结果

跌落试验共进行 1000 次。共有 24 个完成了底部填充的样品,所有样品都通过了 500 次跌落试验。第一次失效发生在第 550 次跌落,失效位置如图 4.19 所示。从图中可以看到,一些位于 PCB 中心附近的拐角焊点发生了失效。失效模式如图 4.20 所示,焊点的开裂位置发生在芯片和焊料的界面。

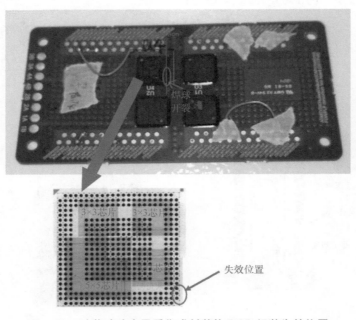

图 4.19 跌落试验中异质集成封装体 PCB 组装失效位置

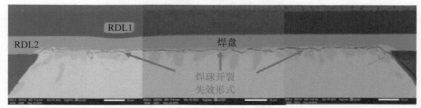

图 4.20 跌落试验后异质集成封装体 PCB 组装的失效模式

4.2.12 总结和建议

本节研究了通过新的 FOWLP 方法实现包含 4 颗芯片异质集成薄封装的设计、材料、工艺和制作。一些重要的结论和建议总结如下 [12, 13]。

1）制备的 RDL 金属最小线宽 / 线距为 5μm，该精度经过了光学显微镜和扫描电子显微镜验证。

2）FOWLP 中大芯片（5mm×5mm）与小芯片（3mm×3mm）间距为 100μm。通过 C-SAM 发现干膜 EMC 和间隔空间中均无空洞存在。

3）选择了一种干膜 EMC，开发了一种干膜 EMC 层压的方法。

4）通过 X 射线和光学显微镜验证了封装体中 4 颗芯片均组装合格。

5）通过 X 射线和光学显微镜验证了封装体的 RDL 层制备合格。

6）完成的异质集成封装体厚度为 300μm（不包括焊球）。这不仅节省了昂贵的 EMC 成本，同时实现了一种超级小型化（薄型）的封装。

7）所提出的组装方法，即利用带有 LTHC 层的临时玻璃支撑片支撑超薄重构晶圆，更好地控制了晶圆的翘曲问题，并为节省 EMC 成本以及实现超级小型化封装提供了可能。

8）所制作的薄的异质集成封装体在进行 PCB 组装后经过了跌落试验。

4.3 扇出型（先上晶且面朝下）板级封装（FOPLP）

本节 FOPLP 的研究采用了与 4.2 节中 FOWLP 研究同样的测试芯片（5mm×5mm 以及 3mm×3mm）[15]。封装体外部尺寸也与 4.2 节一致（10mm×10mm），但是封装体的内部结构发生了改变。与 FOWLP 中金属线宽和线距采用 5μm 和 10μm 不同，FOPLP 中金属线宽和线距选择是 10μm 和 25μm。FOWLP 的制作采用晶圆加工设备，而 FOPLP 的制作采用 PCB 加工设备；FOPLP 的临时支撑片采用了矩形面板。

为了获得非常高的产率、薄型封装以及节省昂贵的 EMC 材料，RDL 的制作采用了一种叫作"统一基板集成封装"（uni-substrate-integrated package，Uni-SIP）的技术[15]。面板的尺寸为 508mm×508mm（见图 4.21），在重构面板上采用干膜 EMC 层压（而非模压液态 EMC）的方式进行模塑。

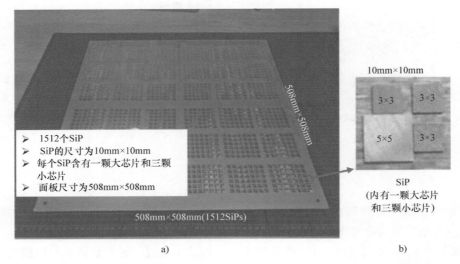

图 4.21 含有 1512 个 SiP 的重构面板（508mm×508mm），每个 SiP 含有四颗芯片

4.3.1 测试封装体的异质集成

图 4.21b 所示的测试封装体的外观与参考文献 [12，13] 中介绍的一样。然而，它们的最大区别在于：①采用干膜 EMC 取代了液态 EMC，② EMC 采用层压的方式而非模压的方式，③该封装体的金属线宽/线距从 20μm 减小到了 10μm，④该封装体使用的面板尺寸从 340mm×340mm 增大到了 508mm×508mm，⑤该封装体采用了一种新的去胶渣和化学沉铜的工艺。该封装体中大芯片和小芯片的间隔距离为 100μm。封装体内有 2 层 RDL。图 4.22a 为芯片和 RDL1 之间的 RDL 示意图，图 4.22b 为 RDL1 和 RDL2 之间的 RDL 示意图，图 4.22c 为 RDL2 和 PCB 之间的 RDL 示意图。图 4.22d 为待测试封装体的轮廓图。这些封装体在图 4.21a 所示的 508mm×508mm 重构面板上制作得到，面板上共有 1512 个异质集成封装体，每个封装体中包含 4 颗芯片。

图 4.23 为测试芯片的截面示意图。从图中总共可以看到 2 层 RDL，RDL1 和 RDL2 的金属层厚度为 10μm。RDL1 的金属线宽/线距为 10μm（与参考文献 [14] 中的 20μm 不同），RDL2 的金属线宽/线距为 25μm，介质层 1、2 和 3 的厚度为 20μm。介质层 1（DL1）的过孔连接了测试芯片的铜焊盘和第一层 RDL（RDL1），其直径为 50μm；RDL1 的焊盘直径为 135μm，它与

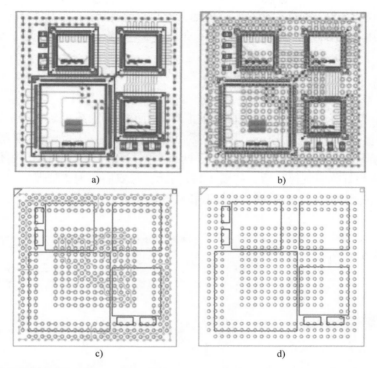

图 4.22 测试封装体，包含一颗 5mm×5mm 芯片和三颗 3mm×3mm 芯片。封装体尺寸为 10mm×10mm。a）芯片和 RDL1 之间的 RDL；b）RDL1 和 RDL2 之间的 RDL；c）RDL2 和 PCB 之间的 RDL；d）待测试封装体的轮廓

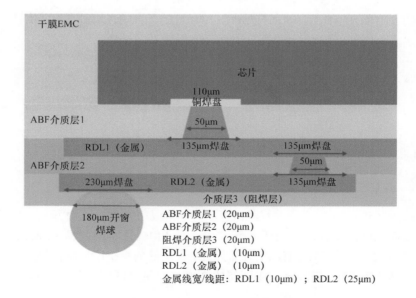

图 4.23 测试封装体的截面图

RDL2 通过一直径为 50μm 的过孔连接。类似地，RDL2 上的焊盘直径为 135μm。最后在 RDL2 上形成直径为 230μm 的焊球铜焊盘，阻焊层（介质层 3）开窗直径为 180μm。焊盘上所植焊球直径为 200μm，节距为 0.4mm。

4.3.2 一种新的 Uni-SIP 工艺

图 4.24 所示为新的 Uni-SIP 工艺。临时支撑片采用了热膨胀系数为 $4\times10^{-6}/℃$ 的有机基板，有机支撑片的厚度为 1mm。采用 1mm 厚的支撑片有如下好处：①在清洗和复用时不容易损坏；②可以增加重构晶圆的刚度，抵抗拾放芯片、两级真空层压以及干膜 EMC 固化过程中引发的弯曲（翘曲）。选择好临时支撑片后，如图 4.24a 所示，将双面热剥离胶带（Nitto Denko 的 RE-VALPHA）粘贴在支撑片上。然后将芯片拾取和放置在胶带上，如图 4.24b 所示。

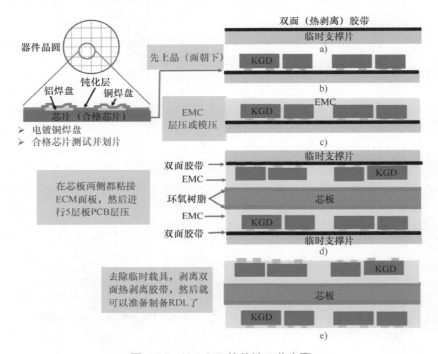

图 4.24 Uni-SIP 的关键工艺步骤

4.3.3 ECM 面板的干膜层压

对于 508mm × 508mm 这样的大尺寸面板，原先液态 EMC 模压的方式难以让 EMC 填满整个面板而没有流痕，而且位于面板角部和中心的 EMC 总厚度差异（total thickness variation，TTV）非常大。干膜 EMC 有以下几点好处：①更好的平整度控制；②整个面板范围内更一致的 TTV；③与 PCB 工艺兼容；④更高的产率；⑤更少的颗粒污染。本研究中采用干膜 EMC。

表 4.1 为干膜 EMC 的材料特性。从图中可以看到，材料成分中 80% 是填料，最大填料尺寸为 5μm。EMC 的杨氏模量为 10GPa，玻璃化转变温度（T_g）为 163℃。如图 4.24c 所示，干膜 EMC 通过一个两级真空层压装置（见图 4.25）层压到支撑片和芯片表面。第一级真空层压（100℃下真空 30s 并加压（0.68MPa）30s）用于增强干膜 EMC 中树脂的粘附力，第二级真空层压用于干膜 EMC 的表面整平（100℃下加压（0.54MPa）60s）。完成后需进行固化，先在 150℃下固化 1h，然后在 180℃下固化 0.5h。图 4.24c 所示的结构为 ECM 面板（本研究中将芯片嵌入在干膜 EMC 中的面板称为 ECM 面板）。

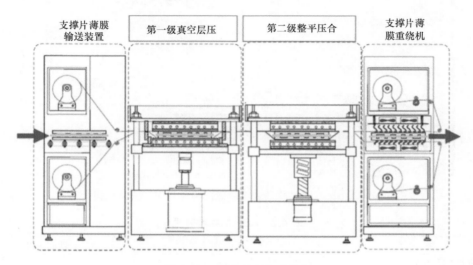

图 4.25　两级真空层压设备

4.3.4　Uni-SIP 结构的层压

下一步操作是在另一块有机芯板支撑片（基板）两面粘贴环氧树脂，然后再在芯板两侧粘贴 ECM 面板，粘贴时 ECM 面板背面与芯板接触，如图 4.24d 所示。接下来再次利用图 4.25 所示的两级真空层压设备，采用同样的工艺条件，进行 5 层板层压。层压完成后，如图 4.24e 所示，去除临时支撑片并剥离胶带，就可以准备在芯片两侧制作 RDL 了。

4.3.5　新 ABF 的层压、激光钻孔、去胶渣

图 4.26 所示为在 Uni-SIP 结构（见图 4.24e）两侧由铜焊盘上制作 RDL 的关键工艺步骤。首先，在面板两面都层压一新的 ABF（见表 4.2）。为了控制面板的翘曲，ABF 和 EMC 材料的选择至关重要。ABF 的热膨胀系数需要与某一个 ECM 面板的等效热膨胀系数匹配。同样，对于需要细金属线宽/线距（≤10μm）的应用，还要求填料尺寸较小。在本研究中，ABF（5μm 厚）的热膨胀系数（15×10^{-6}/℃）明显较小。此外，之前使用的干膜和激光直写技术只能得到有限的 20μm 分辨率，而新的干膜（12μm）搭配更先进的激光直写技术可以在玻璃上将分辨率缩小到 5μm。

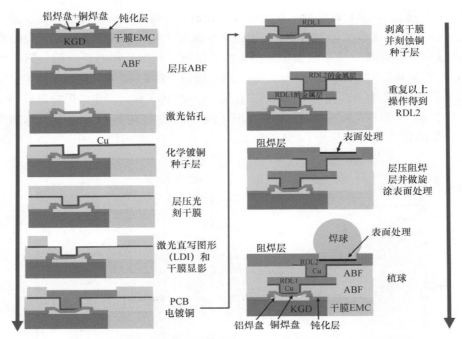

图 4.26 从 Uni-SIP 结构两侧铜焊盘上制备 RDL 的关键工艺步骤

表 4.2 ABF 的材料特性

性 质	老的 ABF	新的 ABF
填料最大直径 /μm	15	5
填料含量（%）	82	80
热膨胀系数（30~150℃）/（10^{-6}/K）	7	15
杨氏模量 /GPa	7	10
介电常数（D_k）	3.2	3.3

如图 4.26 所示，在芯板两面层压好 ABF 后，采用激光在 ABF 钻孔并停止在铜焊盘（这也是之前在器件晶圆上制作铜焊盘的原因）。在金属化之前需要对 ABF 材料进行激光后去胶渣处理，以便为后面的化学沉铜提供合适的表面粗化以增强铜层与 ABF 的粘附力。本研究采用了陶氏化学（DOW）半加成法（semi-additive process，SAP）方案——CIRCUPOSIT 7800 去胶渣工艺（见图 4.27）和 CIRCUPOSIT ADV 8500 化学沉铜工艺（见图 4.28），并优化了其中某些参数。如图 4.29 所示，新的 ABF 表面粗糙度比老的 ABF 要平滑，这有助于在 ABF 上实现细金属线宽/线距的 RDL。

剥离强度是表征积层金属化层性能的关键参数，它与去胶渣工艺后的表面粗糙度有关（见图 4.30）。对该参数进行优化主要考虑三个关键影响因素：溶胀剂浸泡时间、促进剂浸泡时间以及热处理时间。陶氏化学进行了详细的正交试验，得到了图 4.30 所示的剥离强度结果。结果表明，180℃热处理搭配 12min 以上的促进剂浸泡时间可以得到更好的性能表现——剥离强度高于 0.4kgf/cm²。后续的工艺调整基于更好的剥离强度，化学沉铜的平均厚度为 0.65μm。

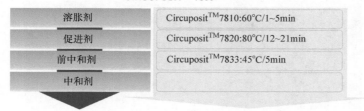

图 4.27　CIRCUPOSIT 7800 去胶渣工艺

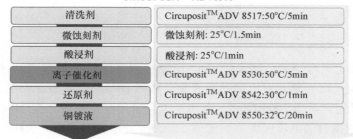

图 4.28　CIRCUPOSIT ADV 8500 化学沉铜工艺

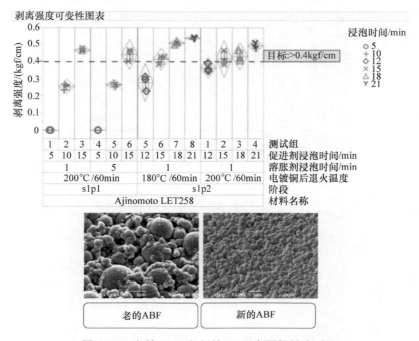

图 4.29　老的 ABF 和新的 ABF 表面粗糙度对比

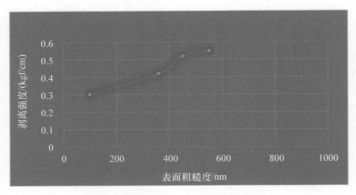

图 4.30 剥离强度结果

4.3.6 激光直写图形和 PCB 镀铜

在面板双面完成化学镀铜后,继续在 Uni-SIP 结构双面层压厚度为 12μm 的光刻胶干膜。层压完成后进行激光直写光刻并对干膜进行显影,激光直写的位置精度为 ±5μm,UV 激光的波长为 365nm,曝光剂量为 200mJ/cm²。电镀铜工艺所用铜镀液成分为 240g/L 的 $CuSO_4$ 加 60g/L 的 H_2SO_4,电镀铜时的电流密度值恒定为 2.0A/dm²。电镀完成后,这些细线条会完全按照预定的光刻图形布局在 ABF 上形成。最后,剥离光刻胶并刻蚀铜种子层就可以得到 RDL1。如图 4.31 截面的顶视图测量所示,制作得到的金属线宽和线距(平均)为 10μm。重复以上

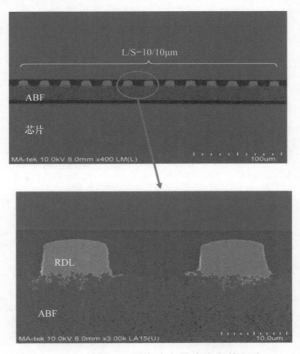

图 4.31 RDL1 的 SEM 图像(金属线宽和线距为 10μm)

步骤制作得到 RDL2。图 4.32 是含有 4 颗芯片的异质集成封装。从图中可以看到，大芯片（5mm×5mm）和小芯片（3mm×3mm）之间的间隔距离为 100μm，且在干膜 EMC 中无空洞。

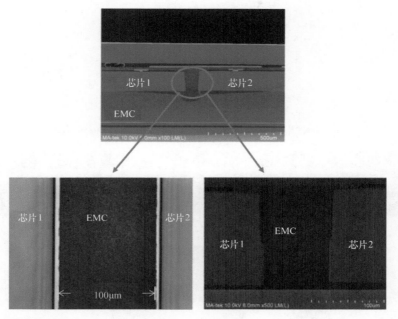

图 4.32　测试封装体的图像显示大芯片和小芯片之间有 100μm 的间距且干膜 EMC 中没有空洞存在

图 4.33 上方为在 508mm×508mm 面板上以先上晶且芯片面朝下 FOPLP 工艺制作的 SiP。图 4.33 下方为 SiP 的 PCB 组装截面图，从图中可以看出 SiP 制作合格。

4.3.7　总结和建议

本节演示了采用 FOPLP 方法实现 4 颗芯片薄型异质集成的可行性，包括设计、材料、工艺和制备。一些重要的结论和建议总结如下[15]：

1）完全依靠 PCB SAP 技术和设备完成了扇出型异质集成封装，RDL1 的线宽和线距为 10μm。

2）基于给定的关键参数，通过两级真空层压设备成功在 ECM 面板上完成了干膜 EMC 层压。

3）成功实现了 5 层 Uni-SIP 结构（2 层 ECM 面板 + 芯板）的层压，成功实现了临时支撑片的解键合和双面胶带的去除。

4）选用了一种热膨胀系数与干膜 EMC 相同的新的 ABF（热膨胀系数为 $15 \times 10^{-6}/℃$），填料尺寸更小（5μm）。

5）通过选用了陶氏化学的 SAP 方案——CIRCUPOSIT 7800 去胶渣工艺搭配 CIRCUPOSIT ADV 8500 化学镀铜工艺，并对部分参数进行了优化，成功得到了光滑的 ABF 表面以及出色的剥离强度。

6）演示了通过 LDI+PCB 电镀铜技术制作得到了线宽和线距 10μm 的 RDL1。

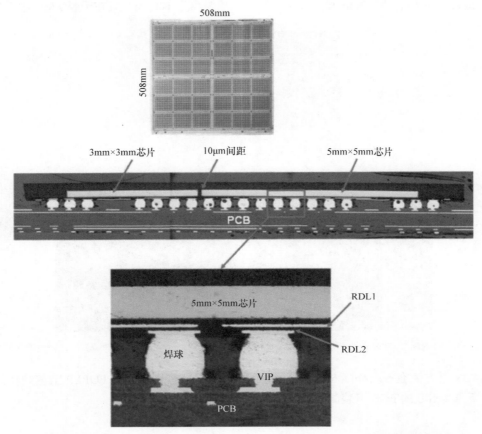

图 4.33 （上）含有 SiP 的 508mm×508mm 的面板；（下）SiP 组装的截面图

4.4 扇出型（先上晶且面朝上）晶圆级封装

4.4.1 测试芯片

图 4.34 为扇出型（先上晶且面朝上）封装的测试芯片。从图中可以看到，在芯片的铝焊盘上电镀了一个铜接触焊盘（凸台）。

4.4.2 工艺流程

制作扇出型（先上晶且面朝上）封装的关键工艺流程如图 4.35 所示，图 4.36 为封装体的截面图。从图中可以看到 3 层最小金属线宽和线距为 5μm 的 RDL。该封装的设计、材料、工艺、制备以及可靠性都已经在参考文献 [83-86] 中说明，本书就不再详细阐述。

第 4 章 扇出型晶圆级 / 板级封装　147

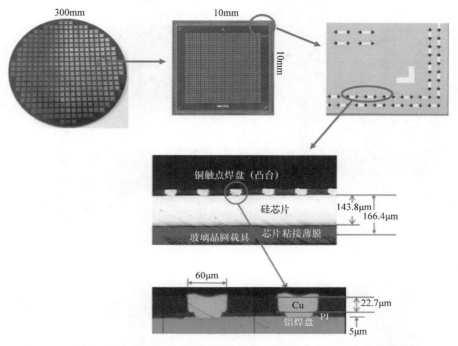

图 4.34　测试芯片。(上) 制造好的芯片的特写镜头；(下) 测试芯片的截面图

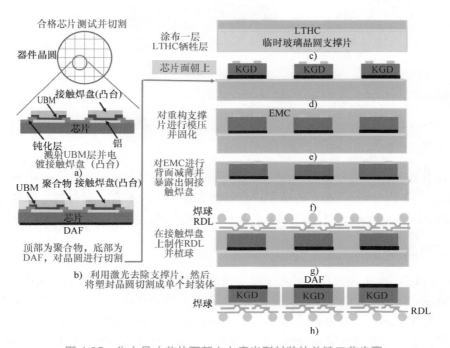

图 4.35　先上晶（芯片面朝上）扇出型封装的关键工艺步骤

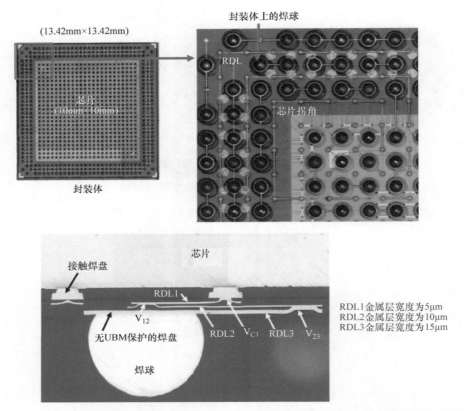

图 4.36 （上）制备好的先上晶且面朝上扇出型封装的特写镜头；（下）测试封装体的截面图

4.5 扇出型（先上晶且面朝上）板级封装

4.5.1 封装结构

对于低互连密度、高产量需求的芯片如电源管理芯片（power management integrated circuit，PMIC），德卡（Deca）提出的图 4.37 所示的 M-Series 技术[80]，通过利用临时面板支撑片可以降低成本。M-Series 是一种先上晶且芯片面朝上的工艺。

4.5.2 工艺流程

Deca 的 M-Series 工艺流程在参考文献 [80] 中有描述，如图 4.38 所示。从图中可以看到，先拾取合格芯片并面朝上放置在一块临时支撑片上，然后对这些芯片的顶部进行连续和全方位的模塑。如图 4.37 所示，在器件晶圆上电镀了铜柱以形成芯片和扇出 RDL 层的互连。这些全方位的模塑为芯片正面的介质层提供了机械保护，又额外充当了 KGD 和 RDL 之间的缓冲层。图 4.39 是典型 M-Series 扇出型先上晶且面朝上封装的典型扫描电镜截面图[80]。

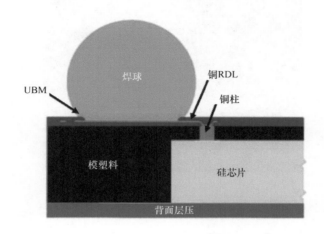

图 4.37 （上）M-Series 先上晶（芯片面朝上）扇出型封装；（下）临时面板支撑片[80]

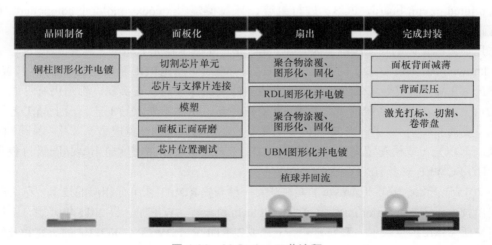

图 4.38 M-Series 工艺流程

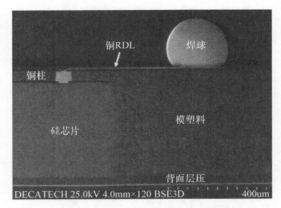

图 4.39　M-Series 扇出型封装截面图

4.6　扇出型（后上晶或先 RDL）晶圆级封装

自 2006 年开始，日本电气电子株式会社（NEC Electronics Corporation，现瑞萨电子株式会社（Renesas Electronics Corporation））一直在开发一种全新的 SMArt 芯片馈电互连转接板（SMArt chip connection with feed through interposer，SMAFTI）封装技术，用于芯片间宽带数据传输[95, 96]、在逻辑器件中集成 3D 堆叠存储颗粒[97-101]、实现晶圆级封装内系统（system-in wafer-level package，SiWLP）[102] 和先 RDL（或后上晶）扇出型晶圆级封装[103]。SMAFTI 所用馈电转接板（feed through interposer，FTI）是一种具有极细线宽和线距 RDL 的薄膜。FTI 的介质层通常采用 SiO_2 或聚合物，RDL 导线层通常采用铜。FTI 技术不仅能支持芯片正下方的 RDL 层，还能对芯片边沿之外的 RDL 提供支撑。在 FTI 底部植上面阵列焊球，可以与印制电路板（printed circuit board，PCB）进行组装；环氧模塑料（epoxy molding compound，EMC）用于埋置芯片并支撑 RDL 和焊球[11, 104]。

采用后上晶工艺的一个关键原因[11, 104] 是为了尽量减少在先上晶工艺制备 RDL 过程中合格芯片（known good die，KGD）的损失[11]。但这只有在后上晶先 RDL 基板能在如芯片 - 基板键合、底部填充和 EMC 模塑等其他工艺前快速完成功能性测试才能真正实现，否则仍有可能因为不合格的后上晶基板导致系统测试之后 KGD 要被丢弃。相比于先上晶工艺，后上晶工艺更加复杂且成本更高[11, 104]。除了制作 RDL，后上晶工艺还需完成晶圆凸点成型、芯片 - 晶圆 / 面板键合以及底部填充（或模塑填充）。因此，先 RDL 工艺通常用于高密度扇出型晶圆级封装或板级封装（HD FOWLP 或者 HD FOPLP）。

自 2015 年开始，安靠（Amkor）就在开发一种与先 RDL 工艺十分相似的技术，称为硅晶圆集成扇出技术（silicon wafer integrated fan-out technology，SWIFT）。自 2015 年开始，新加坡微电子研究所（IME）发表了一系列有关后上晶或先 RDL 的 FOWLP 和 FOPLP 的研究和开发技术论文[107-116]。2016 年，矽品科技（SPIL）展示了第一个窄线宽等离子体增强化学气相沉积（plasma-enhanced chemical vapor deposition，PECVD）SiO_2 与聚合物介质层的 RDL 的混合集成[117]。

2017 年，安靠（Amkor）为其无硅集成模块（silicon-less integrated module，SLIM）发表了类似的混合 RDL 技术 [118, 119]。2017 年，欣兴电子（Unimicron）发表了在 370mm×470mm 面板上最小线宽为 8μm 的先 RDL 技术的论文 [120]。2018 年，三星（Samsung）发表了 2 篇技术论文，1 篇利用先 RDL 技术制备无硅 RDL 并应用于高性能计算（High-performance computing，HPC）[121]，1 篇是应用于先进可移动终端的"肩并肩"系统级封装 [122]。从 2018 年开始，中国台湾"工业技术研究院（ITRI）"发表了关于先 RDL FOWLP 和 FOPLP 的论文 [123-125]。2019 年，日月光（ASE）在它的基板上扇出型芯片（fan-out chip on substrate，FOCoS）技术中采用了先 RDL 工艺，台积电（TSMC）也开始采用先 RDL 工艺制作用于高密度异质集成大封装的 RDL 转接板。2019 和 2020 年，神钢株式会社（Shinko）也开始采用先 RDL 技术制作高密度有机封装基板。2020 年，欣兴电子（Unimicron）发表了 510mm×515mm 面板上最小金属线宽小到 2μm 的先 RDL 技术。截至本书写作时，以上先 RDL FOWLP 和 FOPLP 工艺都还未实现大规模量产。

4.6.1 IME 的先 RDL FOWLP

IME 在扇出型后上晶或者先 RDL 工艺方面已经发表了许多文章 [107-116]。本节会简要介绍 IME 基于圆形（晶圆）临时支撑片的后上晶技术。

4.6.2 测试结构

图 4.40 所示为测试结构。图中共可以看到 3 颗测试芯片：芯片 1（9.0mm×8.0mm）、芯片 2（5.0mm×4.0mm）、芯片 3（3.0mm×2.0mm），它们的凸点节距分别为 120μm、80μm、60μm。测试封装体尺寸为 20mm×20mm，封装体内共有 3 层 RDL，金属布线线宽 / 线距（L/S）分别为 2μm/2μm、5μm/5μm 和 10μm/10μm。封装体共有 2400 个焊球，焊球节距为 400μm。

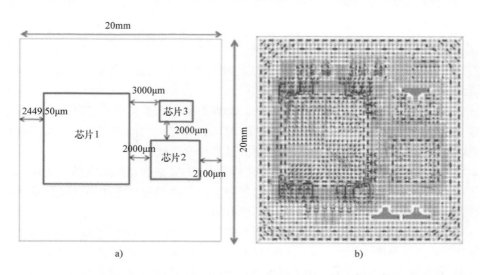

图 4.40 多芯片 FOWLP 布局：a）芯片分布图；b）菊花链和封装级凸点布局 [109]

4.6.3 先 RDL 关键工艺步骤

图 4.41 所示为制作扇出型先 RDL 封装的关键工艺步骤。从图中可以看到，临时支撑片是一块玻璃晶圆，晶圆上有一层作为牺牲层的激光剥离材料。第一步先在玻璃支撑片晶圆背面用 PVD 沉积一层 Ti 层，使得玻璃晶圆背面不透明以便后续夹持工艺操作。同时在临时支撑片正面制作好牺牲剥离层。第二步在带有牺牲剥离层的支撑片表面制作 UBM 以及多层线宽/线距（L/S）分别为 10/10μm、5/5μm、2/2μm 的 RDL。

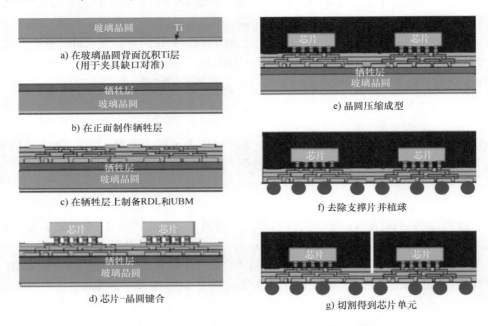

图 4.41 先 RDL FOWLP 制作集成工艺流程 [109]

并行地，在器件晶圆上采用标准 PVD 工艺、电镀铜和焊料工艺完成微凸点成型，之后将晶圆切割得到制作好微凸点的单颗芯片。

第三步，利用芯片-晶圆（chip-on-wafer，CoW）键合技术实现具有微凸点的器件芯片与临时支撑片上的多层 RDL 薄膜的键合。芯片和 RDL 薄膜之间的间隙利用毛细底部填充料进行填充。第四步，对整个晶圆进行模压以实现对测试芯片的包封。

最后将临时晶圆解键合，并清洗牺牲层材料露出 UBM 进行植球。在整个流程中，晶圆模塑放在了最后，这样可以防止在 RDL 制备过程中发生芯片偏移、芯片突出以及晶圆翘曲等问题，同时在临时支撑片上可以成功制作窄节距的 RDL。

图 4.42 为光刻胶（PRA 和 PRB）对线宽/线距分别为 2/2μm 和 5/5μm RDL 的影响。从图中可以看到，PRA（见图 4.42a、b）的侧壁和 PRB（见图 4.42c、d）的侧壁相比近乎垂直。

图 4.43 为光敏介质（PIA 和 PIB）对 RDL 过孔形貌的影响。两种介质材料都在 200℃下固化 1h，且如图 4.43 所示，可以开出直径 3μm 和 5μm 的过孔且侧壁形貌近乎垂直。

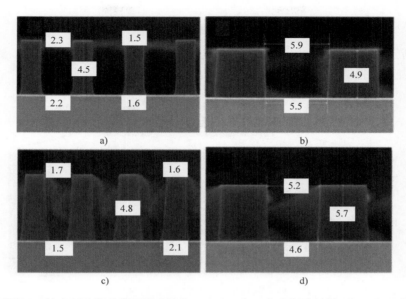

图 4.42 窄节距 RDL 的光刻胶形貌截面图（单位：μm）：a）、b）光刻胶材料 A——RDL 线宽/线距分别为 2μm/2μm 和 5μm/5μm；c）、d）光刻胶材料 B——RDL 线宽/线距分别为 2μm/2μm 和 5μm/5μm[109]

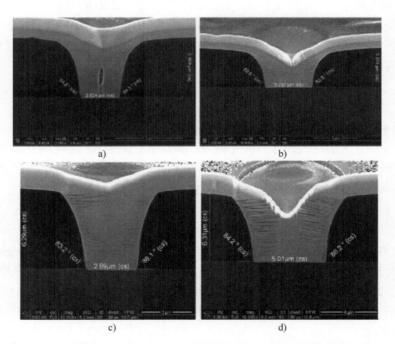

图 4.43 介质层过孔形貌截面图：a）、b）介质层材料 A——过孔开窗直径分别为 3μm 和 5μm；c）、d）介质层材料 B——过孔开窗直径分别为 3μm 和 5μm[109]

4.6.4 先 RDL FOWLP 的 PCB 组装

图 4.44 所示为先 RDL FOWLP 在 PCB 上组装的完整集成图。从图中可以看到，3 层 RDL 制作成功，且芯片 - 晶圆键合和焊点完成得很好。

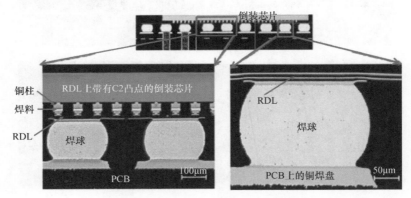

图 4.44　FOWLP 与 PCB 组装截面图[109]

4.7　扇出型（后上晶或先 RDL）板级封装

本节将介绍 3 颗芯片（1 颗大芯片、2 颗小芯片）在 3 层 RDL 基板（金属层的最小线宽/线距可达到 2/2μm）异质集成的材料、工艺、制作和可靠性，RDL 基板是在 515mm × 510mm 的面板上制作的[130, 131]。该结构的物理意义是应用处理器的芯片组，大芯片可以是处理器芯片，小芯片可以是存储器芯片。

4.7.1　测试芯片

图 4.45 所示为测试芯片。大芯片（芯片 1）的尺寸为 10mm × 10mm，小芯片（芯片 2）的尺寸为 7mm × 5mm。大芯片上共有 3760 个焊盘，焊盘节距为 90mm；小芯片上共有 1512 个焊盘，焊盘节距为 60mm。两颗芯片上的微凸点相同：①铜柱的直径和高度分别是 35μm 和 37μm；② Ni 阻挡层厚度为 3μm；③ SnAg 焊料帽厚度为 15μm。

4.7.2　测试封装体

图 4.46 为测试封装体的截面图和俯视图。从图中可以看到，20mm × 20mm 的基板上有 3 颗芯片（1 颗大芯片和 2 颗小芯片）。基板总共包含 3 层 RDL，RDL1 第一层金属层（ML1）的线宽/线距为 2/2μm；RDL2 第二层金属层（ML2）的线宽/线距为 5/5μm；RDL3 第三层金属层（ML3）的线宽/线距为 10/10μm。ML1、ML2、ML3 的厚度分别为 2.5μm、2.5μm 和 8μm。介质层（DL）的厚度 DL01（铜焊盘和 ML1 之间的 DL）、DL12（ML1 和 ML2 之间的 DL）、DL23（ML2 和 ML3 之间的 DL）分别为 5μm、6.5μm 和 6.5μm，如图 4.47a 所示。阻焊层开窗直径和厚度分别为 245μm 和 5μm。

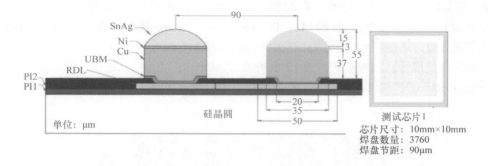

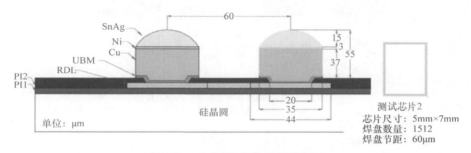

图 4.45 测试芯片的截面示意图

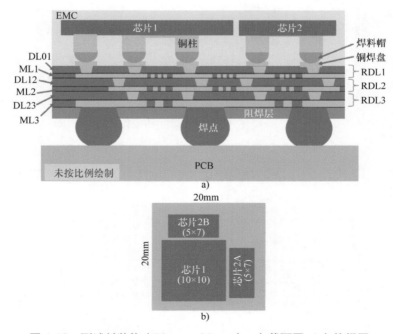

图 4.46 测试封装体（20mm×20mm）：a）截面图；b）俯视图

图 4.47b 是 3 颗芯片 RDL 基板的俯视图,图 4.47c 是从 PCB 焊球面的底视图。从图中可以看到,基板顶部有 6784 个焊盘,焊盘节距分别为 90μm 和 60μm;基板底部有 2780 个焊盘,焊盘节距为 0.35mm。如图 4.48 所示,20mm×20mm 的 RDL 基板是在 515mm×510mm 的面板上制作完成的,从图中可以看到,一次性可以生产 396 块这样的基板(20mm×20mm)。当 RDL 基板制作完成后,面板被切割成 12 根条带(240mm×74mm),每根条带上有 33 个封装基板(20mm×20mm)。后续的组装和工艺步骤均以条带形式完成。

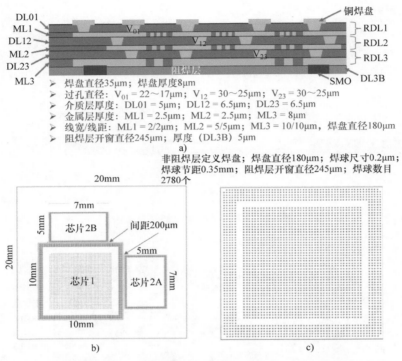

图 4.47　a)三层先 RDL 基板;b)俯视图;c)底视图

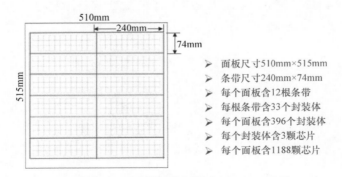

图 4.48　制作先 RDL 基板的面板(515mm×510mm)

4.7.3 异质集成用先 RDL 板级封装

制作先 RDL 基板、表面处理、芯片 - 基板键合、底部填充、环氧模塑料（EMC）模塑、阻焊层开窗（solder resist opening，SRO）、植球以及切割等关键工艺流程示于图 4.49a、b 中。与如图 4.6a、b 所示的参考文献 [130] 中制作 RDL 基板的工艺相比，两者有很大的不同。主要区别在于参考文献 [130] 中，RDL 制作基于 2/2μm 线宽 / 线距，而本工艺的 RDL 制作基于 10/10μm 线宽 / 线距，而且连接不同金属层的过孔直径也不同。另外，本工艺在进行芯片 - 基板键合前不需要将 RDL 基板转移到另一块临时支撑片上。

4.7.4 RDL 基板的制作

首先将可剥离薄膜（牺牲层）采用狭缝式涂布在临时玻璃支撑片（515mm×550mm）上，然后采用狭缝式涂布感光介质（photoimageable dielectric，PID）作为阻焊层（或者钝化层）DL3B，如图 4.49a 所示。然后采用物理气相沉积在支撑片上溅射一层 Ti/Cu 种子层，之后进行涂布光刻胶、激光直写图形（laser direct imaging，LDI）和显影工艺，再用电化学沉积法（electrochemical deposition，ECD）镀铜，最后剥离光刻胶并刻蚀 Ti/Cu 层就得到 10/10μm 的 RDL3 的金属层（ML3）。

接下来继续用狭缝式涂布 PID 并用 LDI 制作获得 RDL3 的介质层（DL23）。然后溅射 Ti/Cu 种子层、狭缝式涂布光刻胶、LDI、显影、ECD 镀铜。之后剥离光刻胶、刻蚀 Ti/Cu 种子层，获得 RDL2 的 5/5μm 的金属层（ML2）。继续狭缝式涂布 PID 并用 LDI 制作得到 RDL2 的介质层（DL12）。重复以上步骤得到 RDL1 的 2/2μm 的金属层（ML1）和介质层（DL01）。再次溅射 Ti/Cu、狭缝式涂布光刻胶、LDI、显影、ECD 镀铜，继续剥离光刻胶、蚀刻 Ti/Cu 得到芯片的键合焊盘（引脚），如图 4.49a 最后一步所示。

接着，将带有 396 个 RDL 基板的玻璃面板（515mm×510mm）切成 12 根条带。每根条带尺寸为 240mm×74mm，每根条带上有 33 个 RDL 基板（20mm×20mm）。图 4.50 为面板、条带以及单颗 RDL 基板。

图 4.51 为 RDL 基板典型截面的 SEM 照片。截面显示了 RDL1、RDL2、RDL3 的 3 个金属层（ML1、ML2、ML3）。图 4.51 下面是放大的 SEM 图像，可以看到 RDL1 的实际线宽为 2.15μm、2.24μm，并非目标设定的 2μm；实际线距为 1.4μm、1.49μm，并非目标设定的 2μm；实际厚度为 1.68μm，也并非目标设定的 2.5μm。这表明该工艺还有提升空间（比如采用更好的 PID、更高精度的 LDI 或者步进式光刻）。相比之下，ML2 的线宽、线距和厚度就好多了，它们分别为 5.14μm、4.77μm 和 2.15μm，目标设定值分别为 5μm、5μm 和 2.5μm。ML3 的线宽和厚度分别为 10.38μm 和 7.49μm，这与目标设定值的 10μm 和 8μm 非常接近。这一结果的出现很好理解，因为线宽 / 线距值越小，制作时的误差越大。

芯片 - 基板键合前 RDL 基板制作的最后一步是铜焊盘的表面处理。在本研究中表面处理采用化学镀钯浸金（electroless palladium immersion gold，EPIG），图 4.52 为 RDL 基板局部位置的表面处理结果。

图 4.49 a)扇出型先 RDL 板级封装的工艺流程(1); b)扇出型先 RDL 板级封装的工艺流程(2)

第 4 章 扇出型晶圆级 / 板级封装

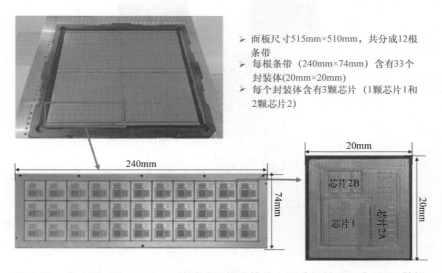

- 面板尺寸515mm×510mm，共分成12根条带
- 每根条带（240mm×74mm）含有33个封装体(20mm×20mm)
- 每个封装体含有3颗芯片（1颗芯片1和2颗芯片2）

图 4.50　在 515mm×510mm 的玻璃面板支撑片上制造好的三层先 RDL 基板

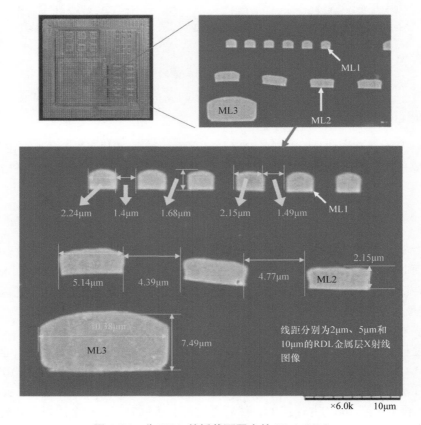

线距分别为2μm、5μm和10μm的RDL金属层X射线图像

图 4.51　先 RDL 基板截面图中的 ML1~ML3

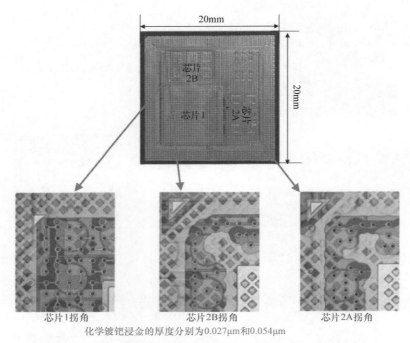

图 4.52　先 RDL 基板键合焊盘的表面处理（EPIG）

4.7.5　晶圆凸点成型

如图 4.45 所示，在测试晶圆上制备 C2（chip connection，芯片互连）凸点。在凸点成型后，将晶圆切割成单颗芯片。每颗测试芯片凸点的结构都包含铜柱、Ni 阻挡层以及 SnAg 焊料帽三个部分。

4.7.6　芯片 - 基板键合

现在开始进行芯片 - 基板的键合。如图 4.49b 所示，由于有临时玻璃支撑片，基板条带都很坚固，表面非常平整，适于键合。同时，由于 2/2μm 线宽 / 线距的 RDL1 金属层 ML1 面朝上，因此不需要在键合前将 RDL 基板条带转移到另一临时支撑片上。这是与现有工艺以及参考文献 [130] 工艺的最大差异。

第一步，在室温下将所有芯片拾取和放置（pick and place，P&P）在条带上。大芯片（芯片 1）P&P 过程中的吸头参数示于图 4.53 中，可以看到温度先从 75℃ 快速上升至 250℃，然后升温到 275℃ 并保持 2.5s，再快速降至 75℃，整个过程中吸头施加的力较小（10N）。小芯片（芯片 2A 和芯片 2B）P&P 过程中的吸头参数也示于图 4.53 中，可以看到温度先从 75℃ 快速上升至 225℃，再上升到 260℃ 并保持 1s，然后再快速下降到 75℃，整个过程中吸头施加的力非常小（5N）。

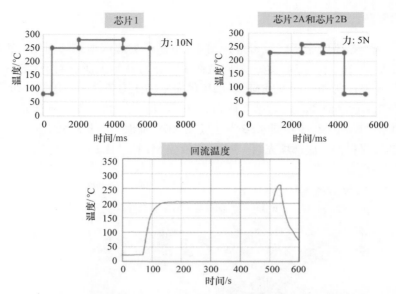

图 4.53 大芯片和小芯片芯片-基板（时间对温度）的键合条件；所有芯片的批量回流条件

所有芯片（3×3 = 99 颗）在 RDL 基板上的 P&P 操作完成后，被放进回流炉中进行所有芯片的批量回流。回流温度曲线如图 4.53 所示。从图中可以看到，对于 SnAg 焊料帽最高温度为 250℃。图 4.54a 为一根含有 33 块 RDL 基板的条带，图 4.54b 显示每个基板上有 3 颗芯片。与参考文献 [1] 中芯片-基板键合是通过热压键合（1 次键合 1 颗芯片）实现的工艺不同，现在这种大批量回流的方式拥有更高的产率。

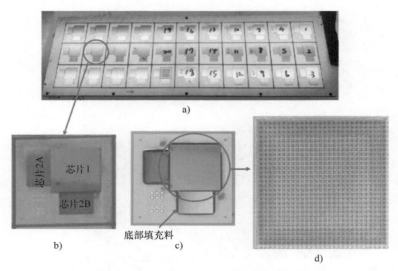

图 4.54 a）经批量回流后含有 33 个异质集成封装体的条带；b）大倍数图片；
c）底部填充后的大倍数图片；d）其中一颗大芯片的 X 射线图像

4.7.7 底部填充和 EMC 模塑

完成芯片-基板条带键合后，就可进行图 4.49b 所示的底部填充和固化。底部填充的固化条件是 165℃下固化 2h，图 4.54c 是一个典型例子。在对条带上所有封装体进行底部填充后，在整个条带上层压厚度为 400μm 的 EMC。层压的温度和压力条件为：①第一阶段，120℃下真空 30s，并压合 30s（5.68MPa）；②第二阶段，100℃下压合 60s（5.58MPa）。图 4.54d 为一颗大芯片的 X 射线图像，图 4.55 和图 4.56 为连接在 RDL 基板上的芯片截面图。从图中可以清楚地看到铜柱、芯片焊料帽、底部填充料、RDL 基板的 ML1、ML2、ML3。

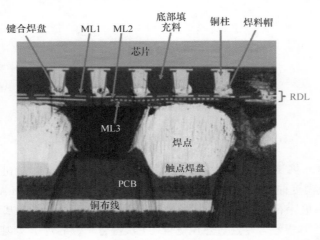

图 4.55　截面图显示芯片-RDL 基板键合、植球以及在 PCB 上的 SMT 回流（1）

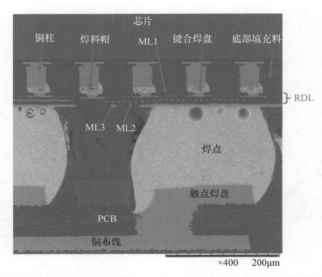

图 4.56　截面图显示芯片-RDL 基板键合、植球以及在 PCB 上的 SMT 回流（2）

4.7.8 面板/条带转移

根据 EMC 的厚度选择是否需要将条带进行转移。如果 EMC 的厚度足够厚（>500μm），即埋置异质集成封装的 EMC 条带足够坚硬，可以抵抗临时支撑片去除后的大翘曲（>1mm），那就没有必要将条带转移到另一块临时支撑片上。否则，就需要将条带转移到另一块支撑片上以确保后续如植球等工艺的进行。本研究先将条带转移到了另一块临时玻璃支撑片上，再移除原先的临时玻璃支撑片，然后进行阻焊层开窗和铜接触焊盘的表面处理，如图 4.49b 所示。在本研究中，为了节省 EMC 材料和实现一个薄外形的异质集成封装，EMC 的厚度仅为 400μm，因此将条带转移到另一个临时玻璃支撑片上是必须的。图 4.57 所示为条带转移到另一块支撑片上，并移除原先的玻璃支撑片的情况，最大翘曲量为 0.7mm。

图 4.57　芯片-基板键合、EMC 模塑后将条带与原先玻璃支撑片解键合并转移到另一块玻璃支撑片上

4.7.9 阻焊层开窗和表面处理

如图 4.49b 所示，在用激光去除第一块临时玻璃支撑片后，可以进行阻焊层开窗和表面处理操作。通过 CO_2 激光 4 次脉冲可以在阻焊层开出直径为 245μm 的窗口。然后如图 4.58 所示，用等离子体刻蚀 PID 460s，化学刻蚀 Ti 层 270s。最后用 EPIG 工艺对焊盘做表面处理，得到 Pd 层和 Au 层的厚度分别为 0.056μm 和 0.069μm。Pd 层和 Au 层的目标厚度值为 0.05μm。

4.7.10 植球、解键合和条带切割

完成阻焊层开窗和 EPIG 表面处理，接下来进行植球。图 4.59 所示为条带上完成的植球（平均焊球高度 154.7μm）及单个焊球的尺寸，焊球材料为 Sn3Ag0.5Cu 合金。在植球完成后，将第二块临时玻璃支撑片也从条带上移除。最后，将条带切割得到单个异质集成封装体。

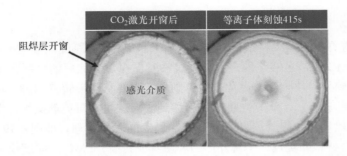

图 4.58 阻焊层开窗、等离子体刻蚀、Ti 层刻蚀、EPIG 表面处理

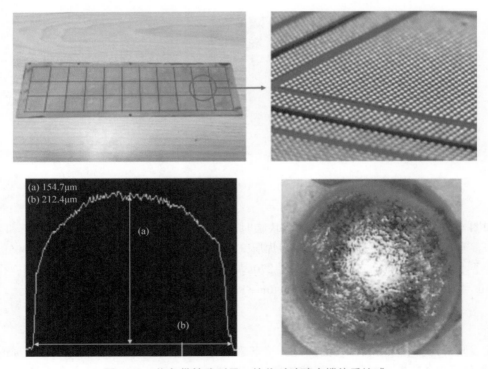

图 4.59 将条带转移到另一块临时玻璃支撑片后植球

4.7.11 先 RDL 板级封装的 PCB 组装

异质集成封装的可靠性评估采用跌落实验进行。

1. 印制电路板

异质集成封装体 PCB 为 FR-4 制作，如图 4.60 所示。板上共有 8 颗封装体，PCB 尺寸为 132mm×77mm×0.992mm，为 8 层板。每个封装体有 2780 个焊盘（焊盘节距为 0.35mm）。焊盘为无阻焊层定义，直径为 0.2mm，表面用有机可焊性保护层进行处理。阻焊层开窗直径为 0.245mm，植球焊盘直径为 0.18mm。

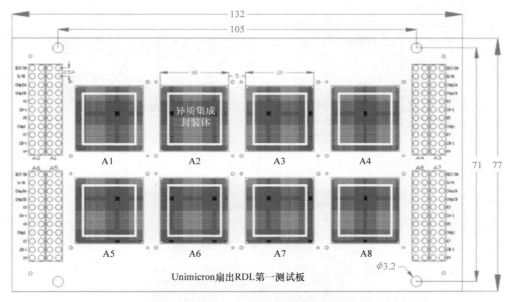

图 4.60 有 8 颗 20mm×20mm 异质集成封装体的可靠性测试 PCB

2. 封装体的表面安装技术（surface mount technology，SMT）组装

图 4.61 为异质集成封装体到 PCB 的表面安装技术[151]组装。图 4.61a 为回流温度曲线，从图中可以看到，回流最高温度为 245℃。图 4.61b 为组装了 8 个异质集成封装体 PCB。图 4.55、图 4.56 和图 4.61c 是典型的 PCB 组装结构截面图，从图中可以清楚地看到芯片、铜柱、焊料帽、EMC、RDL 金属层、焊点和 PCB。

4.7.12 跌落试验结果和失效分析

如图 4.62 所示，冲击（跌落）试验装置根据 JEDEC 标准 ESD22-B111 搭建。装置包括一个跌落塔和一个用于放置 PCB 样品的跌落平台。装置一共有 22 个独立通道，外加 1 个公共通道和 1 个数据采集系统（data acquisition system，DAS）。1500g/ms 跌落谱（1500g，0.5ms 半正弦脉冲）如图 4.63 所示。跌落共进行 30 次，失效的判据定义为测得的样品电阻值达到 1000Ω（见图 4.64a），电阻值低于 1000Ω 就认为没有发生失效（见图 4.64b）。

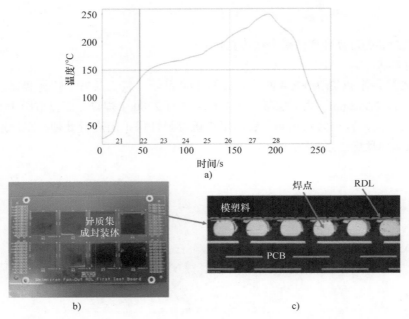

图 4.61　a）SMT 回流温度曲线；b）组装后的测试板；c）PCB 组装结构截面图

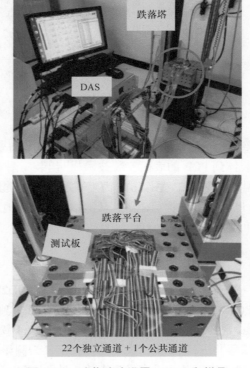

图 4.62　跌落试验设置、DAS 和样品

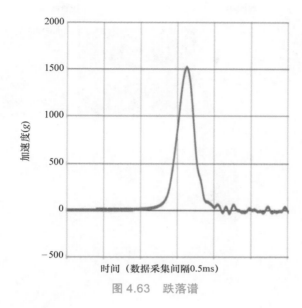

图 4.63 跌落谱

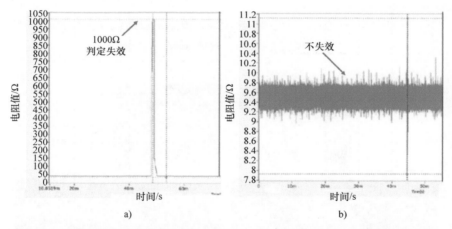

图 4.64 失效判据：a）当电阻值 ≥ 1000Ω 时，判定失效；b）否则判定不失效

在 30 次跌落后，只在第 23 次跌落时发生了一个样品失效，通过 DAS 确定是第二个封装体发生失效（见图 4.65）。如图 4.65 所示，样品的 X 射线图像和截面图像表明失效位置发生在靠近 PCB 中间侧的封装拐角附近焊点。然而，这个特别的焊点一开始就有问题，是一个"枕头效应"（head-in-pillow）的焊点[151]。在进行跌落试验前，DAS 无法将它分辨出来，因此将其视为好焊点。经过第 23 次跌落后，焊点"头"部（head）已经与"枕"部（pillow）完全分离，也就是说焊点完全裂开了。该数据不应计入最终测试结果，应认为 PCB 组装顺利通过了跌落试验。然而，我们对一些未发生失效（电阻值远小于 1000Ω）的样品做了分析，如图 4.64b 所示；结果表明这些样品在焊点处有存在微裂痕，如图 4.66 所示。

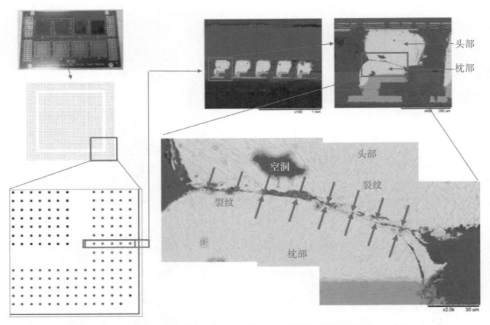

图 4.65 跌落试验中唯一的失效样品的失效位置和失效模式分析。可以看到该样品在测试前质量就不合格

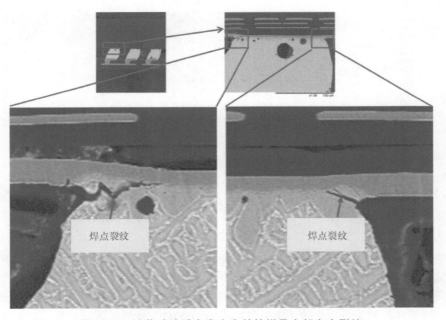

图 4.66 跌落试验后未发生失效的样品内部有小裂纹

4.7.13 热循环试验结果和失效分析

如图 4.67 和图 4.68 所示,异质集成封装体 PCB 组装被放置在热循环试验箱中。试验箱腔体内的温度通过热电偶(在空气中)测量,如图 4.69 所示。从图中可以看到,温度曲线以 50min 为一个周期,在 −55℃ ⇌ 125℃ 之间进行变化。测试箱共有 49 个外接通道,每个通道连接一排对应的外侧焊点。

1. 失效判据

在本研究中,热循环试验的失效判据定义为异质集成封装体与 PCB 组装的菊花链电阻值升高 50%。失效周期(cycle-to-failure)定义为封装体第一个焊点发生失效时所经历的周期。

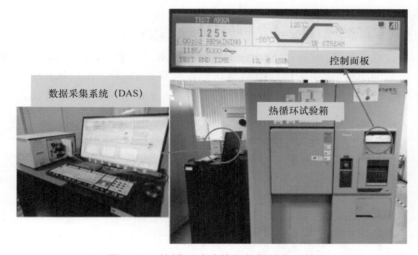

图 4.67 热循环试验箱和数据采集系统

图 4.68 热循环试验箱内带有封装体的测试板

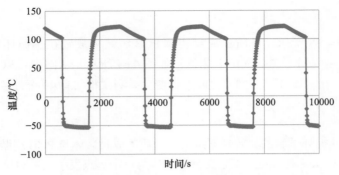

图 4.69　热循环测试温度曲线

2. 中位秩条件下的试验结果

表 4.3 列出了异质集成封装体 PCB 组装的热循环测试结果。从表中可以看到共有 12 个样品发生失效。中位秩由公式 $F(x)=(j-0.3)/(n+0.4)$ 得出，其中 j 为失效发生的序数，n 为样品数量（$n = 49$）。图 4.70 为采用中位秩绘制得到的数据 Weibull 分布图。从图中可以看到，样品的特征寿命（θ）（63.2% 样品失效）为 1702 个周期，Weibull 斜率（β）为 2.28。

表 4.3　中位秩、5% 秩、95% 秩条件下的热循环测试结果

失效序数	失效周期	$F(x)$		
		中位（50%）秩	90% 置信度	
			5% 秩	95% 秩
1	300	1.41	0.1	5.93
2	400	3.43	0.73	9.32
3	552	5.46	1.69	12.3
4	552	7.48	2.84	15.07
5	552	9.51	4.11	17.71
6	552	11.53	5.47	20.27
7	624	13.56	6.9	22.74
8	788	15.58	8.39	25.16
9	872	17.6	9.93	27.54
10	872	19.63	11.51	29.86
11	1003	21.65	13.13	32.15
12	1003	23.68	14.78	34.41

注：样品数量为 49 个。

3. 其他秩条件下的测试结果

对于 90% 的置信度，我们需要 5% 秩和 95% 秩条件下的结果。我们也将其列于表 4.3 中。举个例子，当样品数为 49 且在 5% 秩条件下，$j = 1$ 的样品失效百分比为 0.1%，$j = 2$ 的样品失效百分比为 0.73%。当样品数为 49 且在 95% 秩条件下，$j = 1$ 的样品失效百分比为 5.93%，$j = 2$ 的样品失效百分比为 9.32%。图 4.71 为 90% 置信度下异质集成封装体 PCB 组装焊点的 Weibull 分布图。从图中可以看到，90% 置信度下，10% 的样品发生失效的失效周期在 387~817 个周期之间。

第 4 章 扇出型晶圆级 / 板级封装 171

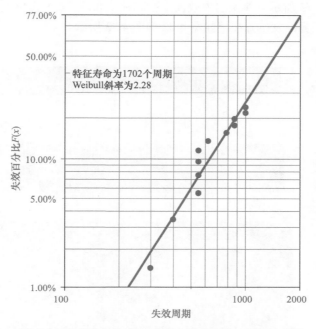

图 4.70 中位秩条件下异质集成焊点的 Weibull 分布图

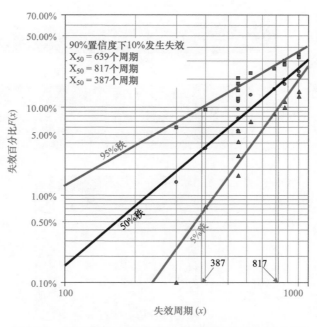

图 4.71 90% 置信度下异质集成焊点的 Weibull 分布图

4.6 面模塑 PLCSP PCB 组装的失效位置和失效模式

如图 4.72 和图 4.73 所示,典型的失效发生在芯片 1 和芯片 2A 拐角下的焊点附近。失效模式是焊料与 RDL 基板界面发生开裂,如图 4.72 和图 4.73 所示。

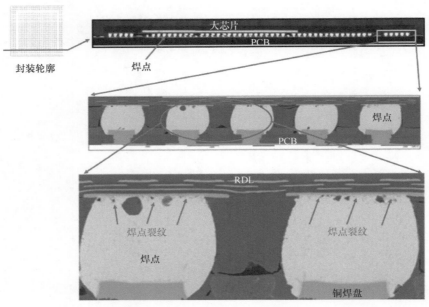

图 4.72　失效位置和失效模式(1)

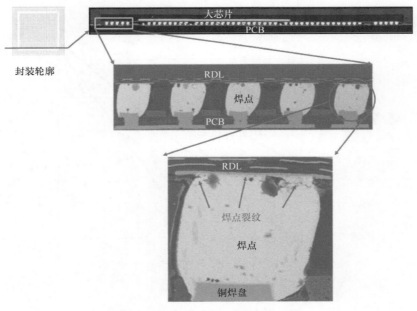

图 4.73　失效位置和失效模式(2)

5. 真 Weibull 斜率

置信度定义为根据样本测试数据给定的区间所得到总体的参数的概率（以 Weibull 分布为例，相应参数为特征寿命和 Weibull 斜率）。可以这么说，置信度的概念适用于测试，而可靠性的概念适用于产品本身！

为预测真实的 Weibull 斜率（β_t），需要估计与样品数量和所需置信度相关的 Weibull 斜率误差。Weibull 斜率的误差（E）取决于失效发生的数量（N）和所要求的置信度（C），具体如下式所示[151]：

$$\frac{1}{\sqrt{\pi}} \int_{-\infty}^{E\sqrt{2N}} e^{-\frac{t^2}{2}} dt = \frac{1+C}{2}$$

此积分不能以闭合形式进行，可近似为下式：

$$\frac{1}{\sqrt{\pi}} \int_{-\infty}^{E\sqrt{2N}} e^{-\frac{t^2}{2}} dt = 1 - Z(x)\left(b_1 t + b_2 t^2 + b_3 t^3 + b_4 t^4 + b_5 t^5\right) + \varepsilon(x)$$

式中，

$$Z(x) = \frac{1}{\sqrt{2\pi}} e^{-\frac{x^2}{2}}$$

$$t = \frac{1}{1+px}$$

$$x = E\sqrt{2N}$$

$$p = 0.2316419$$

$$b_1 = 0.31938153$$

$$b_2 = -0.356563782$$

$$b_3 = 1.781477937$$

$$b_4 = 1.821255978$$

$$b_5 = 1.330274429, \quad |\varepsilon(x)| < 7.5 \times 10^{-8}$$

于是，对于一个给定的 C（置信度）和 N（失效数量），E（误差）可以由上式求得。在本例中，对于 90% 的置信度，即 $C = 0.9$，$N = 12$，我们可以得出 $E = 0.32$。因此，该异质集成焊点的真 Weibull 斜率 β_t 会落在区间（$2.28 - 0.32 \times 2.28$）$\leq \beta_t \leq$（$2.28 + 0.32 \times 2.28$）（即 $1.55 \leq \beta_t \leq 3.01$）。

6. 线性加速模型

对于无铅焊料合金，有很多线性加速模型[151]。它们基本都基于以下类型[152]：①保持时间和最大温度[153]；②频率和最大温度[154]；③频率和平均温度[155]。本研究采用由 Lall 等[154] 提出的一个模型：

$$\alpha = \left(\frac{\Delta T_t}{\Delta T_o}\right)^{2.3} \left(\frac{f_o}{f_t}\right)^{2.3} \exp\left[4562\left(\frac{1}{T_{max,o}} - \frac{1}{T_{max,t}}\right)\right]$$

式中，$T_{max,t}$、f_t、ΔT_t 和 $T_{max,o}$、f_o、ΔT_o 分别是循环测试/服役中最大温度（K）、温度循环频率、温度范围（℃）。在本例中，$\Delta T_t = 180℃$，$T_{max,t} = 125℃$，$f_t = 28.8$ 个周期/天，且我们知道焊点可以经受住高达 300 个周期（见表 4.3 和图 4.70）。然而，我们还需要知道这 300 个周期的值是否已经足够满足 5 年的服役条件（每天经历 0~80℃ 的一个周期变化，即 $\Delta T_o = 80℃$、$T_{max,o} = 80℃$、$f_o = 1$ 个周期/天）。

$$\alpha = \left(\frac{180}{80}\right)^{2.3} \left(\frac{1}{28.8}\right)^{2.3} \exp\left[4562\left(\frac{1}{273+80} - \frac{1}{273+125}\right)\right]$$

$$\alpha = 10.76$$

该焊点会经历 10.76×300 个周期/天，也就代表在实际服役条件下可以经历 3228÷365 = 8.84 年 >5 年。

4.7.14 热循环仿真

参考文献 [130] 对异质集成封装体 PCB 组装进行了仿真。图 4.74 为不同位置焊点的累积蠕变应变随时间变化的曲线。从图中可以看到，最大累积蠕变应变发生在芯片 1 下方的拐角焊点 A，第二大累积蠕变应变发生在芯片 2A 下方的拐角焊点 F。也就是说，任何焊点的失效都

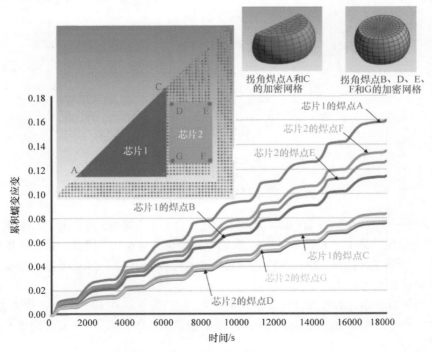

图 4.74 不同位置焊点累积蠕变应变随时间的变化

应率先发生在该位置。该结论得到了图 4.72 和图 4.73 所示的热循环试验结果的验证。

图 4.75 和图 4.76 分别是焊点 A 和焊点 F 在 450s（85℃）和 2250s（-40℃）下的累积蠕变应变。从图中可以看到，85℃和-40℃的焊点最大累积蠕变应变都发生在焊料和 RDL 基板的界面。因此，焊点的失效模式是焊料与 RDL 界面开裂。该结论同样得到热循环试验结果的验证（见图 4.72 和图 4.73）。

4.7.15 总结和建议

一些重要的结论和建议总结如下：

1）对 3 颗芯片异质集成封装体 PCB 组装进行了热循环试验。试验共有 49 条测试通道，每条通道对应一排外侧焊点。测试条件是 50min 内完成 -55℃ ⇌ 125℃的一个周期变化。

2）失效判据定义异质集成封装体 PCB 组装的菊花链电阻值上升 50%。某一个通道第一次检测出失效时所经历的周期定义为异质集成封装的失效周期。

3）在中位秩条件下，PCB 组装的焊点 Weibull 斜率以及特征寿命分别是 2.28 和 1702 个周期。

4）在 90% 的置信度下，真 Weibull 斜率分布为 $1.55 \leq \beta_t \leq 3.01$，真 10% 寿命为 387 个周期 $\leq x_t \leq$ 817 个周期。

5）异质集成封装体 PCB 组装的典型失效位置发生在芯片 1 的拐角下方（A）和芯片 2A 的拐角下方（F）。

6）异质集成封装体 PCB 组装的典型失效模式是在焊点和 RDL 基板的互连界面发生开裂。

7）在服役条件（每天经历 0~80℃的一个周期变化）下，所研究的 3 颗芯片异质集成封装的焊点寿命可以达到 8.84 年。

8）通过一个异质集成封装体 PCB 组装的非线性、时间和温度相关的 3D 有限元模型验证了实际热循环测试过程中的失效位置和失效模式。

9）更多有关扇出型封装的信息，请见参考文献 [16，104]。

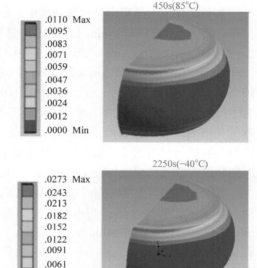

图 4.75　450s（85℃）和 2250s（-40℃）时焊点 A 的累积蠕变应变

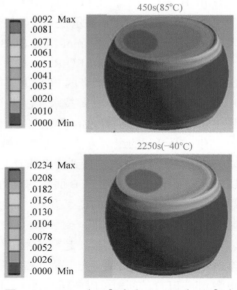

图 4.76　450s（85℃）和 2250s（-40℃）时焊点 F 的累积蠕变应变

4.8 Mini-LED RGB 显示器的扇出型板级封装

本节研究先上晶扇出型板级封装工艺制作 mini-LED RGB 显示器的可行性。重点在于设计、材料、工艺、制造以及 mini-LED RGB 显示器封装在 PCB 上的可靠性。mini-LED 规格如下：红色（R）LED 为 125μm×250μm×100μm，绿色（G）LED 为 130μm×270μm×100μm，蓝色（B）LED 为 130μm×270μm×100μm。RGB mini-LED 阵列的间距为 80μm，像素之间的间距也约为 80μm，像素节距为 625μm。制作 RDL 基板的玻璃面板尺寸为 515mm×510mm×1.1mm。为了提升 PCB 组装的良率，mini-LED 阵列以 4 个 1 组（2×2 像素点）的形式组装为表面安装器件（surface mount device，SMD），即每个 SMD 包括 12 个 R、G、B mini-LED。为对 mini-LED 封装进行跌落试验，我们还设计制造了一块 132mm×77mm 的 PCB，并利用非线性、温度、时间相关有限元模型对 mini-LED SMD PCB 组装进行热循环模拟。

发光二极管（light emitting diode，LED）是一种在电流经过时会发光的特殊半导体器件。当电流通时，器件变亮；当电流断时，器件变暗。传统 LED 的平均尺寸为 1mm[156]，而一个 mini-LED 的尺寸在 75~300μm 之间[157]，一个 micro-LED 则 ≤ 75μm[157]。自从 2020 年 4 月微星科技（Micro-Star International，MSI）出货的 Creator 17 游戏笔记本电脑上开始配备 mini-LED 背光液晶显示器（liquid-crystal display，LCD）[158]，mini-LED 显示技术开始得到大量关注。例如，苹果（Apple）[159] 准备在 2021 年第一季度发布它的第一款带有 mini-LED 背光 LCD 屏幕的 iPad Pro，并准备在 2021 年第二季度开始大规模量产；三星电子（Samsung Electronics）[160] 计划在 2021 年出货 200 万台 mini-LED 背光 LCD 电视机。

目前至少有两种 mini-LED 显示技术：一种是 mini-LED 背光 LCD 技术，另一种是 mini-LED RGB 显示技术。在 LCD 技术中，每个颜色透镜定义一个子像素，通过控制颜色透镜来产生图像；mini-LED RGB 显示技术中，每个 mini-LED 自身就是一个子像素，通过控制 mini-LED 来产生图像。本研究关注 mini-LED RGB 显示技术。

目前大部分的 mini-LED RGB 显示技术的组装方式是先从 LED 晶圆上拾取单颗的 mini-LED，然后放置到印有焊膏的 PCB 上进行 SMT 回流[151]。这是一种非常老的板上 mini-LED（芯片）（COB）技术[161]。这种技术存在两个不足：一方面 mini-LED 间的节距会受到 PCB 的限制，另一方面 PCB 电镀铜焊盘的平整性问题会导致组装的良率不高。

有一种可行的组装方法是利用塑料球栅阵列（plastic ball grid array，PBGA）封装，它可以将部分 mini-LED 先连接到有机基板上，然后再通过焊球连接到 PCB。这样能稍微减小 mini-LED 的节距，也可消除 PCB 面临的压力。但是这种方法还是无法消除有机封装基板上电镀铜焊盘平整度的问题带来的良率损失，而且这种方法还会带来其他成本的增加：① LED 晶圆凸点制作的成本或者 LED 晶圆/封装基板模板印刷焊膏的成本；②封装基板的成本。

为了得到非常高密度（超窄节距）和低成本（最小良率损失）的 mini-LED RGB 显示封装，本研究采用一种扇出型（先上晶且芯片面朝下）组装工艺[1-10]。由于扇出技术天然的优势[16, 104]，单颗 mini-LED 的间距可以做到 <50μm。同时，扇出再布线层（redistribution-layer，RDL）的存在可以让 mini-LED 不需要与 PCB 上的电镀铜焊盘直接接触。与 PBGA 方法相比，免除了封装基板以及 LED 与基板之间焊料凸点成型的成本和工艺，而且还能因扇出型封装带来封装小外形的优

势。当然，扇出型技术需要临时支撑片和制作 RDL。

至少有两种 RDL 制作用临时支撑片：圆形（晶圆）[12, 13] 和矩形（面板）[14, 15]。为了能获得高产率，在本研究中，采用矩形面板作为临时支撑片。

为 mini-LED 封装的跌落试验设计并制造了 PCB，还利用 3 维非线性温度、时间相关有限元模型对 mini-LED RGB 显示 SMD 与 PCB 组装的热循环试验进行了模拟。

4.8.1 测试 mini-LED

图 4.77 所示为测试 mini-LED，其尺寸如下：红色（R）LED 为 125μm×250μm×100μm，绿色（G）LED 为 130μm×270μm×100μm，蓝色（B）LED 为 130μm×270μm×100μm。这些 mini-LED 的 P、N 极焊盘采用 Cr/Au 进行表面处理，绿色和蓝色 mini-LED 的焊盘尺寸为 96μm×63μm，红色 mini-LED 的焊盘尺寸为 90μm×60μm。为了提高 SMT 组装的良率，将 12 个单独的 mini-LED 按照 "4 个 1 组" 形成一个 SMD，如图 4.78a 所示。从图中可以看到，每个 RGB mini-LED 的间距为 80μm，像素间距也约为 80μm，像素节距为 625μm。Mini-LED 之间通过图 4.78c 所示的菊花链连接，图 4.78d 为 SMD 底部的焊盘分布图[162]。

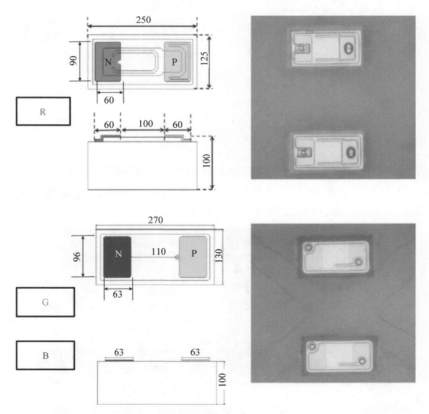

图 4.77　测试 LED：红色（R）125μm×250μm×100μm，绿色（G）130μm×270μm×100μm，蓝色（B）130μm×270μm×100μm

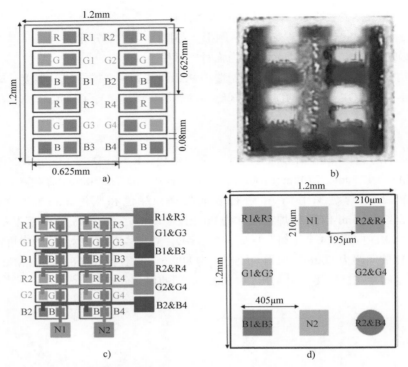

图 4.78 SMD：a）每个 SMD 包含 4 个像素（2×2），共 12 个 R、G、B mini-LED；b）点亮后的 SMD；c）SMD 的菊花链布局；d）SMD 底部布局

4.8.2 测试 mini-LED RGB 显示器的 SMD 封装

图 4.79 是测试 mini-LED RGB 显示器的 SMD 封装的截面示意图。从图中可以看到，mini-LED 通过扇出型先上晶且面朝下工艺进行封装。如图 4.80 所示，封装体共包括 2 层 RDL，RDL1 线宽/线距为 35/35μm，RDL2 线宽/线距为 35/50μm。

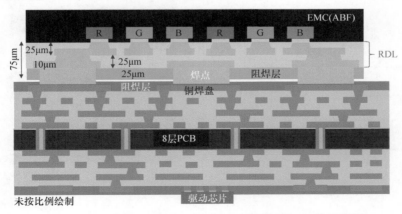

图 4.79 测试 mini-LED RGB SMD 封装 PCB 组装截面图

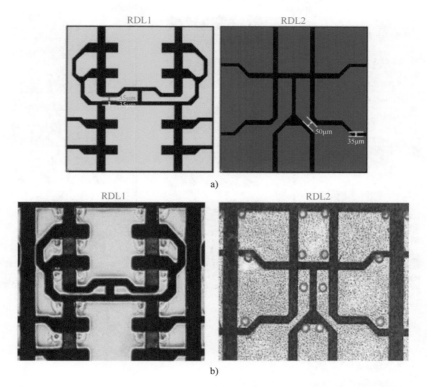

图 4.80　RDL：a）RDL1 和 RDL2 的布局；b）制造好的 RDL1 和 RDL2

4.8.3　RDL 和 mini-LED RGB SMD 制造

图 4.81 为采用扇出（先上晶且面朝下）工艺制造 mini-LED RGB 显示封装 RDL 的关键工艺步骤[12, 13]。

1. 拾取和放置 Mini-LED

先在 515mm×510mm×1.1mm 的临时玻璃支撑片上用物理气相沉积溅射 Ti 层，然后用液体光刻胶（liquid photoresist，LPR）光刻实现 LED 对齐标记、刻蚀并去除光刻胶、粘贴好 Nitto Revalpha 热剥离胶带，然后进行图 4.81 所示的 mini-LED 的拾取和放置（pick and place，P&P）。每块 515mm×510mm 的面板上可以放置 629712 个 mini-LED。为了节省 mini-LED，在临时支撑片放置了 74064 个 mini-LED。该步骤采用图 4.82 所示的 ASM AD420 高产率射片机完成，图 4.83 为拾取和放置了 mini-LED 的临时面板。

2. ABF 模塑

完成 mini-LED 的 P&P 后，进行图 4.81 所示的模塑操作。采用的环氧模塑料为 ABF（Ajinomoto build-up film）。ABF 的层压分两阶段进行：①第一阶段，在 120℃下真空 30s，并压合 30s（0.68MPa）；②第二阶段，在 100℃的温度下压平 60s（0.58MPa）。图 4.84 是模塑后的面板。

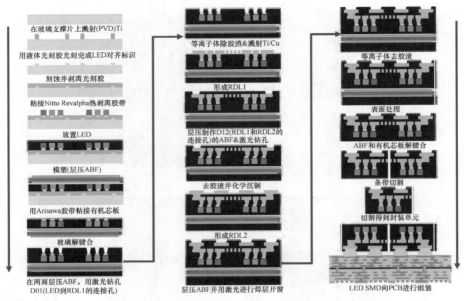

图 4.81 mini-LED 扇出型板级封装的关键工艺步骤

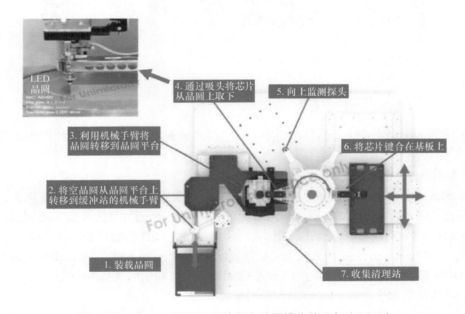

图 4.82 对 mini-LED 进行拾取和放置操作的设备（ASM）

3. 第二块支撑片粘接

为了节省 ABF 材料同时实现薄型封装，之前的工艺仅用了 150μm 厚的 ABF。因此，在解键合临时玻璃支撑片前，需要如图 4.81 所示先在 ABF 背面用 Arisawa 胶带粘接有机芯板。

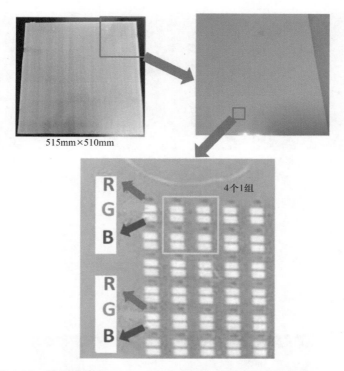

图 4.83　放置在粘有 Revalpha 胶带的临时玻璃面板上的 mini-LED

4. 第一块临时玻璃支撑片解键合

临时玻璃支撑片解键合温度为 170℃。在该温度下可以非常容易地去除玻璃支撑片并剥离胶带。然后，用化学方法清洁表面。

5. RDL 制作

首先在重构面板双面层压 ABF。然后依次进行激光钻孔、等离子体去胶渣并溅射 Ti/Cu 层（因为 mini-LED 做了 Au 表面处理），如图 4.85 所示。通过干膜层压、激光直写图形（laser direct imaging，LDI）和显影、电镀铜、干膜剥离、Ti/Cu 层刻蚀得到图 4.80b 所示的第一层 RDL（RDL1）。RDL1 的介质层为 ABF，金属导线层为铜。

图 4.84　环氧模塑料（ABF）模塑

为制作 RDL2，再次层压 ABF 并进行激光钻孔。然后，依次经过去胶渣、化学沉铜、干膜层压、LDI 和显影、电镀铜、干膜剥离和种子层刻蚀得到图 4.80b 所示的第二层 RDL（RDL2）。

6. 阻焊层开窗以及表面处理

继续层压 ABF，用激光进行阻焊层开窗，并用等离子体去胶渣。去胶渣完成后，如图 4.86 所示用化学镀钯浸金（electroless palladium immersion gold，EPIG）法对焊盘进行表面处理。

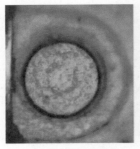 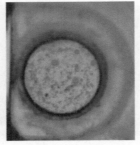 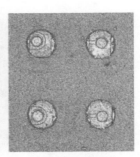

等离子体处理前的通孔　　等离子体处理后的通孔　　溅射后的通孔

图 4.85　等离子体去胶渣并在 ABF 通孔中溅射 Ti/Cu 层

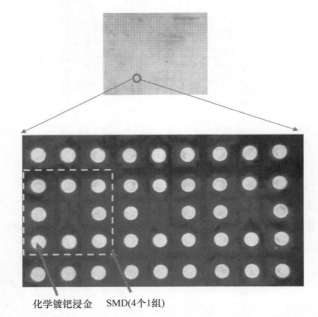

化学镀钯浸金　　SMD(4个1组)

图 4.86　通过激光钻孔进行阻焊层开窗并用 EPIG 进行表面处理

7. 背面减薄以及面板切割

最后通过背面减薄去除掉一定的 ABF 和有机芯板，将重构面板划片后得到含 2 层 RDL 的单颗 mini-LED RGB 显示 SMD，如图 4.87 所示。

4.8.4　PCB 组装

1. PCB 布局

图 4.88 为用于支撑 mini-LED RGB 显示 SMD 用作跌落试验和热循环试验的 PCB（132mm × 77mm）布局。该 PCB 是 8 层板，其上可以有规格不同的 mini-LED RGB 显示 SMD，如 4 个 1 组、16 个 1 组等。

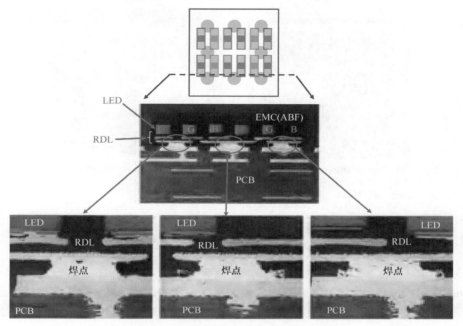

图 4.87 mini-LED SMD PCB 组装界面图

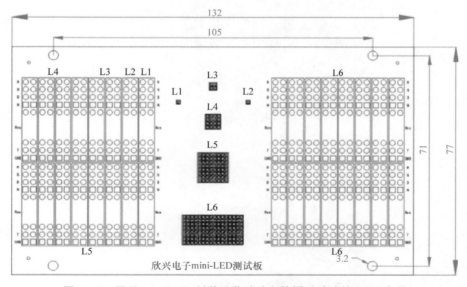

图 4.88 用于 mini-LED 封装跌落试验和热循环试验的 PCB 布局

2. PCB 上模板印刷焊膏

如图 4.89a 所示，采用一开窗直径 170μm 的不锈钢模板在 PCB 上印刷 Sn3Ag0.5Cu 焊膏，图 4.89b 和图 4.90a 为回流后得到的焊球。

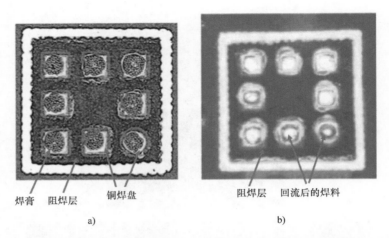

图 4.89　a）印刷在 PCB 上的焊膏；b）回流后 PCB 上的焊膏

3. SMT 组装

将 mini-LED RGB 显示 SMD 拾取和放置在已有焊球和助焊剂的 PCB 上。然后进行标准的 SMT 回流工艺。图 4.87 为一个 mini-LED RGB 显示 SMD PCB 组装的典型截面。为观察到 mini-LED，截面是在边缘切割得到而非由中间切割。图 4.90b 为最终组装体的 X 射线图像。

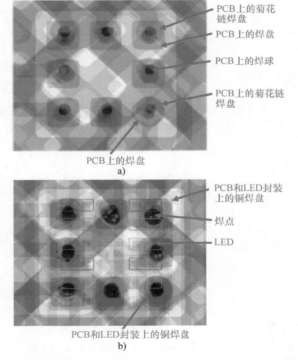

图 4.90　a）焊料回流后的 PCB X 射线图像；b）mini-LED SMD PCB 组装的 X 射线图像

4.8.5 跌落试验

1. 试验装置

对 mini-LED RGB 显示 SMD PCB 组装进行跌落试验。跌落试验装置按照 JEDEC 标准 JESD22-B111 搭建，如图 4.91 所示。在经过 10 多次尝试后，确定了能产生如图 4.92a 所示 1522g/ms 跌落谱 (1522g，0.5ms 半正弦脉冲) 的高度 (648mm)，试验样品数量 22 个。

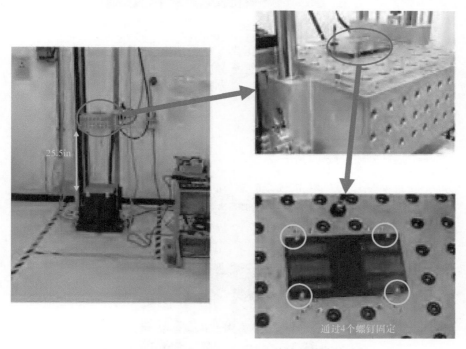

图 4.91　mini-LED SMD PCB 组装的跌落试验装置

2. 试验结果

在经过 100 次跌落后，SMD 点亮后仍与图 4.78b 所示一样好，没有发生失效。但是通过放大照片观察测试样品的截面，可以看到 SMD 其中一根铜金属线几乎断开了（仅勉强连接在一起），如图 4.93 所示。

4.8.6 热循环仿真

1. 结构和动力学边界条件

对 mini-LED RGB 显示 SMD PCB 组装结构进行分析。如图 4.94 所示，由于结构在 x 轴和 y 轴有对称性，可以仅对结构的 1/4 进行建模。图 4.95 为有限元模型的不同视图。模型的动力学边界条件如下：①在 yz 平面上没有位移（u_x），绕 y 轴没有旋转；②在 xz 平面上没有位移（u_y），绕 x 轴没有旋转；③PCB 底部中心在 z 方向上固定。表 4.4 为模型的材料参数。

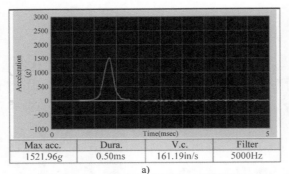

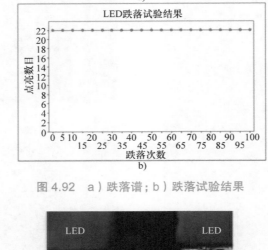

图 4.92　a）跌落谱；b）跌落试验结果

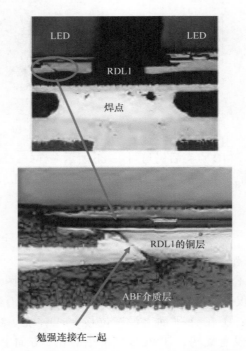

图 4.93　100 次跌落试验后的样品。RDL1 的铜层勉强连接在一起

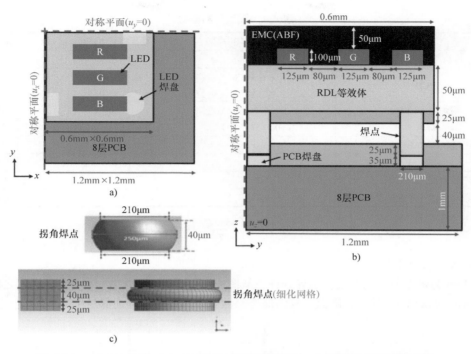

图 4.94　a）1/4 结构；b）1/4 组装结构的截面图；c）拐角焊点的细化网格

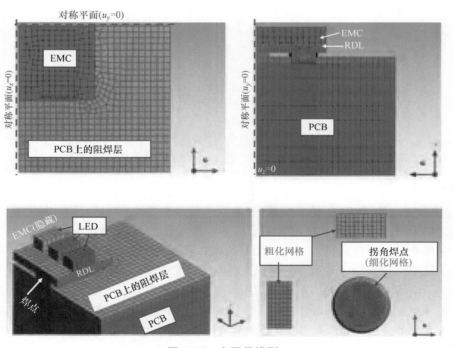

图 4.95　有限元模型

表 4.4 模型材料参数

材料	热膨胀系数 /(10^{-6}/℃)	杨氏模量 /GPa	泊松比
铜	16.3	121	0.34
PCB	$\alpha_x = \alpha_y = 18$ $\alpha_z = 70$	$E_x = E_y = 22$ $E_z = 10$	0.28
阻焊层（LED）	7.0	7.0	0.3
焊料	$21+0.017T$（℃）	$49-0.07T$（℃）	0.3
EMC（ABF）	7.0	7.0	0.3
AlGaInP	5.0	103	0.31
阻焊层（PCB）	39	4.1	0.3
再布线层（等效体）	9.79	41.2	0.312

注：焊料连续性方程为 $\dfrac{d\varepsilon}{dt} = 500000\left[\sinh(0.01\sigma)\right]^5 \exp\left(-\dfrac{5800}{T}\right)$，其中 T 单位为 K，σ 单位为 MPa。

2. 动力学边界条件

图 4.96 为模型的运动、温度边界条件。从图中可以看到，温度在 $-40℃ \leftrightarrows 85℃$ 之间变化。循环的时间为 60min，升温、降温、高温保持、低温保持的时间分别都是 15min。

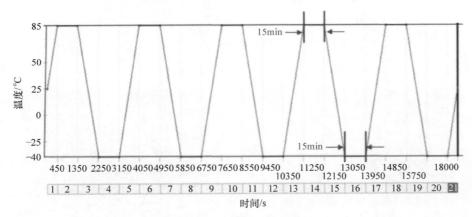

图 4.96 温度边界条件

3. 模拟结果

图 4.97 为 mini-LED RGB 显示 SMD PCB 组装的形变形状（颜色云图）和无形变形状（黑线）。从图 4.97b 中可以看到，在 85℃时，PCB 膨胀量比 SMD 更大，整个结构呈现凹面变形（笑脸）。从图 4.97c 中可以看到，在 -40℃时，PCB 收缩量比 SMD 更大，整个结构呈现凸面变形（哭脸）。这些结果均表明位移、旋转以及温度边界都描述正确。

图 4.98 为不同时间拐角焊点的最大累积蠕变应变分布图，其中图 4.98a 是 450s（85℃）时的焊点，图 4.98b 是 2250s（-40℃）时的焊点。从图中可以看到，累积蠕变应变的最大值仅出现在局部非常小的区域，其他大部分区域的数值都非常小。图 4.99 为拐角焊点最大累积蠕变应变随时间变化的曲线。同样，每个周期的最大累积蠕变应变量都非常小（小于 1%）。

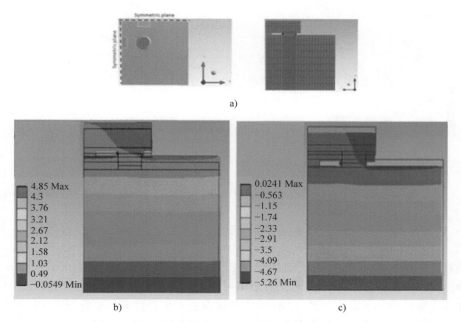

图 4.97 a）结构的俯视图和截面图；b）450s（85℃）形变后的形状（彩色云图）；
c）2250s（-40℃）形变后的形状（彩色云图）

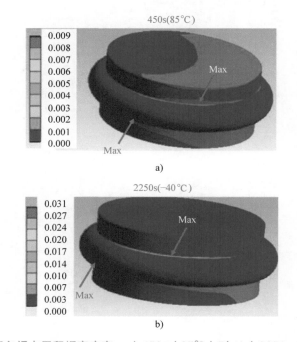

图 4.98 拐角焊点累积蠕变应变：a）450s（85℃）时；b）2250s（-40℃）时

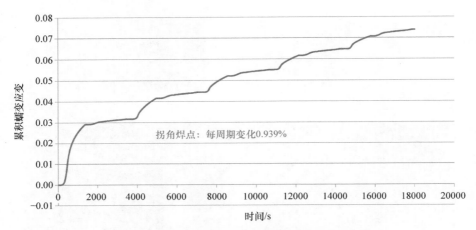

图 4.99　拐角焊点最大累积蠕变应变随时间变化图

图 4.100 为拐角焊点最大蠕变应变能量密度分布云图，其中图 4.100a 为 450s（85℃），图 4.100b 为 2250s（-40℃）。同样地，从图中可以看到，最大值仅出现在局部非常小的区域，其他大部分区域的值都非常小。图 4.101 为拐角焊点最大累积蠕变应变能量密度随时间变化图。同样地，从图中可以看到，每个周期蠕变应变能量密度都非常小（小于 0.3MPa）。

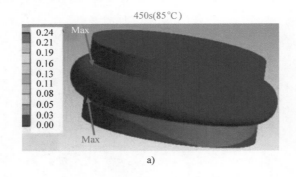

a)

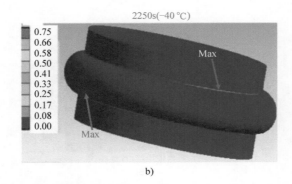

b)

图 4.100　拐角焊点最大累积蠕变应变能量密度：a）450s（85℃）；b）2250s（-40℃）

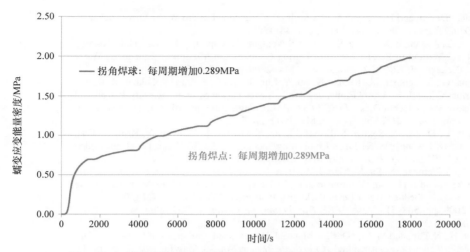

图 4.101 拐角焊点最大累积蠕变应变能量密度随时间变化曲线

4.8.7 总结和建议

一些重要的结论和建议总结如下：

1）本节演示了先上晶扇出型板级封装工艺制作 mini-LED RGB 显示 SMD 的设计、材料、工艺、制造以及 PCB 组装的可靠性和可行性。

2）该封装方法不仅产率和组装良率高，而且还具有低成本的潜在优势。

3）采用层压 ABF 的方法取代传统 EMC 压模，实现了薄型化 SMD 封装。

4）mini-LED RGB 显示 SMD PCB 组装的可靠性通过跌落试验进行了评估，在经过 100 次跌落后仍没有发生失效。

5）通过热循环仿真的方法演示了 mini-LED RGB 显示 SMD 的 PCB 组装的可靠性。从仿真中可以发现，每个周期的最大累积蠕变应变和蠕变应变能量密度都非常小，不会引起明显的可靠性问题。

参 考 文 献

1. Hedler, H., T. Meyer, and B. Vasquez, "Transfer wafer level packaging," *US Patent 6,727,576*, filed on Oct. 31, 2001; patented on April 27, 2004.
2. Lau, J. H., "Patent Issues of Fan-Out Wafer/Panel-Level Packaging", *Chip Scale Review*, Vol. 19, November/December 2015, pp. 42–46.
3. Brunnbauer, M., E. Furgut, G. Beer, T. Meyer, H. Hedler, J. Belonio, E. Nomura, K. Kiuchi, and K. Kobayashi, "An Embedded Device Technology Based on a Molded Reconfigured Wafer", *IEEE/ECTC Proceedings*, May 2006, pp. 547–551.
4. Brunnbauer, M., E. Furgut, G. Beer, and T. Meyer, "Embedded Wafer Level Ball Grid Array (eWLB)", *IEEE/EPTC Proceedings,* May 2006, pp. 1–5.
5. Keser, B., C. Amrine, T. Duong, O. Fay, S. Hayes, G. Leal, W. Lytle, D. Mitchell, and R. Wenzel, "The Redistributed Chip Package: A Breakthrough for Advanced Packaging", *Proceedings of IEEE/ECTC*, May 2007, pp. 286–291.
6. Kripesh, V., V. Rao, A. Kumar, G. Sharma, K. Houe, X. Zhang, K. Mong, N. Khan, and J.

H. Lau, "Design and Development of a Multi-Die Embedded Micro Wafer Level Package", *IEEE/ECTC Proceedings*, May 2008, pp. 1544–1549.
7. Khong, C., A. Kumar, X. Zhang, S. Gaurav, S. Vempati, V. Kripesh, J. H. Lau, and D. Kwong, "A Novel Method to Predict Die Shift During Compression Molding in Embedded Wafer Level Package", *IEEE/ECTC Proceedings*, May 2009, pp. 535–541.
8. Sharma, G., S. Vempati, A. Kumar, N. Su, Y. Lim, K. Houe, S. Lim, V. Sekhar, R. Rajoo, V. Kripesh, and J. H. Lau, "Embedded Wafer Level Packages with Laterally Placed and Vertically Stacked Thin Dies", *IEEE/ECTC Proceedings*, 2009, pp. 1537–1543. Also, *IEEE Transactions on CPMT*, Vol. 1, No. 5, May 2011, pp. 52–59.
9. Kumar, A., D. Xia, V. Sekhar, S. Lim, C. Keng, S. Gaurav, S. Vempati, V. Kripesh, J. H. Lau, and D. Kwong, "Wafer Level Embedding Technology for 3D Wafer Level Embedded Package", *IEEE/ECTC Proceedings*, May 2009, pp. 1289–1296.
10. Lim, Y., S. Vempati, N. Su, X. Xiao, J. Zhou, A. Kumar, P. Thaw, S. Gaurav, T. Lim, S. Liu, V. Kripesh, and J. H. Lau, "Demonstration of High Quality and Low Loss Millimeter Wave Passives on Embedded Wafer Level Packaging Platform (EMWLP)", *IEEE/ECTC Proceedings*, 2009, pp. 508–515. Also, *IEEE Transactions on Advanced Packaging*, Vol. 33, 2010, pp. 1061–1071.
11. Lau, J. H., N. Fan, and M. Li, "Design, Material, Process, and Equipment of Embedded Fan-Out Wafer/Panel-Level Packaging", *Chip Scale Review*, Vol. 20, May/June 2016, pp. 38–44.
12. Lau, J. H., M. Li, M. Li, T. Chen, I. Xu, X. Qing, Z. Cheng, N. Fan, E. Kuah, Z. Li, K. Tan, Y. Cheung, E. Ng, P. Lo, K. Wu, J. Hao, S. Koh, R. Jiang, X. Cao, R. Beica, S. Lim, N. Lee, C. Ko, H. Yang, Y. Chen, M. Tao, J. Lo, and R. Lee, "Fan-Out Wafer-Level Packaging for Heterogeneous Integration", *IEEE Transactions on CPMT*, 2018, September 2018, pp. 1544–1560.
13. Lau, J. H., M. Li, Y. Lei, M. Li, I. Xu, T. Chen, Q. Yong, Z. Cheng, K. Wu, P. Lo, Z. Li, K. Tan, Y. Cheung, N. Fan, E. Kuah, C. Xi, J. Ran, R. Beica, S. Lim, N. Lee, C. Ko, H. Yang, Y. Chen, M. Tao, J. Lo, and R. Lee, "Reliability of Fan-Out Wafer-Level Heterogeneous Integration", *IMAPS Transactions, Journal of Microelectronics and Electronic Packaging*, Vol. 15, Issue: 4, October 2018, pp. 148–162.
14. Ko, CT, H. Yang, J. H. Lau, M. Li, M. Li, C. Lin, J. W. Lin, T. Chen, I. Xu, C. Chang, J. Pan, H. Wu, Q. Yong, N. Fan, E. Kuah, Z. Li, K. Tan, Y. Cheung, E. Ng, K. Wu, J. Hao, R. Beica, M. Lin, Y. Chen, Z. Cheng, S. Koh, R. Jiang, X. Cao, S. Lim, N. Lee, M. Tao, J. Lo, and R. Lee, "Chip-First Fan-Out Panel-Level Packaging for Heterogeneous Integration", *IEEE Transactions on CPMT*, September 2018, pp. 1561–1572.
15. Ko, C. T., H. Yang, J. H. Lau, M. Li, M. Li, C. Lin, J. Lin, C. Chang, J. Pan, H. Wu, Y. Chen, T. Chen, I. Xu, P. Lo, N. Fan, E. Kuah, Z. Li, K. Tan, C. Lin, R. Beica, M. Lin, C. Xi, S. Lim, N. Lee, M. Tao, J. Lo, and R. Lee, "Design, Materials, Process, and Fabrication of Fan-Out Panel-Level Heterogeneous Integration", *IMAPS Transactions, Journal of Microelectronics and Electronic Packaging*, Vol. 15, Issue: 4, October 2018, pp. 141–147.
16. Lau, J. H., "Recent Advances and Trends in Fan-Out Wafer/Panel-Level Packaging", *ASME Transactions, Journal of Electronic Packaging*, Vol. 141, December 2019, pp. 1–27.
17. Lau, J. H., "Recent Advances and Trends in Heterogeneous Integrations", *IMAPS Transactions, Journal of Microelectronics and Electronic Packaging*, Vol. 16, April 2019, pp. 45–77.
18. Kurita, Y., T. Kimura, K. Shibuya, H. Kobayashi, F. Kawashiro, N. Motohashi, and M. Kawano, "Fan-out wafer-level packaging with highly flexible design capabilities," *IEEE/ESTC Proceedings*, May 2010, pp. 1–6.
19. Motohashi, N., T. Kimura, K. Mineo, Y. Yamada, T. Nishiyama, K. Shibuya, H. Kobayashi, Y. Kurita, and M. Kawano, "System in wafer-level package technology with RDL-first process," *IEEE/ECTC Proceedings*, May 2011, pp. 59–64.
20. Yoon, S., J. Caparas, Y. Lin, and P. Marimuthu, "Advanced Low Profile PoP Solution with Embedded Wafer Level PoP (eWLB-PoP) Technology", *IEEE/ECTC Proceedings*, May 2012, pp. 1250–1254.
21. Tseng, C., Liu, C., Wu, C., and D. Yu, "InFO (Wafer Level Integrated Fan-Out) Technology", *IEEE/ECTC Proceedings*, May 2016, pp. 1–6.
22. Hsieh, C., Wu, C., and D. Yu, "Analysis and Comparison of Thermal Performance of Advanced Packaging Technologies for State-of-the-Art Mobile Applications", *IEEE/ECTC Proceedings*,

May 2016, pp. 1430–1438.
23. Yoon, S., P. Tang, R. Emigh, Y. Lin, P. Marimuthu, and R. Pendse, "Fanout Flipchip eWLB (Embedded Wafer Level Ball Grid Array) Technology as 2.5D Packaging Solutions", *IEEE/ECTC Proceedings*, May 2013, pp. 1855–1860.
24. Lin, Y., W. Lai, C. Kao, J. Lou, P. Yang, C. Wang, and C. Hseih, "Wafer Warpage Experiments and Simulation for Fan-out Chip on Substrate", *IEEE/ECTC Proceedings*, May 2016, pp. 13–18.
25. Chen, N., T. Hsieh, J. Jinn, P. Chang, F. Huang, J. Xiao, A. Chou, and B. Lin, "A Novel System in Package with Fan-out WLP for high speed SERDES application", *IEEE/ECTC Proceedings*, May 2016, pp. 1495–1501.
26. Chang, H., D. Chang, K. Liu, H. Hsu, R. Tai, H. Hunag, Y. Lai, C. Lu, C. Lin, and S. Chu, "Development and Characterization of New Generation Panel Fan-Out (PFO) Packaging Technology", *IEEE/ECTC Proceedings*, May 2014, pp. 947–951.
27. Liu, H., Y. Liu, J. Ji, J. Liao, A. Chen, Y. Chen, N. Kao, and Y. Lai, "Warpage Characterization of Panel Fab-out (P-FO) Package", *IEEE/ECTC Proceedings*, May 2014, pp. 1750–1754.
28. Braun, T., S. Raatz, S. Voges, R. Kahle, V. Bader, J. Bauer, K. Becker, T. Thomas, R. Aschenbrenner, and K. Lang, "Large Area Compression Molding for Fan-out Panel Level Packing", *IEEE/ECTC Proceedings*, May 2015, pp. 1077–1083.
29. Che, F., D. Ho, M. Ding, X. Zhang, "Modeling and design solutions to overcome warpage challenge for fanout wafer level packaging (FO-WLP) technology," *IEEE/EPTC Proceedings*, May 2015, pp. 2–4.
30. Che, F., D. Ho, M. Ding, D. MinWoopp, "Study on Process Induced Wafer Level Warpage of Fan-Out Wafer Level Packaging", *IEEE/ECTC Proceedings*, May 2016, pp. 1879–1885.
31. Hsu, I., C. Chen, S. Lin, T. Yu, M. Hsieh, K. Kang, S. Yoon, "Fine-Pitch Interconnection and Highly Integrated Assembly Packaging with FOMIP (Fan-out Mediatek Innovation Package) Technology", *IEEE/ECTC Proceedings*, May 2020, pp. 867–872.
32. Lai, W., P. Yang, I. Hu, T. Liao, K. Chen, D. Tarng, and C. Hung, "A Comparative Study of 2.5D and Fan-out Chip on Substrate : Chip First and Chip Last", *IEEE/ECTC Proceedings*, May 2020, pp. 354–360.
33. Julien, B., D. Fabrice. K. Tadashi, B. Pieter, K. Koen, P. Alain, M. Andy, P. Arnita, B. Gerald, and B. Eric, "Development of compression molding process for Fan-Out wafer level packaging", *IEEE/ECTC Proceedings*, May 2020, pp. 1965–1972.
34. Lee, K., Y. Lim, S. Chow, K. Chen, W. Choi, and S. Yoon, "Study of Board Level Reliability of eWLB (embedded wafer level BGA) for 0.35 mm Ball Pitch *IEEE/ECTC Proceedings*, May 2019, pp. 1165–1169.
35. Wu, D., R. Dahlbäck, E. Öjefors and M. Carlsson, F. Lim, Y. Lim, A. Oo, W. Choi, and S. Yoon, "Advanced Wafer Level PKG solutions for 60 GHz WiGig (802.11ad) Telecom Infrastructure", *IEEE/ECTC Proceedings*, May 2019, pp. 968–971.
36. Fowler, M., J. Massey, T. Braun, S. Voges, R. Gernhardt, and M. Wohrmann, "Investigation and Methods Using Various Release and Thermoplastic Bonding Materials to Reduce Die Shift and Wafer Warpage for eWLB Chip-First Processes", *IEEE/ECTC Proceedings*, May 2019, pp. 363–369.
37. Theuss, H., C. Geissler, F. Muehlbauer, C. Waechter, T. Kilger, J. Wagner, T. Fischer, U. Bartl, S. Helbig, A. Sigl, D. Maier, B. Goller, M. Vobl, M. Herrmann, J. Lodermeyer, and U. Krumbein, and A. Dehe, "A MEMS Microphone in a FOWLP", *IEEE/ECTC Proceedings*, May 2019, pp. 855–860.
38. Huang, C., T. Hsieh, P. Pan, M. Jhong, C. Wang, and S. Hsieh, "Comparative Study on Electrical Performance of eWLB, M-Series and Fan-Out Chip Last", *IEEE/ECTC Proceedings*, May 2018, pp. 1324–1329.
39. Ha, J., Y. Yu, and K. Cho, "Solder Joint Reliability of Double sided Assembled PLP Package", *IEEE/EPTC Proceedings,* December 2020, pp. 408–412.
40. Mei, S., T. Lim, X. Peng, C. Chong, and S. Bhattacharya, "FOWLP RF Passive Circuit Designs for 77 GHz MIMO radar applications", *IEEE/EPTC Proceedings,* December 2020, pp. 445–448.
41. Zhang, X., B. Lau, H. Chen, Y. Han, M. Jong, S. Lim, S. Lim, X. Wang, Y. Andriani, and S. Liu, "Board Level Solder Joint Reliability Design and Analysis of FOWLP", *IEEE/EPTC Proceedings,* December 2020, pp. 316–320.

42. Ho, S., S. Boon, L. Long, H. Yao, C. Choong, S. Lim, T. Lim, and C. Chong, "Double Mold Antenna in Package for 77 GHz Automotive Radar", *IEEE/EPTC Proceedings,* December 2020, pp. 257–261.
43. Jeon, Y., and R. Kumarasamy, "Impact of Package Inductance on Stability of mm-Wave Power Amplifiers", *IEEE/EPTC Proceedings*, December 2020, pp. 255–256.
44. Han, Y., T. Chai, and T. Lim, "Investigation of Thermal Performance of Antenna in Package for Automotive Radar System", *IEEE/EPTC Proceedings,* December 2020, pp. 246–250.
45. Bhardwaj, S., S. Sayeed, J. Camara, D. Vital, P. Raj, "Reconfigurable mmWave Flexible Packages with Ultra-thin Fan-Out Embedded Tunable Ceramic IPDs", *IMAPS Proceedings*, October 2019, pp. 1.1–1.4.
46. Hdizadeh, R., A. Laitinen, N. Kuusniemi, V. Blaschke, D. Molinero, E. O'Toole, and M. Pinheiro, "Low-Density Fan-Out Heterogeneous Integration of MEMS Tunable Capacitor and RF SOI Switch", *IMAPS Proceedings*, October 2019, pp. 5.1–5.5.
47. Ostholt, R., R. Santos, N. Ambrosius, D. Dunker, and J.Delrue, "Passive Die Alignment in Glass Embedded Fan-Out Packaging", *IMAPS Proceedings*, October 2019, pp. 7.1–7.5.
48. Ali, B., and M. Marshall, "Automated Optical Inspection (AOI) for FOPLP with Simultaneous Die Placement Metrology", *IMAPS Proceedings*, October 2019, pp. 8.1–8.8.
49. Ogura, N., S. Ravichandran, T. Shi, A. Watanabe, S. Yamada, M. Kathaperumal, and R. Tummala, "First Demonstration of Ultra-Thin Glass Panel Embedded (GPE) Package with Sheet Type Epoxy Molding Compound for 5G/mm-Wave Applications", *IMAPS Proceedings*, October 2019, pp. 9.1–9.7.
50. Yoon, S., Y. Lin, S. Gaurav, Y. Jin, V. Ganesh, T. Meyer, C. Marimuthu, X. Baraton, and A. Bahr, "Mechanical Characterization of Next Generation eWLB (embedded Wafer Level BGA) Packaging", *IEEE/ECTC Proceedings*, May 2011, pp. 441–446.
51. Jin, Y., J. Teysseyre, X. Baraton, S. Yoon, Y. Lin, and P. Marimuthu, "Development and Characterization of Next Generation eWLB (embedded Wafer Level BGA) Packaging", *IEEE/ECTC Proceedings*, May 2012, pp. 1388–1393.
52. Osenbach, J., S. Emerich, L. Golick, S. Cate, M. Chan, S. Yoon, Y. Lin, and K. Wong, "Development of Exposed Die Large Body to Die Size Ratio Wafer Level Package Technology", *IEEE/ECTC Proceedings*, May 2014, pp. 952–955.
53. Lin, Y., E. Chong, M. Chan, K. Lim, and S. Yoon, "WLCSP + and eWLCSP in Flex-Line: Innovative Wafer Level Package Manufacturing", *IEEE/ECTC Proceedings*, May 2015, pp. 865–870.
54. Lin, Y., C. Kang, L. Chua, W. Choi, and S. Yoon, "Advanced 3D eWLB-PoP (embedded Wafer Level Ball Grid Array - Package on Package) Technology", *IEEE/ECTC Proceedings*, May 2016, pp. 1772–1777.
55. Chen, K., L. Chua, W. Choi, S. Chow, and S. Yoon, "28 nm CPI (Chip/Package Interactions) in Large Size eWLB (Embedded Wafer Level BGA) Fan-Out Wafer Level Packages", *IEEE/ECTC Proceedings*, May 2017, pp. 581–586.
56. Yap, D., K. Wong, L. Petit, R. Antonicelli, and S. Yoon, "Reliability of eWLB (embedded wafer level BGA) for Automotive Radar Applications", *IEEE/ECTC Proceedings*, May 2017, pp. 1473–1479.
57. Braun, T., T. Nguyen, S. Voges, M. Wöhrmann, R. Gernhardt, K. Becker, I. Ndip D. Freimund, M. Ramelow, K. Lang, D. Schwantuschke, E. Ture, M. Pretl, S. Engels, "Fan-out Wafer Level Packaging of GaN Components for RF Applications", *IEEE/ECTC Proceedings,* May 2020, pp. 7–13.
58. Braun, T., K. Becker, O. Hoelck, S. Voges, R. Kahle, P. Graap, M. Wöhrmann, R. Aschenbrenner, M. Dreissigacker, M. Schneider-Ramelow, K. Lang, "Fan-out Wafer Level Packaging - A Platform for Advanced Sensor Packaging", *IEEE/ECTC Proceedings,* May 2019, pp. 861–867.
59. Woehrmann, M., H. Hichri, R. Gernhardt, K. Hauck, T. Braun, M. Toepper, M. Arendt, K. Lang, "Innovative Excimer Laser Dual Damascene Process for ultra-fine line multi-layer Routing with 10 μm Pitch Micro-Vias for Wafer Level and Panel Level Packaging", *IEEE/ECTC Proceedings*, May 2017, pp. 872–877.
60. Braun, T., S. Raatz, U. Maass, M. van Dijk, H. Walter, O. Holck, K.-F. Becker, M. Topper, R. Aschenbrenner, M. Wohrmann, S. Voges, M. Huhn, K.-D. Lang, M. Wietstruck, R. Scholz, A. Mai, and M. Kaynak, "Development of a Multi-Project Fan-Out Wafer Level Packaging

Platform", *IEEE/ECTC Proceedings,* May 2017, pp. 1–7.
61. Braun, T., K.-F. Becker, S. Raatz, M. Minkus, V. Bader, J. Bauer, R. Aschenbrenner R. Kahle, L. Georgi, S. Voges, M. Wohrmann, K.-D. Lang, "Foldable Fan-out Wafer Level Packaging", *EEE/ECTC Proceedings,* May 2016, pp. 19–24.
62. Braun, T., K.-F. Becker, S. Voges, J. Bauer, R. Kahle, V. Bader, T. Thomas, R. Aschenbrenner, K.-D. Lang, "24" × 18" Fan-out Panel Level Packing", *EEE/ECTC Proceedings,* May 2014, pp. 940–946.
63. Braun, T., K.-F. Becker, S. Voges, T. Thomas, R. Kahle, J. Bauer, R. Aschenbrenner, K.-D. Lang, "From Wafer Level to Panel Level Mold Embedding", *EEE/ECTC Proceedings,* May 2013, pp. 1235–1242.
64. Braun, T., K.-F. Becker, S. Voges, T. Thomas, R. Kahle, V. Bader, J. Bauer, K. Piefke, R. Krüger, R. Aschenbrenner, K.-D. Lang, "Through Mold Vias for Stacking of Mold Embedded Packages", *EEE/ECTC Proceedings,* May 2011, pp. 48–54.
65. Braun, T., K.-F. Becker, L. Böttcher, J. Bauer, T. Thomas, M. Koch, R. Kahle, A. Ostmann, R. Aschenbrenner, H. Reichl, M. Bründel, J.F. Haag, and U. Scholz, "Large Area Embedding for Heterogeneous System Integration", *EEE/ECTC Proceedings,* May 2010, pp. 550–556.
66. Chiu, T., J. Wu, W. Liu, C. Liu, D. Chen, M. Shih, and D. Tarng, "A Mechanics Model for the Moisture Induced Delamination in Fan-Out Wafer-Level Package", *IEEE/ECTC Proceedings*, May 2020, pp. 1205–1211.
67. Poe, B., "An Innovative Application of Fan-Out Packaging for Test & Measurement-Grade Products"", *IWLPC Proceedings*, October 2018, pp. 1.1–1.6.
68. Hadizadeh,R., A. Laitinen, D. Molinero, N. Pereira, and M. Pinheiro, "Wafer-Level Fan-Out for High-Performance, Low-cost Packaging of Monolithic RF MEMS/CMOS", *IWLPC Proceedings*, October 2018, pp. 2.1–2.6.
69. Lianto, P., C. Tan, Q. Peng, A. Jumat, X. Dai, K. Fung, G. See, S. Chong, S. Ho, S. Soh, S. Lim, H. Chua, A. Haron, H. Lee, M. Zhang, Z. Ko, Y. San, and H. Leong, "Fine-Pitch RDL Integration for Fan-Out Wafer-Level Packaging", *IEEE/ECTC Proceedings*, May 2020, pp. 1126–1131.
70. Ma, S, C. Wang, F. Zheng, D. Yu, H. Xie, X. Yang, L. Ma, P. Li , W. Liu , J. Yu , J. Goodelle, "Development of Wafer Level Process for the Fabrication of Advanced Capacitive Fingerprint Sensors Using Embedded Silicon Fan-Out (eSiFO®) Technology", *IEEE/ECTC Proceedings*, May 2019, pp. 28–34.
71. Cho, J., J. Paul, S. Capecchi, F. Kuechenmeister, T. Cheng, "Experiment of 22FDX® Chip Board Interaction (CBI) in Wafer Level Packaging Fan-Out (WLPFO)", *IEEE/ECTC Proceedings*, May 2019, pp. 910–916.
72. Weichart, J., J. Weichart, A. Erhart, K. Viehweger, "Preconditioning Technologies for Sputtered Seed Layers in FOPLP", *IEEE/ECTC Proceedings*, May 2019, pp. 1833–1841.
73. Liu, C., et al., "High-performance integrated fan-out wafer level packaging (InFO-WLP): Technology and system integration," *Proc. IEEE Int. Electron Devices Meeting*, December 2012, pp. 323–326.
74. Chen, S., D. Yu, et al., "High-performance inductors for integrated fan-out wafer level packaging (InFO-WLP)," *Symp. on VLSI Technol.*, June 2013, pp. T46–T47.
75. Tsai, C., et al., "Array antenna integrated fan-out wafer level packaging (InFO-WLP) for millimeter wave system applications," *Proc. IEEE Int. Electron Devices Meeting*, June 2013, pp. 25.1.1–25.1.4.
76. Yu, D., "New system-in-package (SIP) integration technologies," *Proc. Custom Integrated Circuits Conf.*, September 2014, pp. 1–6.
77. Yu, D., "A new integration technology platform: integrated fan-out wafer-level-packaging for mobile applications," *Symp. on VLSI Technol.*, June 2015, pp. T46–T47.
78. Tsai, C., *et al.*, "High performance passive devices for millimeter wave system integration on integrated fan-out (InFO) wafer level packaging technology," in *Proc. International Electron Devices Meeting*, Dec. 2015, pp. 25.2.1–25.2.4.
79. Wang, C., *et al.*, "Power saving and noise reduction of 28 nm CMOS RF system integration using integrated fan-out wafer level packaging (InFO-WLP) technology," in *Proc. International 3D Systems Integration Conference*, Aug. 2015, pp. TS6.3.1–TS6.3.4.
80. Rogers, B. D. Sanchez, C. Bishop, C. Sandstrom, C. Scanlan and T. Olson, "Chips "Face-up" Panelization Approach for Fan-Out Packaging", *Proceedings of IWLPC*, October 2015,

pp. 1–8.
81. Wang, C., and D. Yu, "Signal and Power Integrity Analysis on Integrated Fan-out PoP (InFO_PoP) Technology for Next Generation Mobile Applications", *IEEE/ECTC Proceedings*, May 2016, pp. 380–385.
82. Hsu, C., C Tsai, J. Hsieh, K. Yee, C. Wang, and D. Yu, "High Performance Chip-Partitioned Millimeter Wave Passive Devices on Smooth and Fine Pitch InFO RDL", *IEEE/ECTC Proceedings*, May 2017, pp. 254–259.
83. Lau, J. H., M. Li, D. Tian, N. Fan, E. Kuah, K. Wu, M. Li, J. Hao, Y. Cheung, Z. Li, K. Tan, R. Beica, T. Taylor, CT Lo, H. Yang, Y. Chen, S. Lim, NC Lee, J. Ran, X. Cao, S. Koh, and Q. Young, "Warpage and Thermal Characterization of Fan-Out Wafer-Level Packaging", *IEEE Transactions on CPMT*, Vol. 7, Issue 10, October 2017, pp. 1729–1738.
84. Lau, J. H., M. Li, N. Fan, E. Kuah, Z. Li, K. Tan, T. Chen, I. Xu, M. Li, Y. Cheung, K. Wu, J. Hao, R. Beica, T. Taylor, C. Ko, H. Yang, Y. Chen, S. Lim, N. Lee, J. Ran, K. Wee, Q. Yong, C. Xi, M. Tao, J. Lo, and R. Lee, "Fan-Out Wafer-Level Packaging (FOWLP) of Large Chip with Multiple Redistribution Layers (RDLs)", *IMAPS Transactions, Journal of Microelectronics and Electronic Packaging*, Vol. 14, Issue: 4, October 2017, pp. 123–131.
85. Lau, J. H., M. Li, Q. Li, I. Xu, T. Chen, Z. Li, K. Tan, X. Qing, C. Zhang, K. Wee. R. Beica, C. Ko, S. Lim, N. Fan, E. Kuah, K. Wu, Y. Cheung, E. Ng, X. Cao, J. Ran, H. Yang, Y. Chen, N. Lee, M. Tao, J. Lo, and R. Lee, "Design, Materials, Process, and Fabrication of Fan-Out Wafer-Level Packaging", *IEEE Transactions on CPMT*. Vol. 8, Issue 6, June, 2018, pp. 991–1002.
86. Lau, J. H., M. Li, D. Tian, N. Fan, E. Kuah, K. Wu, M. Li, J. Hao, K. Cheung, Z. Li, K. Tan, R. Beica, C. Ko, Y. Chen, S. Lim, N. Lee, K. Wee, J. Ran, and C. Xi, "Warpage Measurements and Characterizations of FOWLP with Large Chips and Multiple RDLs", *IEEE Transactions on CPMT*, Vol. 8, Issue 10, October 2018, pp. 1729–1737.
87. Wang, C., T. Tang, C. Lin, C. Hsu, J. Hsieh, C. Tsai, K. Wu, H. Pu, and D. Yu, "InFO_AiP Technology for High Performance and Compact 5G Millimeter Wave System Integration", *IEEE/ECTC Proceedings*, May 2018, pp. 202–207.
88. Yu, C., L. Yen, C. Hsieh, J. Hsieh, V. Chang, C. Hsieh, C. Liu, C. Wang, K. Yee, and D. Yu, "High Performance, High Density RDL for Advanced Packaging", *IEEE/ECTC Proceedings*, May 2018, pp. 587–593.
89. Su, A., T. Ku, C. Tsai, K. Yee, and D.s Yu, "3D-MiM (MUST-in-MUST) Technology for Advanced System Integration", *IEEE/ECTC Proceedings*, May 2019, pp. 1–6.
90. Wang, C., J. Hsieh, V. Chang, S. Huang, T. Ko, H. Pu, and D. Yu, "Signal Integrity of Submicron InFO Heterogeneous Integration for High Performance Computing Applications", *IEEE/ECTC Proceedings*, May 2019, pp. 688–694.
91. Chen, F., M. Chen, W. Chiou, D. Yu, "System on Integrated Chips (SoIC^{TM}) for 3D Heterogeneous Integration" *IEEE/ECTC Proceedings*, May 2019, pp. 594–599.
92. Hou, S., K. Tsai. M. Wu, M. Ku, P. Tsao, and L. Chu, "Board level Reliability Investigation of FO-WLP Package", *IEEE/ECTC Proceedings*, May 2018, pp. 904–910.
93. Chun, S., T. Kuo, H. Tsai, C. Liu, C. Wang, J. Hsieh, T. Lin, T. Ku, D. Yu, "InFO_SoW (System-on-Wafer) for High Performance Computing", *IEEE/ECTC Proceedings*, May 2020, pp. 1–6.
94. Ko, T., H. Pu, Y. Chiang, H. Kuo, C. Wang, C. Liu, and D. Yu, "Applications and Reliability Study of InFO_UHD (Ultra-High-Density) Technology", *IEEE/ECTC Proceedings*, May 2020, pp. 1120–1125.
95. Kurita, Y., K. Soejima, K. Kikuchi, M. Takahashi, M. Tago, M. Koike, "A Novel "SMAFTI" Package for Inter-Chip Wide-Band Data Transfer", *IEEE/ECTC Proceedings*, May 2006, pp. 289–297.
96. Kawano, M., S. Uchiyama, Y. Egawa, N. Takahashi, Y. Kurita, K. Soejima, "A 3D Packaging Technology for 4 Gbit Stacked DRAM with 3 Gbps Data Transfer", *IEEE/IEMT Proceedings*, May 2006, pp. 581–584.
97. Kurita, Y., S. Matsui, N. Takahashi, K. Soejima, M. Komuro, M. Itou, "A 3D Stacked Memory Integrated on a Logic Device Using SMAFTI Technology", *IEEE/ECTC Proceedings*, May 2007, pp. 821–829.
98. Kawano, M., N. Takahashi, Y. Kurita, K. Soejima, M. Komuro, and S. Matsui, "A 3-D Packaging Technology for Stacked DRAM with 3 Gb/s Data Transfer", *IEEE Transactions on*

Electron Devices, 55 (7), 2008, pp. 1614–1620.
99. Motohashi, N., Y. Kurita, K. Soejima, Y. Tsuchiya, and M. Kawano, "SMAFTI Package with Planarized Multilayer Interconnects", 2009, *IEEE/ECTC Proceedings,* May 2009, pp. 599–606.
100. Kurita, M., S. Matsui, N. Takahashi, K. Soejima, M. Komuro, M. Itou, "Vertical Integration of Stacked DRAM and High-Speed Logic Device Using SMAFTI Technology", *IEEE Transactions on Advanced Packaging*, May 2009, pp. 657–665.
101. Kurita, Y., N. Motohashi, S. Matsui, K. Soejima, S. Amakawa, K. Masu, "SMAFTI Packaging Technology for New Interconnect Hierarchy", *Proceedings of IITC*, June 2009, pp. 220–222.
102. Kurita, Y., T. Kimura, K. Shibuya, H. Kobayashi, F. Kawashiro, N. Motohashi, "Fan-Out Wafer Level Packaging with Highly Flexible Design Capabilities", *Proceedings of the Electronics System Integration Technology Conferences*, 2010, pp. 1–6.
103. Motohashi, N., T. Kimura, K. Mineo, Y. Yamada, T. Nishiyama, K. Shibuya, "System in a Wafer Level Package Technology with RDL-First Process", *IEEE/ECTC Proceedings*, May 2011, pp. 59–64.
104. Lau, J. H., *Fan-Out Wafer-Level Packaging*, Springer, New York, 2018.
105. Lau, J. H., *Heterogeneous Integrations*, Springer, New York, 2019.
106. Huemoeller, R., and C. Zwenger, "Silicon wafer integrated fan-out technology," *Chip Scale Review*, Mar/Apr 2015, pp. 34–37.
107. Bu, L., F. Che, M. Ding, S. Chong, and X. Zhang, "Mechanism of Moldable Underfill (MUF) Process for Fan-Out Wafer Level Packaging", *IEEE/EPTC Proceedings*, May 2015, pp. 1–7.
108. Che, F., D. Ho, M. Ding, and D. Woo, "Study on Process Induced Wafer Level Warpage of Fan-Out Wafer Level Packaging", *IEEE/ECTC Proceedings*, May 2016, pp. 1879–1885.
109. Rao, V., C. Chong, D. Ho, D. Zhi, C, Choong, S. Lim, D. Ismael, and Y. Liang, "Development of High Density Fan Out Wafer Level Package (HD FOWLP) with Multi-layer Fine Pitch RDL for Mobile Applications", *IEEE/ECTC Proceedings*, May 2016, pp. 1522–1529.
110. Chen, Z., F. Che, M. Ding, D. Ho, T. Chai, V. Rao, "Drop Impact Reliability Test and Failure Analysis for Large Size High Density FOWLP Package on Package", May 2017, *IEEE/ECTC Proceedings*, 2017, pp. 1196–1203.
111. Lim, T., and D. Ho, "Electrical design for the development of FOWLP for HBM integration", *IEEE/ECTC Proceedings*, May 2018, pp. 2136–2142.
112. Ho, S., H. Hsiao, S. Lim, C. Choong, S. Lim, and C. Chong, "High Density RDL build-up on FO-WLP using RDL-first Approach", *IEEE/EPTC Proceedings*, December 2019, pp. 23–27.
113. Boon, S., D. Wee, R. Salahuddin, and R. Singh, "Magnetic Inductor Integration in FO-WLP using RDL-first Approach", *IEEE/EPTC Proceedings*, December 2019, pp. 18–22.
114. Hsiao, H., S. Ho, S. S. Lim, W. Ching, C. Choong, S. Lim, H. Hong, and C. Chong, "Ultrathin FO Package-on-Package for Mobile Application", *IEEE/ECTC Proceedings*, May 2019, pp. 21–27.
115. Lin, B., F. Che, V. Rao, and X. Zhang, "Mechanism of Moldable Underfill (MUF) Process for RDL-1st Fan-Out Panel Level Packaging (FOPLP)", *IEEE/ECTC Proceedings*, May 2019, pp. 1152–1158.
116. Sekhar, V., V. Rao, F. Che, C. Choong, and K. Yamamoto, "RDL-1st Fan-Out Panel Level Packaging (FOPLP) for Heterogeneous and Economical Packaging", *IEEE/ECTC Proceedings*, May 2019, pp. 2126–2133.
117. Ma, M., S. Chen, P.I. Wu, A. Huang, C.H. Lu, A. Chen, C. Liu, and S. Peng, "The Development and the Integration of the 5 μm to 1 μm Half Pitches Wafer Level Cu Redistribution Layers", *IEEE/ECTC Proceedings,* May 2016, pp. 1509–1614.
118. Kim, Y., J. Bae, M. Chang, A. Jo, J. Kim, S. Park, D. Hiner, M. Kelly, and W. Do, "SLIM™, High Density Wafer Level Fan-out Package Development with Submicron RDL", *IEEE/ECTC Proceedings*, December 2017, pp. 18–13.
119. Hiner, D., M. Kolbehdari, M. Kelly, Y. Kim, W. Do, J. Bae, M. Chang, and A. Jo, "SLIM™ Advanced Fan-out Packaging for High Performance Multi-die Solutions", *IEEE/ECTC Proceedings*, May 2017, pp. 575–580.
120. Lin, B., C. Ko, W. Ho, C. Kuo, K. Chen, Y. Chen, and T. Tseng, "A Comprehensive Study on Stress and Warpage by Design, Simulation and Fabrication of RDL-First Panel Level Fan-Out Technology for Advanced Package", *IEEE/ECTC Proceedings,* May 2017, pp. 1413–1418.
121. Suk, K., S. Lee, J. Youn, K. Lee, H. Kim, S. Lee, P. Kim, D. Kim, D. Oh, and J. Byun, "Low Cost Si-less RDL Interposer Package for High Performance Computing Applications",

IEEE/ECTC Proceedings, May 2018, pp. 64–69.
122. Hwang, T., D. Oh, E. Song, K. Kim, J. Kim, and S. Lee, "Study of Advanced Fan-Out Packages for Mobile Applications", *IEEE/ECTC Proceedings,* May 2018, pp. 343–348.
123. Lee, C, J. Su, X. Liu, Q. Wu, J. Lin, P. Lin, C. Ko, Y. Chen, W. Shen, T. Kou, S. Huang, A. Lin, Y. Lin, and K. Chen, "Optimization of laser release process for throughput enhancement of fan-out wafer level Packaging", *IEEE/ECTC Proceedings,* May 2018, pp. 1818–1823.
124. Cheng, W., C. Yang, J. Lin, W. Chen, T. Wang, and Y. Lee, "Evaluation of Chip-last Fan-out Panel Level Packaging with G2.5 LCD Facility using FlexUPTM and Mechanical De-bonding Technologies", *IEEE/ECTC Proceedings,* May 2018, pp. 386–391.
125. Cheng, S., C. Yang, W. Cheng, S. Cheng, W. Chen, H. Lai, T. Wang, and Y. Lee, "Application of Fan-Out Panel Level Packaging Techniques for Flexible Hybrid Electronics Systems", *IEEE/ECTC Proceedings,* May 2019, pp. 1877–1822.
126. Chang, K., C. Huang, H. Kuo, M. Jhong, T. Hsieh, M. Hung, and C. Wan, "Ultra High Density IO Fan-Out Design Optimization with Signal Integrity and Power Integrity", *IEEE/ECTC Proceedings,* May 2019, pp. 41–46.
127. Lin, Y., M. Yew, M. Liu, S. Chen, T., Lai, P. Kavle, C. Lin, T. Fang, C. Chen, C. Yu, K. Lee, C. Hsu, P. Lin, F. Hsu, and S. Jeng, "Multilayer RDL Interposer for Heterogeneous Device and Module Integration", *EEE/ECTC Proceedings,* 2019, pp. 931–936.
128. Miki, S., H. Taneda, N. Kobayashi, K. Oi, K. Nagai, and T. Koyama, "Development of 2.3D High Density Organic Package using Low Temperature Bonding Process with Sn-Bi Solder *IEEE/ECTC Proceedings,* May 2019, pp. 1599–1604.
129. Murayama, K., S. Miki, H. Sugahara, K. Oi, "Electro-migration evaluation between organic interposer and build-up substrate on 2.3D organic package", *IEEE/ECTC Proceedings,* May 2020, pp. 716–722.
130. Lau, J. H., C. Ko, K. Yang, C. Peng, T. Xia, B. Lin, J. J. Chen, P. Huang, H. Liu, T. Tseng, E. Lin, and L. Chang, "Panel-Level Fan-Out RDL-First Packaging for Heterogeneous Integration" *IEEE Transactions on CPMT,* Vol. 10, No. 7, July 2020, pp. 1125–1137.
131. Lau, J. H., C. Ko, T. Peng, K. Yang, T. Xia, P. Lin, J. Chen, P. Huang, T. Tseng, E. Lin, L. Chang, C. Lin, and W. Lu, "Chip-Last (RDL-First) Fan-Out Panel-Level Packaging (FOPLP) for Heterogeneous Integration", *IMAPS Transactions, Journal of Microelectronics and Electronic Packaging,* Vol. 17, No. 3, October 2020, pp. 89–98.
132. Takahashi, N., Y. Susumago, S. Lee, Y. Miwa, H. Kino, T. Tanaka, T. Fukushima, "RDL-first Flexible FOWLP Technology with Dielets Embedded in Hydrogel", *IEEE/ECTC Proceedings,* May 2020, pp. 811–816.
133. Scott, G., J. Bae, K. Yang, W. Ki, N. Whitchurch, M. Kelly, C. Zwenger, J. Jeon, "Heterogeneous Integration Using Organic Interposer Technology", *IEEE/ECTC Proceedings,* May 2020, pp. 885–892.
134. Son, S., D. Khim, S. Yun, J. Park, E. Jeong, J. Yi, J. Yoo, K. Yang, M. Yi, S. Lee, W. Do, and J. Khim, "A New RDL-First PoP Fan-Out Wafer-Level Package Process with Chip-to-Wafer Bonding Technology", *IEEE/ECTC Proceedings,* May 2020, pp. 1910–1915.
135. Mok, I., J. Bae, W. Ki, H. Yoo, S. Ryu, S. Kim, G. Jung, T. Hwang, and W. Do, "Wafer Level Void-Free Molded Underfill for High-Density Fan-out Packages", *IEEE/EPTC Proceedings,* December 2020, pp. 419–424.
136. Chong, S., V. Rao, K. Yamamoto, S. Lim, and S. Huang, "Development of RDL-1st Fan-Out Panel-Level Packaging (FO-PLP) on 550 mm × 650 mm size panels", *IEEE/EPTC Proceedings,* December 2020, pp. 425–429.
137. Rotaru, M., and K. Li, "Electrical characterization and design of hyper-dense interconnect on HD-FOWLP for die to die connectivity for AI and ML accelerator applications", *IEEE/EPTC Proceedings,* December 2020, pp. 430–434.
138. Lim, S., N. Jaafar, S. Chong, S. Lim, and T. Chai, "Development of wafer level solder ball placement process for RDL-first", FOWLP *IEEE/EPTC Proceedings,* December 2020, pp. 435–439.
139. Chai, T., D. Ho, S. Chong, H. Hsiao, S. Soh, S. Lim, S. Lim, E. Wai, B. Lau, W. Seit, G. Lau, T. Phua, K. Lim, S.Lim, Y. Ye, "Fan-Out Wafer Level Packaging Development Line", *IEEE/EPTC Proceedings,* December 2020, pp. 440–444.
140. Boon, S., W. Ho, S. Boon, S. Lim, R. Singh, and S. Raju, "Fan-Out Packaging with Thin-film Inductors", *IEEE/EPTC Proceedings,* December 2020, pp. 449–452.

141. Ji, L., T. Chai, G. See, and P. Suo, "Modelling and prediction on process dependent wafer warpage for FOWLP technology using finite element analysis and statistical approach", *IEEE/EPTC Proceedings,* December 2020, pp. 386–393.
142. Sayeed, S., D. Wilding, J. Camara, D. Vital, S. Bhardwaj, and P. Raj, "Deformable Interconnects with Embedded Devices in Flexible Fan-Out Packages", *IMAPS Proceedings*, October 2019, pp. 8.1–8.6.
143. Boulanger, R., J. Hander, and R. Moon, "Innovative Panel Plating for Heterogeneous Integration", *IMAPS Proceedings*, October 2019, pp. 8.7–8.11.
144. Fang, J., M. Huang, H. Tu, W. Lu, P. Yang, "A Production-worthy Fan-Out Solution – ASE FOCoS Chip Last", *IEEE/ECTC Proceedings*, May 2020, pp. 290–295.
145. Lin, J., C. Chung, C. Lin, A. Liao, Y. Lu, J. Chen, D. Ng, "Scalable Chiplet package using Fan-Out Embedded Bridge", *IEEE/ECTC Proceedings*, May 2020, pp. 14–18.
146. Wang, T., H. Lai, Y. Chung, C. Feng, L. Chang, J. Yang, T. Yu, S. Yan, Y. Lee, and S. Chiu, "Functional RDL of FOPLP by Using LTPS-TFT Technology for ESD protection Application", *IEEE/ECTC Proceedings*, May 2020, pp. 25–30.
147. Chong, S., E. Ching, S. Lim, S. Boon, T. Chai, "Demonstration of Vertically Integrated POP using FOWLP Approach", *IEEE/ECTC Proceedings*, May 2020, pp. 873–878.
148. Podpod, A., A. Phommahaxay, P. Bex, J. Slabbekoorn, J. Bertheau, A. Salahouelhadj, E. Sleeckx, A. Miller, G. Beyer, E. Beyne, A. Guerrero, K. Yess, K. Arnold, "Advances in Temporary Carrier Technology for High-Density Fan-Out Device Build-up", *IEEE/ECTC Proceedings*, May 2019, pp. 340–345.
149. Elmogi, A., A. Desmet, J. Missinne, H. Ramon, J. Lambrecht, P. Heyn, M. Pantouvaki, J. Campenhout, J. Bauwelinck, and G. Steenberge, "Adaptive Patterning of Optical and Electrical Fan-out for photonic chip packaging", *IEEE/ECTC Proceedings*, May 2019, pp. 1757–1763.
150. Chen, D., I. Hu, K. Chen, M. Shih, D. Tarng, D. Huang, J. On, "Material and Structure Design Optimization for Panel-Level Fan-Out Packaging", *IEEE/ECTC Proceedings*, May 2019, pp. 1710–1715.
151. Lau, J. H., and N. C. Lee, *Assembly and Reliability of Lead-Free Solder Joint*, Springer, New York, 2020.
152. Lau, J. H., State of the Art of Lead-Free Solder Joint Reliability, *ASME Transactions, Journal of Electronic Packaging,* June 2021, Vol. 143, pp. 1–21.
153. Pan, N., G. Henshall, F. Billaut, S. Dai, M. Strum, R. Lewis, E. Benedetto, and J. Rayner, "An Acceleration Model for Sn-Ag-Cu Solder Joint Reliability under Various Thermal Cycle Conditions", *SMTA International Conference Proceedings,* September 2005, pp. 876–883.
154. Lall, P., A. Shirgaokar, and D. Arunachalam, "Norris-Landzberg Acceleration Factor and Goldmann Constants for SAC305 Lead-Free Electronics", *ASME Transactions, Journal of Electronic Packaging*, Vol. 134, September 2012, pp. 1–8.
155. Osterman, M., "Modeling Temperature Cycle Fatigue Life of Select SAC Solders", *SMTA International Conference,* September 2018.
156. Lau, J. H., R. Lee, M. Yuen, and P. Chan, "3D LED and IC Wafer Level Packaging", *Journal of Microelectronics International,* Vol. 27, Issue 2, 2010, pp. 98–105.
157. Chen, Y., "Latest Mini LED and Micro LED Definition and Technology Analysis", *LED Inside*, July 2020.
158. Han, S., "MSI to Launch Gaming Notebook Backlit by Lextar-Made Mini-LED", *Digitimes*, April 2020.
159. Chen, M., "Taiwan Suppliers to Play Major Role in Apple Mini-LED Offerings in 2021", *Digitimes,* November 2020.
160. Yoon, G., "Samsung Electronics Preparing for Mass-Production of Mini-LED TVs", *Korea IT News*, November 2020.
161. Lau, J. H., *Chip On Board Technologies for Multichip Modules*, Van Nostrand Reinhold, New York, 1994.
162. Lau, J. H., C. T. Ko, C. Lin, T. Tseng, K. Yang, T. Xia, P. Lin, C. Peng, E. Lin, L. Chang, N. Liu, S. Chiu, and Z. N. Lee, "Design, Materials, Process, Fabrication, and Reliability of Mini-LED RGB Display by Fan-Out Panel-Level Packaging", *IEEE/ECTC Proceedings*, May 2021.

第 5 章

2D、2.1D 和 2.3D IC 集成

5.1 引言

本章讨论 2D、2.1D 和 2.3D IC 集成技术。在 2D IC 集成中，会介绍引线键合、倒装芯片、引线键合-倒装芯片混合、先上晶扇出、后上晶扇出和混合 RDL 的后上晶扇出。在 2.1D IC 集成中，会讨论积层封装基板上的薄膜层，还会讨论积层基板嵌入式互连桥、扇出型环氧模塑料（EMC）嵌入式互连桥。在 2.3D IC 集成中，有两种类型，分别是积层封装基板上的无芯板有机转接板和积层封装基板上的扇出型（包括先上晶和后上晶）RDL 转接板。2D、2.1D 和 2.3D IC 集成中不包含硅通孔（through-silicon via，TSV），硅通孔相关技术将在第 6 章讨论。

5.2 2D IC 集成——引线键合

图 5.1 为一个封装基板或印制电路板（printed circuit board，PCB）上多芯片引线键合的例子。从图中可以看出芯片在同一块基板上肩并肩地排布。

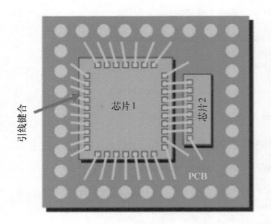

图 5.1 采用引线键合的多芯片

5.3　2D IC 集成——倒装芯片

图 5.2 为一个多芯片通过倒装芯片技术先连接到封装基板然后连接到 PCB 的例子。从图中可以看出，芯片是倒装的，在同一块封装基板上肩并肩地排布。

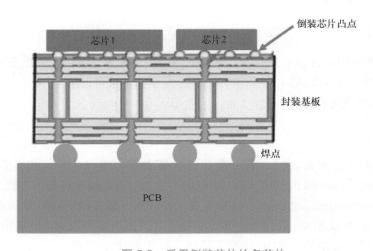

图 5.2　采用倒装芯片的多芯片

5.4　2D IC 集成——引线键合和倒装芯片

图 5.3 为一个多芯片同时采用引线键合和倒装芯片工艺的示意图。从图中可以看到，IC 芯片通过焊球凸点以倒装的形式安装到封装基板上，同时在基板上还通过引线键合连接了微机电系统（micro-electro-mechanical system，MEMS）芯片。

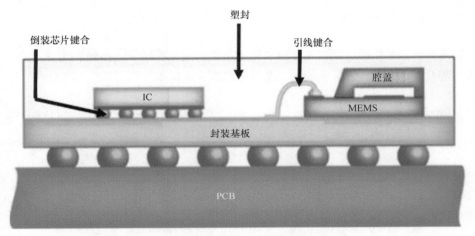

图 5.3　采用引线键合和倒装芯片的多芯片

5.5　RDL

如本书第 4 章所介绍的，扇出型技术的核心是 RDL。至少有三种不同的扇出型 RDL，分别是有机 RDL、无机 RDL 和混合 RDL。

5.5.1　有机 RDL

有机 RDL[1-3] 是指用聚合物（无论是否光敏）做介质层和电化学镀沉铜以及刻蚀形成金属导线层所得到的 RDL，它可以用 4.2.7 节介绍的先上晶工艺，也可以用 4.7.4 节介绍的后上晶工艺。有机 RDL 的金属线宽 / 线距一般 ≥ 2μm（后上晶）和 ≥ 5μm（先上晶）。

5.5.2　无机 RDL

对于线宽 / 线距 < 2μm 的 RDL，其介质层通过等离子体增强化学气相沉积（plasma enhanced chemical vapor deposition，PECVD）来制作，金属层通过铜大马士革电镀工艺和化学机械抛光（chemical-mechanical polishing，CMP）来制作，这是最传统的半导体后道工艺。该工艺的介质层材料为 SiO_2 或 SiN，通过电化学方法在整片晶圆表面沉积铜，然后采用 CMP 去除多余的铜层和种子层，最终得到 RDL 的铜导线层。参考文献 [1-3] 展示了该工艺的关键步骤。通过 PECVD、铜大马士革电镀工艺和 CMP 制作的 RDL 称为无机 RDL。无机 RDL 的线宽 / 线距可以达到亚微米。

5.5.3　混合 RDL

无机 RDL 和有机 RDL 的组合称为混合 RDL。参考文献 [1-3] 中描述其关键工艺步骤如下：①首先在玻璃临时支撑片上涂布一层牺牲层；②第一层 RDL 采用 PECVD 制作 SiO_2 介质层，采用铜大马士革电镀工艺和 CMP 制作导线层（无机 RDL）；③其余的 RDL 层采用聚合物（无论是否光敏）、电镀铜和刻蚀的方法制备（有机 RDL）。混合 RDL 的金属线宽/线距最小可以达到亚微米。

5.6 2D IC 集成——扇出型（先上晶）

采用扇出型（先上晶）的 2D IC 集成封装有很多例子[4-16]，本节给出其中 2 个案例。

5.6.1 HTC 的 Desire 606W

图 5.4 所示为扇出型封装早期在手机中的应用。它是 HTC 在 2013 年出货的 Desire 606W 手机中的基带调制解调器应用处理器芯片组。从图中可以看到，调制解调器芯片与应用处理器芯片通过扇出型（先上晶且面朝下）的方式肩并肩封装在了一起。芯片通过 2 层 RDL 扇出，并通过焊球连接到了 PCB 上。RDL 制作的关键工艺是采用有机 RDL 方法，芯片在重构晶圆上的组装方法为第 2 章中所介绍的表面安装技术（surface mount technology，SMT）。

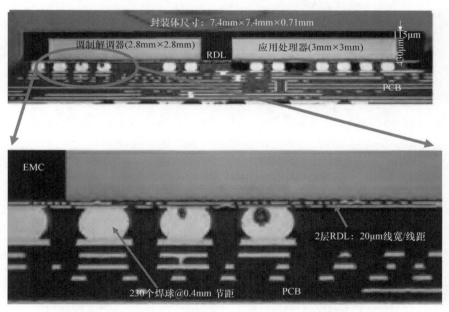

图 5.4 HTC 采用扇出型（先上晶）技术的多芯片封装

5.6.2 4 颗芯片异质集成

图 5.5 和图 5.6 所示为采用扇出型封装技术实现的 4 颗芯片异质集成封装[11-16]。图中大芯片可以是应用处理器芯片，较小的芯片可以是存储芯片，通过先上晶且面朝下扇出型封装肩并肩排布和粘接。有 2 层 RDL，采用有机 RDL 方式制作，最小金属线宽/线距可以小到 5μm。

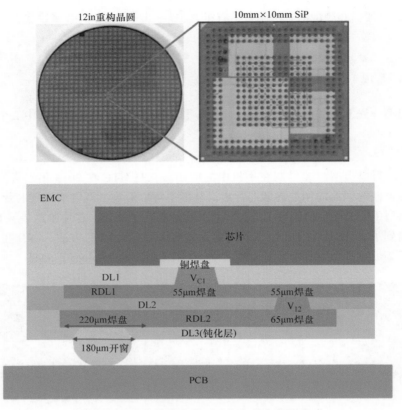

图 5.5 4 颗芯片的扇出型（先上晶）异质集成

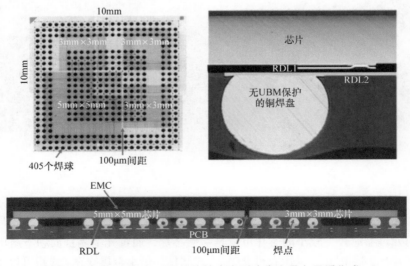

图 5.6 在 PCB 上 4 颗芯片的扇出型（先上晶）异质集成

5.7 2D IC 集成——扇出型（后上晶）

采用扇出型（后上晶）的 2D IC 集成封装例子很多，本节给出其中 5 个案例。如 4.7.4 节所述，在扇出型后上晶（先 RDL）工艺中，RDL 通常先在一块临时玻璃支撑片上制作完成。

5.7.1 IME 的后上晶扇出型封装

图 5.7 和图 5.8 为采用后上晶（或先 RDL）扇出型封装实现的两颗芯片异质集成[17-26]。如 4.7.4 节所述，首先制作 RDL 基板，并行地完成晶圆 C2（铜柱和焊料帽）凸点成型、合格芯片测试以及划片。然后进行芯片 - 晶圆（RDL 基板）的键合、底部填充以及模塑。最后，对临时支撑片解键合、植球、将重构晶圆切割成单颗封装体。

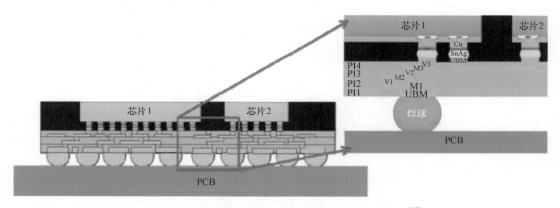

图 5.7 采用扇出型（后上晶）工艺的多芯片封装[19]

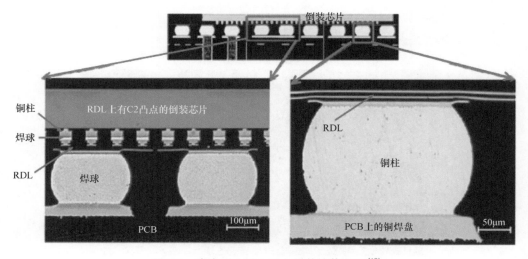

图 5.8 扇出型（后上晶）封装的截面图[19]

5.7.2 Amkor 的 SWIFT

图 5.9 和图 5.10 所示为安靠（Amkor）的硅晶圆集成扇出技术（Silicon wafer integrated fan-out technology，SWIFT）[27-32]，其工艺流程（有机 RDL）与 IME 的非常相似。从图中可以看到，芯片先通过带有焊料帽的铜柱连接到 RDL 基板，然后模组再通过焊球连接到 PCB。

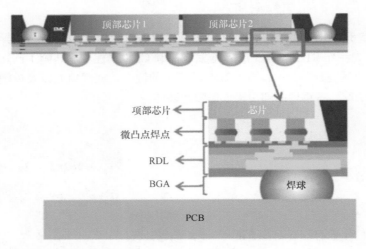

图 5.9　Amkor 的 SWIFT[31]

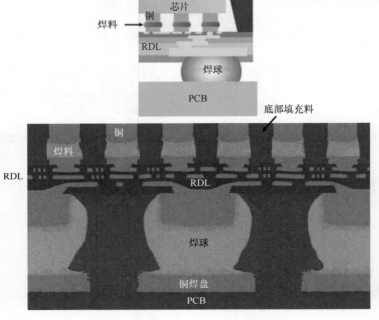

图 5.10　SWIFT 的 SEM 图像[31]

5.7.3 Amkor 的 SLIM

图 5.11 和图 5.12 为安靠（Amkor）的无硅集成模组（silicon-less integrated module，SLIM）[27-32]，SLIM 与 SWIFT 最大的不同是 SLIM 采用了混合 RDL。为了减小金属线宽/线距（减小到亚微米），先制作无机 RDL 后制作有机 RDL 来完成混合 RDL 制作。图 5.12 所示为通过半导体工艺和设备制备的线宽/线距为 0.5μm 的金属线（RDL1）（无机 RDL 方法）以及由聚合物和电化学沉铜（有机 RDL 方法）制得的 RDL2 和 RDL3。

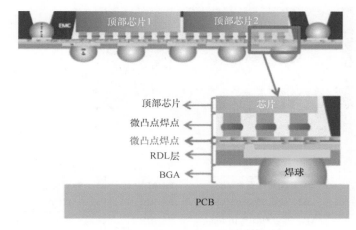

图 5.11　Amkor 的 SLIM[30]

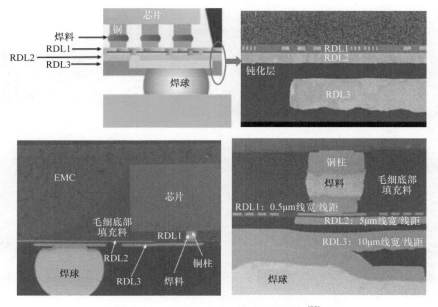

图 5.12　Amkor SLIM 的 SEM 图像[30]

5.7.4 矽品科技的混合 RDL 扇出

2016 年,矽品科技(SPIL)首次演示了在混合 RDL 上实现后上晶扇出型封装的可行性[33],如图 5.13 和图 5.14 所示。对于第一层 RDL(窄金属线宽/线距),采用 PECVD 制备 SiO_2 层,采用铜双大马士革电镀和 CMP 制备线宽/线距为 2/2μm 的铜金属层(无机 RDL);然后采用聚合物制作介质层,通过电化学沉积铜和刻蚀制作 RDL2 和 RDL3(有机 RDL)。

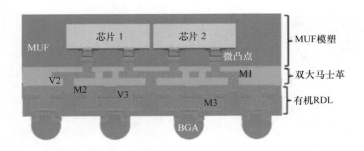

层	方案(thk)	介质层
M1:线宽/线距为 2/2μm	FS-RDL1	SiO_2/SiN_x
M2:线宽/线距为 5/5μm	BS-RDL2	PBO
M3:线宽/线距为 10/10μm (连接 BGA 球)	BS-RDL3	PBO

图 5.13　矽品科技的混合 RDL[33]

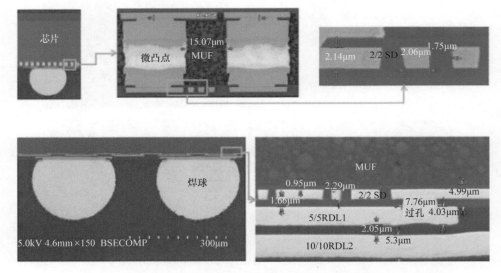

图 5.14　矽品科技的混合 RDL 的 SEM 图像[33]

5.7.5　欣兴电子的扇出型后上晶工艺

图 5.15 和图 5.16 为一颗大芯片（可以是应用处理器芯片）和两颗小芯片（可以是存储芯片）在先 RDL 基板上的异质集成，RDL 基板的最小金属线宽/线距为 2/2μm[34, 35]。从图 5.16 可以看到，大芯片先通过带有焊料帽的铜柱连接到窄金属线宽/线距的 RDL 基板上，然后再通过焊球连接到 PCB 上。

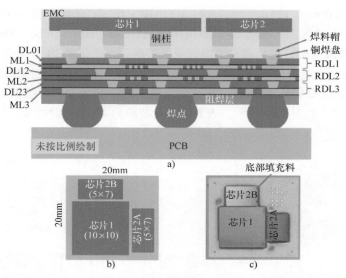

图 5.15　欣兴电子的在扇出型先 RDL 基板上制作的 3 颗芯片异质集成封装

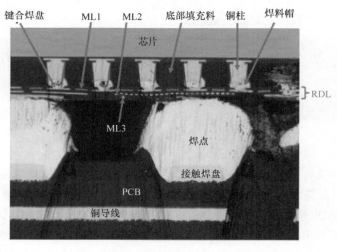

图 5.16　3 颗芯片异质集成 PCB 组装的截面图

5.8 2.1D IC 集成

2.1D IC 集成是指在积层封装基板上制备窄金属线宽/线距薄膜层，本节将简要介绍 8 个案例。

5.8.1 Shinko 的 i-THOP

图 5.17、图 5.18 和图 5.19[36, 37] 所示为神钢（Shinko）的集成薄膜高密度有机封装（integrated thin film high density organic package，i-THOP）技术，Shinko 在 2013 年提出在封装基板的积层之上再制备薄膜层。图 5.17 所示为用于高性能应用的 Shinko i-THOP 基板[36]。它是一个 4+（2-2-3）测试结构，它代表有 2 层金属层芯板、在底部 PCB 一侧有 3 层积层布线、在顶部（芯片）一侧有 2 层积层布线，第一个数字"4"代表在顶部积层布线的表面有 4 层薄膜铜线层（RDL）。薄膜铜 RDL 的厚度、线宽以及线距均小到 2μm，薄膜铜 RDL 层间通过直径为 10μm 的过孔连接，如图 5.17 所示。表面铜焊盘节距为 40μm，铜焊盘直径为 25μm，高度为 10～12μm。

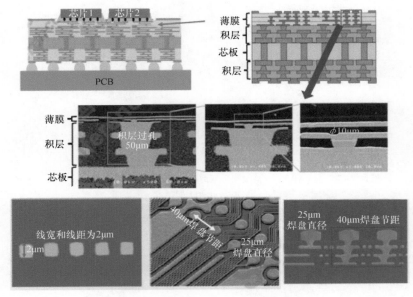

图 5.17 Shinko 的 i-THOP[37]

2014 年，Shinko 演示了超窄节距倒装芯片[37]可以成功实现在 i-THOP 基板上的组装。图 5.18 为两块芯片通过 2 层金属线宽/线距为 2μm 的薄膜 RDL 进行横向通信的示意图，该薄膜层制作在一块 1-2-2 的有机积层基板上，也就是 2+（1-2-2）基板。图 5.19 所示为测试芯片上节距 40μm 的微凸点（铜柱 + Ni + SnAg）和节距 40μm 的倒装芯片键合焊盘（直径 25μm）。图 5.19 为倒装芯片在参数条件优化后组装的截面图，从图中可以看到组装的所有焊点质量良好[37]。

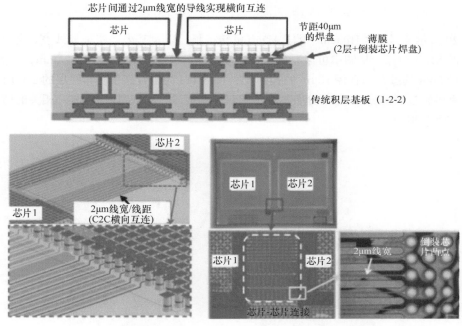

图 5.18 Shinko 含 2 层薄膜的 i-THOP 基板[37]

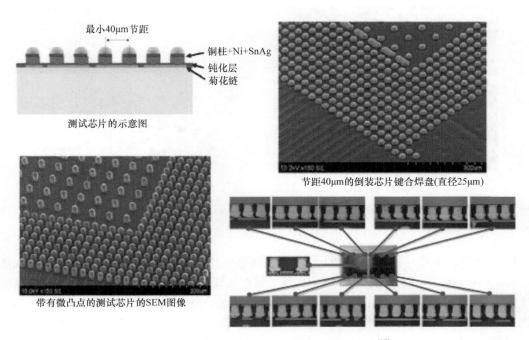

图 5.19 Shinko 的 i-THOP 测试结构和组装[37]

5.8.2 日立的 2.1D 有机转接板

图 5.20 所示为日立（Hitachi）的 2.1D 有机转接板，以及其内部窄金属线宽/线距 RDL 和过孔图形的 SEM 图像。从图中可以看到，转接板走线的线宽/线距达到了 2/2μm，从图中还可以看到直径 5μm 的过孔和直径 10μm 的孔盘[38]。图 5.21a 为含有高密度布线层的 2.1D 封装布局示意图（有机转接板部分），该部分叠加在传统封装基板（见图 5.21b）上方。相关的仿真结果，请见参考文献 [38]。

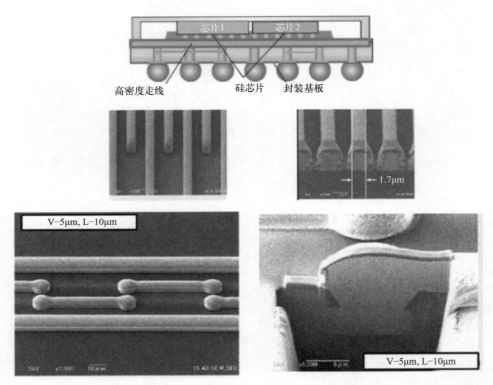

图 5.20　日立的 2.1D 结构以及线宽和线距图像[38]

5.8.3 日月光的 2.1D 有机转接板

图 5.22 所示为日月光（ASE）的 2.1D 有机转接板[39]。可以看到，高密度基板是在大面板（512mm×408mm）上制作的，包含 4 个区域（每个区域尺寸为 256mm×204mm）。每个区域上有 3 根条带（240mm×63.5mm），每根条带分 16 个单元，每个单元（23mm×23mm）可以支持 2 颗芯片。图 5.23a 为不同视角下的高密度薄膜层图案，图 5.23b 为在 2.1D 有机转接板上的倒装芯片组装。

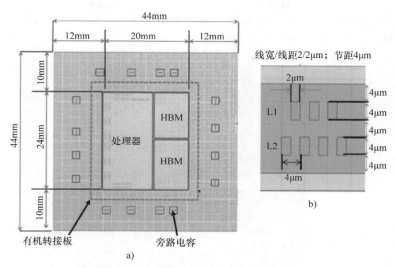

图 5.21 用于分析 PDN 阻抗的结构[38]

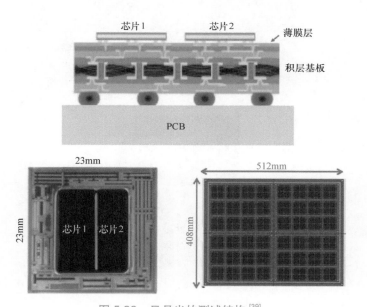

图 5.22 日月光的测试结构[39]

5.8.4 矽品科技的 2.1D 有机转接板

图 5.24 所示为矽品科技（SPIL）的 2.1D 有机转接板[40]。封装体的尺寸为 45mm×45mm× 2.9mm，基板的厚度为 1.2mm（4 层薄膜和 2/2/3 基板）；芯片尺寸为 11.8mm×20.8mm× 0.78mm；最小的金属线宽/线距为 2/2μm，微焊盘节距最小为 40μm。图 5.25 为 2.1D 有机转接

板封装的俯视图和底视图,图 5.25 还给出了微凸点的尺寸以及倒装芯片与薄膜层组装的图像。相关仿真和可靠性试验结果,请见参考文献 [40]。

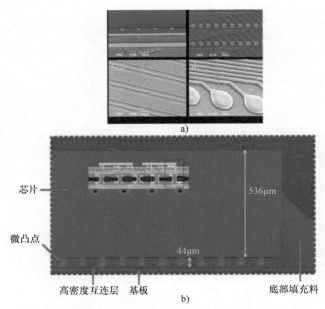

图 5.23 薄膜层的 SEM 图像以及组装结构的截面图[39]

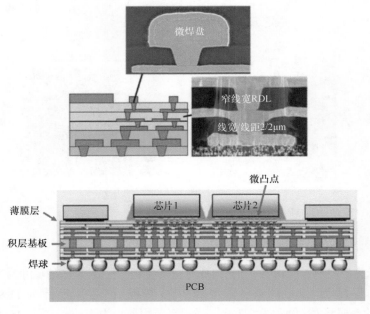

图 5.24 薄膜层的 SEM 图像以及结构(矽品科技)的截面图[40]

图 5.25　薄膜层上芯片（含微凸点和底部填充料）的 SEM 图像[40]

5.8.5　长电科技的 uFOS

图 5.26 所示为长电科技（JCET）的 2.1D 有机转接板，称为超结构有机基板（ultra format organic substrate，uFOS）[41]，图 5.27 为 uFOS 的关键工艺步骤和 SEM 图像，可以看到在无芯板封装基板上制作有金属线宽/线距为 2/2μm 的布线层。为了缓解无芯板封装基板所带来的翘曲问题，制作时在基板最后一层引入了嵌入式加固（embedded stiffness，e-STF）。如图 5.26 所示，即通过将金属条嵌入到半固化片中来加强基板整体的抗弯曲能力[41]。

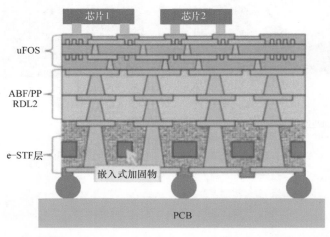

图 5.26　含有薄膜层（uFOS）以及嵌入式加固（e-STF）层的结构示意图[41]

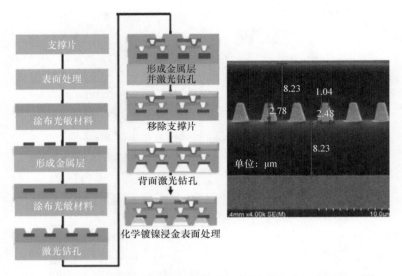

图 5.27　uFOS 工艺流程以及薄膜层的 SEM 图像（长电科技）[41]

5.8.6　英特尔的 EMIB

图 5.28 所示为英特尔（Intel）的 Agilex 现场可编程门阵列（field programmable gate array，FPGA）模组，可以看到 FPGA 上有两种焊料凸点（C4 凸点和 C2 微凸点）。FPGA 和其他芯片组装在含有嵌入式多芯片互连桥（embedded multi-die interconnect bridge，EMIB）的积层封装基板上。EMIB 是一片无功能的、带有窄金属线宽/线距 RDL 的硅片，可以实现芯片水平方向的互连[42, 43]。

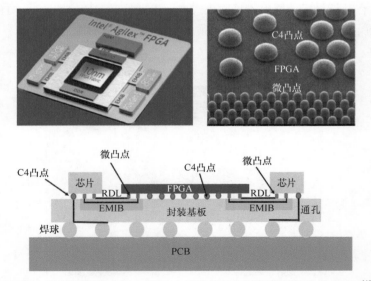

图 5.28　英特尔在其 Agilex FPGA 模组的积层基板顶部埋入的 EMIB[43]

5.8.7 应用材料的互连桥

2017 年 12 月 8 日，应用材料（Applied Materials）提出了一种采用扇出型封装在 EMC 中埋入互连桥的方案，如图 5.29 所示[44]。从图中可以看到，芯片的电路通过 RDL 扇出并通过互连桥水平连接，垂直方向上通过模塑通孔（through molding via，TMV）进行互连[44]。

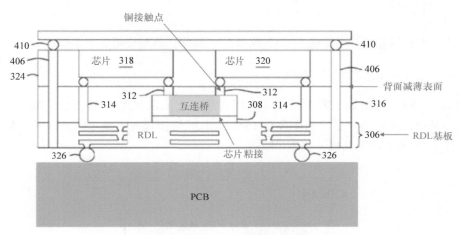

图 5.29 Applied Materials 在 EMC 中埋入的互连桥

5.8.8 台积电的 LSI

在台积电（TSMC）的年度技术讨论会上（2020 年 8 月 25 日），台积电宣布了它的局部硅互连（local Si interconnect，LSI）技术，如图 5.30 所示。从图中可以看到，LSI 技术与 Applied Materials 的方案非常相似。唯一的明显不同是，LSI 的转接板也许不仅仅是一片含 RDL 的无 TSV 转接板，还可能包括 TSV 甚至 CMOS 器件。

5.9 2.3D IC 集成

2.3D IC 集成是指在积层封装基板表面制作一个无芯板有机/无机转接板，无芯板基板（转接板）在过去 10 年受到大量关注。与传统积层封装基板的最大不同在于无芯板基板的所有布线层都是积层的。无芯板基板的优势在于：①由于去除了芯板，成本更低；②由于去除了芯板，具有更高密度的布线能力；③因为具有良好的高速传输特性，其电性能更好；④显然更小的封装外形。但是另一方面，无芯板基板也有缺点：①因为没有芯板，其翘曲更大；②更容易发生分层开裂；③无芯板基板刚性下降会引起焊点良率下降；④需要新的制造设备。2010 年，索尼制作了第一

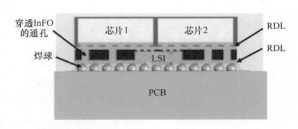

图 5.30 台积电埋入 EMC 的 LSI 互连桥

款无芯板基板，用于其 PlayStation3 产品的 Cell 处理器。

制作无芯板有机/无机转接板至少有 3 种方法，分别是传统的半加成工艺（semi-additive process,SAP）/PCB 方法制作有机转接板、扇出型先上晶法制造有机转接板、扇出型后上晶（先RDL）法制造有机/无机转接板。

5.10 采用 SAP/PCB 法的 2.3D IC 集成

本节中，有机转接板采用传统的 SAP/PCB 工艺制作，简要介绍 2 个案例。

5.10.1 Shinko 的无芯板有机转接板

2012 年，神钢（Shinko）提出采用无芯板封装基板来取代 TSV 转接板，如图 5.31 所示。制作无芯板基板的成本当然会比制作含有 TSV 和 RDL 转接板的更低（因为后者需要用到半导体加工设备），但是无芯板基板的翘曲是问题。

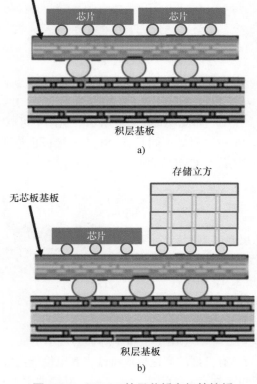

图 5.31　Shinko 的无芯板有机转接板

5.10.2 思科的有机转接板

图 5.32 为思科（Cisco）采用一块窄节距窄线宽互连的大有机转接板（无 TSV 转接板）设计和实现的异质集成封装[45]。有机转接板的尺寸为 38mm×30mm×0.4mm，正反面的最小线宽、线距以及厚度都一致，分别为 6μm、6μm、10μm。这是一块 10 层高密度转接板（基板），过孔直径为 20μm。制作该有机转接板的主要工艺步骤与制作有机积层基板一样，包括①芯板电镀通孔（plating through-hole，PTH）的产生和填充；②芯板电路制作；③采用 SAP 工艺在芯板两面制作积层布线。

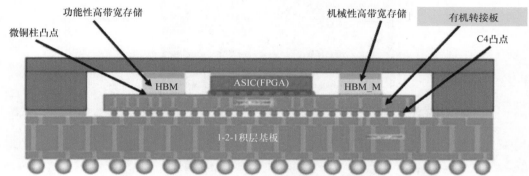

特征	有机转接板
铜布线(介质层)	SAP工艺(有机)
正面焊盘尺寸/节距	30/55μm
正面布线线宽/线距/厚度(最小值)	6/6/10μm
布线层数	10
布线层过孔尺寸	20μm
电镀通孔尺寸/节距/深度	57/150/200μm
背面焊盘尺寸/节距	100/150μm
背面布线线宽/线距/厚度(最小值)	6/6/10μm

图 5.32 思科的有机转接板[45]

在有机转接板顶部安装了 1 颗高性能的专用集成电路（application-specific IC，ASIC）芯片（尺寸为 19.1mm×24mm×0.75mm）与 4 颗由动态随机存储器（dynamic random access memory，DRAM）堆叠形成的高性能存储器（high-bandwidth memory，HBM）。3D HBM 芯片堆叠尺寸为 5.5mm×7.7mm×0.48mm，包括 1 颗底部缓冲芯片和 4 颗 DRAM 核，4 颗 DRAM 核通过 TSV 和窄节距带有焊料帽的微凸点实现垂直互连。有机转接板正面的焊盘尺寸为 30μm，节距为 55μm。图 5.33 为制造好的有机转接板俯视图以及 HBM 芯片堆叠体与有机转接板间的高质量焊点截面[45]。

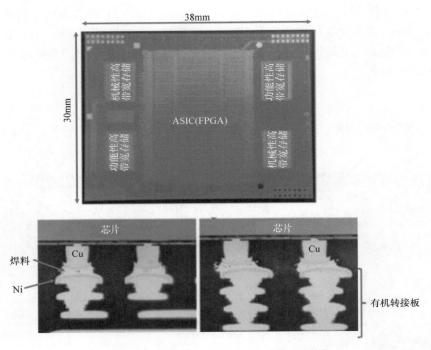

图 5.33　制造的有机转接板及其截面图[45]

5.11　采用扇出型（先上晶）技术的 2.3D IC 集成

本节介绍采用扇出型（先上晶）技术制作无芯板有机转接板上的 2.3D IC 集成，共有 4 个案例。

5.11.1　星科金朋的 2.3D eWLB

在 ECTC2013 会议上，星科金朋（Statschippac）提出采用扇出型倒装芯片（fan-out flip chip，FOFC）- 埋入式晶圆级球栅阵列（embedded wafer level ball grid array，eWLB）来制作芯片的 RDL，用于芯片间的大部分横向通信，如图 5.34 所示。根据专利申请书中的描述（US 9484319 B2，2011 年 12 月 23 日提交），其目的是想用 RDL（无芯板有机转接板）来取代 TSV 转接板、微凸点和底部填充料，该专利在 2016 年 11 月 1 日获得授权，如图 5.35 所示[46]。

联发科（Mediatek）在 2009 年 2 月 12 日也提交了它们的专利申请（US 7838975 B2）并在 2010 年 11 月 23 日获得授权[48]。与星科金朋不同，联发科的目的不是取代 TSV 转接板，而是将 RDL 转接板用于超窄节距的倒装芯片焊盘，这样可以使用低成本的封装基板，如图 5.36 所示。这两份专利的结构看起来相似。

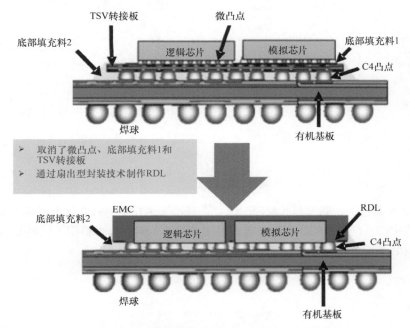

图 5.34　星科金朋的扇出型（先上晶）有机转接板[47]

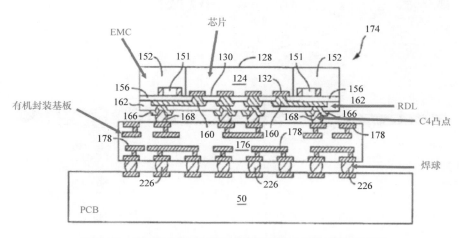

图 5.35　星科金朋关于扇出型（先上晶）有机基板的专利

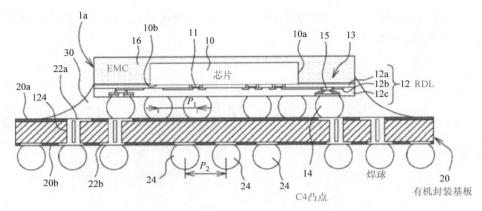

图 5.36 联发科关于扇出型（先上晶）有机转接板的专利

5.11.2 联发科的扇出型（先上晶）技术

在 ECTC2016 会议上，联发科（Mediatek）[49] 提出了采用 FOWLP 技术制作类似无 TSV 转接板的 RDL，如图 5.37 和图 5.38 所示。它们采用微凸点（铜柱＋焊料帽）代替 C4 凸点将底部 RDL 与 6-2-6 封装基板进行互连。

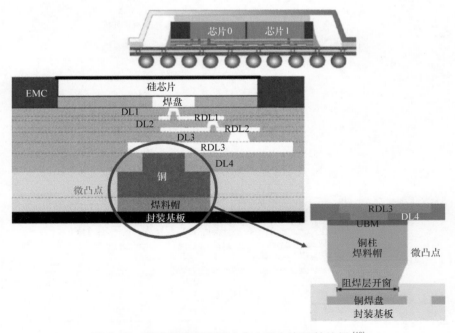

图 5.37 联发科的扇出型（先上晶）有机转接板[49]

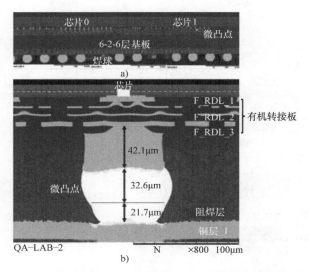

图 5.38 联发科的扇出型（先上晶）有机转接板的截面 SEM 图像[49]

5.11.3 日月光的 FOCoS（先上晶）

在 ECTC2016 会议上，日月光（ASE）提出使用扇出型晶圆级封装（FOWLP）技术（在临时晶圆支撑片上先上晶且芯片面朝下，然后采用模压完成模塑）来制作芯片大部分横向通信所需的 RDL，如图 5.39 和图 5.40 所示，该技术称为扇出型晶圆级基板上芯片（fan-out wafer-level chip-on-substrate，FOCoS）。该技术不需要 TSV 转接板、芯片晶圆凸点成型、助焊剂涂覆、芯片 - 晶圆键合、清洗、底部填充料滴涂和固化。底部 RDL 与封装基板之间通过凸点下金属化层（under-bump metallurgy，UBM）和 C4 焊料凸点连接。日月光的这项技术与星科金朋比较相似。

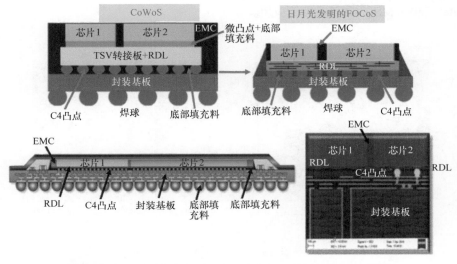

图 5.39 日月光的扇出型（先上晶）有机基板（FOCoS）[50]

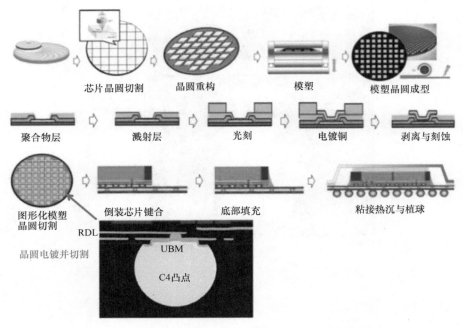

图 5.40 日月光的先上晶 FOCoS 关键工艺步骤[50]

5.11.4 台积电的 InFO_oS 和 InFO_MS

图 5.41 所示为台积电（TSMC）的异质集成扇出型 RDL 基板示意图。图 5.41a 为基板上集成扇出封装（integrated fan-out on substrate，InFO_oS），该技术取消了微凸点、底部填充和含 RDL 的 TSV 转接板。图 5.41b 为基板上存储器集成扇出封装（integrated fan-out with memory on substrate，InFO_MS），用于高性能应用。

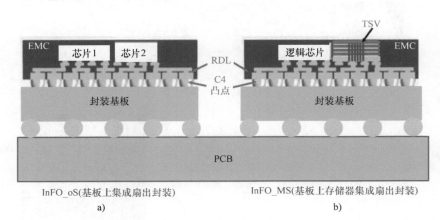

图 5.41 台积电的 a）InFO_oS；b）InFO_MS

5.12 采用扇出型（后上晶）技术的 2.3D IC 集成

本节介绍采用扇出型后上晶（或者先 RDL）技术制作无芯板有机/无机转接板上的 2.3D IC 集成，共有 6 个案例。

5.12.1 矽品科技的 NTI

在 ECTC2016 会议上，矽品科技（SPIL）提出了采用无芯板无机转接板进行 2.3D IC 集成的无硅通孔转接板（non-TSV interposer，NTI）[51, 52]。它们先使用 65nm 工艺在晶圆上制作最小节距达到 0.4μm 的 RDL。然后它们在 RDL 上进行芯片 - 晶圆键合、在芯片和 RDL 转接板之间进行底部填充、用 EMC 对芯片进行模塑。剩下的主要工艺步骤如图 5.42 所示，图 5.43 所示为最后得到的组装体截面图。从图中可以看到，芯片通过微凸点（铜柱 + 焊料帽）粘接到了无机转接板（RDL）上，RDL 转接板再通过 C4 凸点连接到了积层封装基板上。

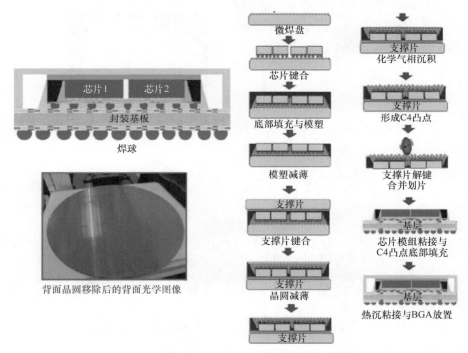

图 5.42 矽品科技的扇出型（先 RDL）无机转接板[52]

5.12.2 三星的无硅 RDL 转接板

在 ECTC2018 会议上，三星（Samsung）也提出利用后上晶或者先 RDL 的 FOWLP 技术来去除在高性能计算异质集成应用中的 TSV 转接板[53, 54]。首先在裸玻璃上制作 RDL，玻璃既可以是晶圆形状也可以是面板形状。与此并行，完成逻辑芯片和 HBM 芯片的晶圆凸点成型。然后依次进行以下步骤：涂覆助焊剂、芯片-晶圆或芯片-面板键合、清洗、底部填充和固化，再进行 EMC 的模压。之后，对 EMC、芯片和 HBM 进行背面减薄并完成 C4 凸点成型（见图 5.44）。这些步骤完成后，将整个模块粘接到积层封装基板上，最后植球并装好散热片。三星将最终制得的结构称作无硅 RDL 转接板。

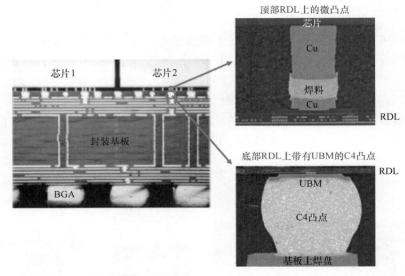

图 5.43 组装截面以及 RDL、微凸点和 C4 凸点的 SEM 图像[52]

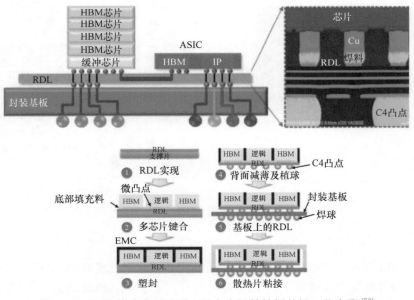

图 5.44 三星的扇出型（后上晶）有机转接板关键工艺步骤[53]

图 5.45 为三星的测试结构。RDL 转接板的尺寸为 55mm×55mm,有 5 层 RDL,包括键合层、信号层、地层。三星表示有机转接板模组中 C4 焊点的热循环测试性能(见图 5.46)要好于 TSV 转接板模组(见图 5.47)[53],这是由于硅转接板与积层封装基板之间的热膨胀系数不匹配要大于与有机转接板之间的热膨胀系数不匹配。

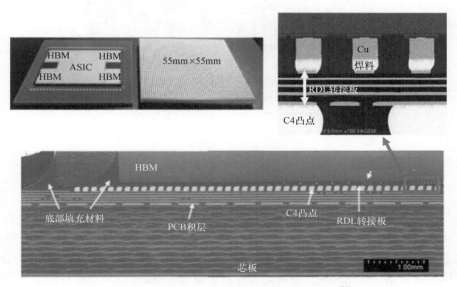

图 5.45 三星的测试结构和截面 SEM 图像[53]

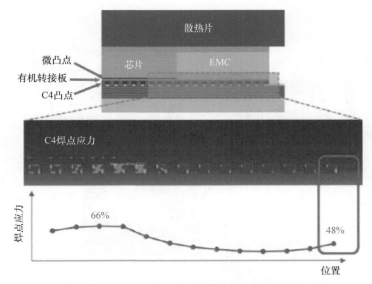

图 5.46 与有机转接板之间焊点的应力[53]

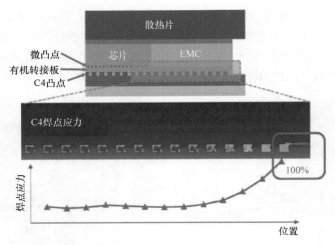

图 5.47　与硅转接板之间焊点的应力[53]

5.12.3　日月光的 FOCoS（后上晶）

图 5.48 和图 5.49 为日月光（ASE）采用扇出型后上晶技术的 FOCoS[55-57]。首先在一片临时玻璃支撑片上制作好 RDL 转接板，从图 5.49 可以看出转接板上至少有 4 层 RDL，即包含堆叠过孔也包含非堆叠过孔。并行地，完成晶圆的微凸点成型。然后，依次进行芯片-RDL 晶圆键合、底部填充和模塑。模塑完成后将临时支撑片解键合，并进行 C4 凸点植球，再切割得到单颗封装模组。最后，将封装模组粘贴到积层封装基板上。

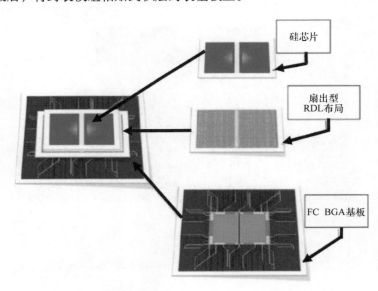

图 5.48　日月光的扇出型（先 RDL）有机转接板[55]

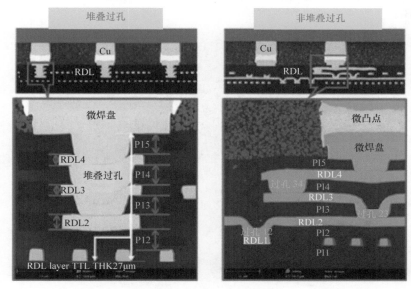

图 5.49 日月光的含堆叠/非堆叠过孔的有机转接板（后上晶）SEM 图像[55]

5.12.4 台积电的多层 RDL 转接板

图 5.50 和图 5.51 为台积电（TSMC）用于异质器件和模块集成的多层 RDL 转接板[58]。该转接板的基本制作工艺与其他文献报道的类似[51-57]：①首先制作 RDL 基板（转接板）；②晶圆凸点成型并切割成单颗芯片；③芯片通过微凸点和底部填充连接到有机/无机 RDL 转接板上；④ RDL 转接板通过 C4 凸点和底部填充连接到积层封装基板上；⑤封装基板通过 BGA 焊球连接到 PCB 上。图 5.51 展示了组装的一些图片，从图中可以看到，芯片和 DRAM 通过微凸点连接到了多层 RDL 转接板上，然后转接板又通过 C4 凸点连接到了封装基板上。RDL 转接板内有 6 层 RDL，RDL 层间通过交错孔、两层堆叠孔或四层堆叠孔连接。

5.12.5 Shinko 的 2.3D 有机转接板

图 5.52、图 5.53 和图 5.54 为神钢（Shinko）用于高性能计算应用的 2.3D 无芯板有机转接板[59, 60]。从图 5.52 中可以看到，Shinko 采用非导电薄膜（non-conductive film，NCF）作为有机转接板和积层封装基板之间的底部填充料。在有机转接板和积层封装基板的互连中，采用了 Sn-Bi 焊料合金取代 SnAgCu。在制作 RDL 有机转接板时，采用了临时支撑片和刚性层，这样在芯片-面板键合前就不用再将转接板转移到另一块临时支撑片上。图 5.53 为有机转接板（薄膜层）的 SEM 图像，从图中可以看到金属的线宽/线距为 2/2μm，层间有堆叠过孔。并行地，完成晶圆凸点成型（铜柱 + 焊料帽）和划片。大芯片的凸点节距为 40μm，小芯片的凸点节距为 55μm（小芯片）。芯片-面板的互连通过热压键合（thermocompression bonding，TCB）实现，如图 5.53 所示。图 5.54 为有机转接板和积层封装基板的组装图像，从图中可以看到互连采用

的材料为 Sn-Bi。

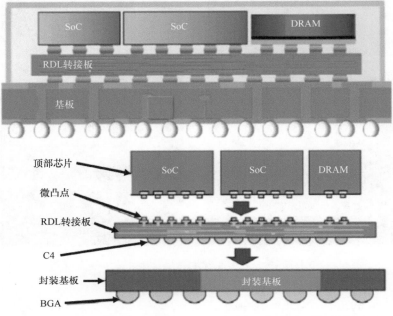

图 5.50 台积电的扇出型（后上晶）有机 RDL 转接板[58]

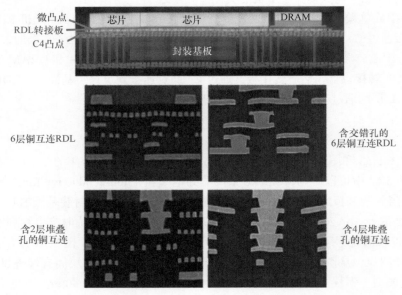

图 5.51 台积电的含交错孔和堆叠孔的后上晶有机 RDL 转接板的 SEM 图像[58]

第 5 章 2D、2.1D 和 2.3D IC 集成 231

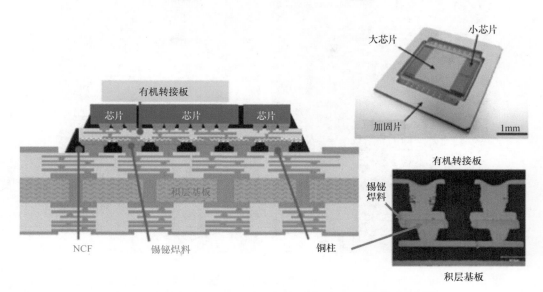

图 5.52 神钢（Shinko）的扇出型（后上晶）有机转接板[59]

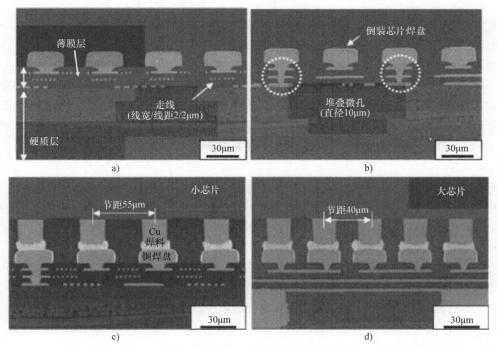

图 5.53 SnAgCu 焊料合金实现的芯片 - 有机基板键合的 SEM 图像[59]

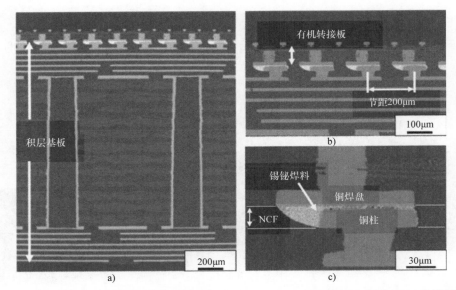

图 5.54 锡铋焊料合金和 NCF 实现的有机转接板 - 积层封装基板键合的 SEM 图像[59]

5.12.6 欣兴电子的 2.3D RDL 转接板

本节介绍扇出型板级后上晶封装用于异质集成的研究,重点放在设计、材料、工艺和制造,包含:①实现 1 颗大芯片(10mm×10mm)和 1 颗小芯片(5mm×5mm)在最小节距 50μm 时的异质集成;②在 515mm×510mm 的临时玻璃支撑片上制作窄线宽/线距的先 RDL 基板;③在 515mm×510mm 的临时玻璃支撑片上制作常规的积层封装基板;④在积层基板顶部焊接 RDL 基板实现混合基板(hybrid substrate);⑤实现芯片与混合基板的键合以及底部填充。可靠性评估以热 - 机械仿真的形式进行,包括采用了非线性温度、时间相关的有限元模型对 2 颗芯片在混合基板上封装的结构进行热循环试验分析[61]。

1. 测试芯片

图 5.55 所示为测试芯片。从图中可以看到大芯片的尺寸为 10mm×10mm×150μm,共有 3592 个按面阵列菊花链排布的焊盘;小芯片的尺寸为 5mm×5mm×150μm,有 1072 个按面阵列菊花链排布的焊盘。两颗芯片的外围周边排布的焊盘节距为 50μm,所有内圈焊盘节距都为 200μm。

大芯片和小芯片的晶圆凸点成型工艺都是采用标准物理气相沉积法(physical vapor deposition,PVD)制作种子层,然后采用电化学沉积法(electrochemical deposition,ECD)制作铜柱和焊料,然后将晶圆切割得到单颗完成凸点成型的芯片。

图 5.55 为两颗芯片的截面图。从图中可以看到铜焊盘的尺寸为 40μm×40μm,Ti/Cu(0.1/0.2μm)UBM 焊盘的尺寸为(直径)32μm,钝化层(PI2)开窗直径为 20μm;铜柱的直径为 32μm,高度为 22μm;SnAg 焊料帽的高度为 15μm,Ni 阻挡层厚度为 3μm。

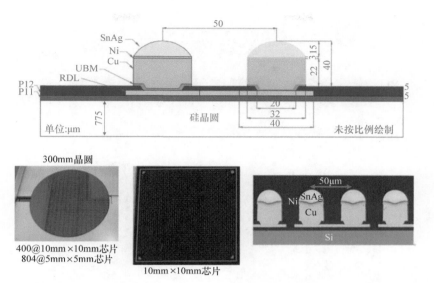

图 5.55 测试芯片：尺寸分别为 10mm×10mm 和 5mm×5mm，凸点节距为 50μm

2. 异质集成测试封装体

图 5.56a 为测试封装体的示意图。从图中可以看到共有 2 颗芯片进行了异质集成，包括一颗大芯片（芯片 1，10mm×10mm）和一颗小芯片（芯片 2，5mm×5mm）。大芯片与小芯片之间的间距为 100μm。芯片键合在混合基板顶部，混合基板包括窄金属线宽/线距的 RDL 基板（20mm×15mm×32μm）和积层基板（23mm×23mm×1.3mm），如图 5.57 所示。

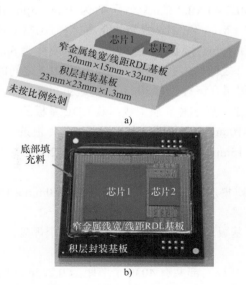

图 5.56 a) 测试封装体的示意图；b) 组装后的测试封装体实物图。芯片在混合基板上键合，混合基板含窄金属线宽/线距的 RDL 基板（20mm×15mm）和积层基板（23mm×23mm）

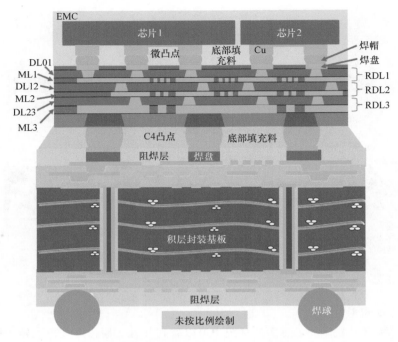

图 5.57 后上晶异质集成测试封装体的截面示意图

RDL 基板包括 3 层 RDL，每层 RDL 包含一层介质层（DL）和一层金属层（ML）。如图 5.58a 所示，RDL1 的 ML1、RDL2 的 ML2、RDL3 的 ML3 线宽/线距分别为 2/2μm、5/5μm、10/10μm。ML1、ML2 和 ML3 的厚度分别为 3μm、5μm 和 5μm。DL01（铜焊盘和 ML1 之间的介质层）、DL12（ML1 和 ML2 之间的介质层）、DL23（ML2 和 ML3 之间的介质层）的厚度分别为 3μm、6μm 和 4μm。焊盘的直径为 36μm，厚度为 6μm。V_{01}（焊盘和 ML1 之间的过孔）、V_{12}（ML1 和 ML2 之间的过孔）、V_{23}（ML2 和 ML3 之间的过孔）的开窗直径分别为 10μm、20μm 和 18μm。阻焊层开窗直径为 80μm，厚度为 2μm。

图 5.58b 所示为用于制造 2 颗芯片异质集成封装 RDL 基板的临时面板，从图中可以看到该玻璃面板尺寸为 515mm×510mm。该面板分为 4 根条带，每根条带（257.5mm×255mm）上有 154 颗 RDL 基板（20mm×15mm）。因此，可以一次完成 1232 颗芯片在 616 个异质集成封装体中的 RDL 基板。

图 5.59 为 2 颗芯片异质集成测试封装体 RDL 基板（20mm×15mm）的俯视图和底视图。从图中可以看到：

1）图 5.59a 显示 RDL 基板顶部共有 3592+1072=4664 个焊盘，这些焊盘均用于芯片-基板键合。

2）图 5.59c 显示 RDL 基板底部共有 4039 个节距为 0.225mm 的焊盘（焊球直径 105μm）。这些焊盘为表面安装焊盘（有阻焊层定义），阻焊层开窗直径为 80μm。这些焊盘上所植焊球为 Sn3Ag0.5Cu 无铅 C4 焊料凸点，焊球直径为 80μm。

3. 混合基板制造

图 5.60 所示为先 RDL 基板、积层基板以及混合基板制作的关键工艺步骤。

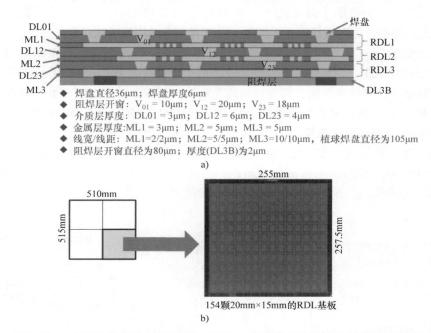

- 焊盘直径36μm；焊盘厚度6μm
- 阻焊层开窗：$V_{01}=10μm$；$V_{12}=20μm$；$V_{23}=18μm$
- 介质层厚度：$DL01=3μm$；$DL12=6μm$；$DL23=4μm$
- 金属层厚度：$ML1=3μm$；$ML2=5μm$；$ML3=5μm$
- 线宽/线距：$ML1=2/2μm$；$ML2=5/5μm$；$ML3=10/10μm$，植球焊盘直径为105μm
- 阻焊层开窗直径为80μm；厚度(DL3B)为2μm

a)

图 5.58　a) 窄金属线宽/线距 RDL 基板；b) 制造 RDL 基板和组装用的面板

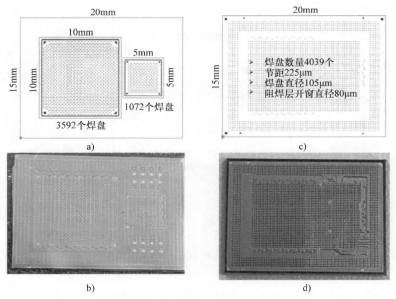

图 5.59　a) RDL 基板的俯视示意图；b) 实际制造的 RDL 基板俯视图；c) RDL 基板的底视示意图；d) 实际制造的 RDL 基板底视图

(1) 先 RDL 基板制造

为提高产率，先 RDL 基板是在 515mm×510mm 的面板上制作的。首先，在临时玻璃支撑片上狭缝式涂布一层牺牲层（剥离胶带），然后通过 PVD 溅射一层 Ti/Cu 种子层，再用光刻胶和激光直写图形显影，之后进行电化学沉铜、剥离光刻胶后得到接触焊盘。

完成以上步骤后，再次狭缝式涂布一层感光介质（photoimageable dielectric，PID），通过激光直写图形得到 RDL1 的第一层介质层（DL01）。然后溅射 Ti/Cu 种子层、涂覆光刻胶、激光直写图形和显影、电化学沉铜，再进行剥离光刻胶、刻蚀种子层得到 RDL1 的第一层金属层（ML1）。重复以上步骤，继续得到 RDL2 的 DL12 和 ML2、RDL3 的 DL23 和 ML3。图 5.59b 为制造好的 RDL 基板顶部的图像。

ML3 制作后，狭缝式涂布一层 PID 并采用 LDI 完成钝化层（阻焊层），再进行化学镀镍钯浸金（electroless nickel electroless palladium immersion gold，ENEPIG）表面处理。至此窄线宽/线距 RDL 基板制作完成，图 5.59d 为 RDL 基板底部的图像。

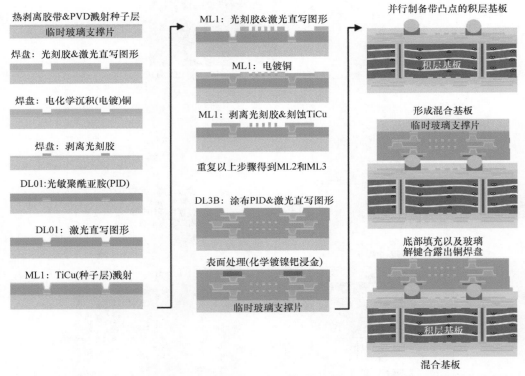

图 5.60 制造先 RDL 混合基板（窄金属线宽/线距 RDL 基板 + 积层基板）的关键工艺步骤

图 5.61 为典型的 RDL 基板截面 SEM 图像，截面图中包含了 ML1、ML2、ML3 以及接触焊盘。通过对 SEM 图像进行放大，可以看到对于 RDL1，实际线宽为 2.07μm、2.08μm 而非 2μm，实

际线距为 1.74μm、1.81μm 而非 2μm，实际厚度为 2.06μm、2.17μm 而非 3μm；对于 RDL2，实际线宽为 5.6μm、6.03μm 而非 5μm，实际线距为 3.91μm、4.5μm 而非 5μm，实际厚度为 4.35μm、4.48μm 而非 5μm；对于 RDL3，实际线宽为 10.24μm、10.69μm 而非 10μm，实际线距为 9.46μm、10.18μm 而非 10μm，实际厚度为 3.39μm、5.11μm 而非 5μm。因此，如果使用更好的 PID 材料、采用更高分辨率的 LDI 或步进式光刻机，制作工艺还有提升的空间。

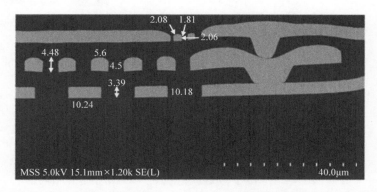

图 5.61　RDL 基板的截面 SEM 图像

（2）积层封装基板制造

如图 5.62a 所示，积层封装基板（23mm×23mm×1.3mm）是 2-2-2 结构。每颗基板上有 4039 个节距为 0.225mm 的焊盘，基板是在一片 510mm×510mm 的面板（见图 5.62b）上采用传统工艺制作的。图 5.62c 为一颗典型的 2-2-2 积层基板截面图像，图 5.62d、e 分别是制造好的积层封装基板面板的俯视图和底视图。

图 5.63 所示为单颗积层封装基板。图 5.63a、b 分别为封装体的俯视示意图以及实物图，图 5.63c、d 分别为制造好的封装体底部示意图以及实物图。可以看出封装基板的顶部与 RDL 基板的底部焊盘（见图 5.59c、d）互相匹配。积层基板的底部有 475 个节距为 1mm 的焊盘。焊盘直径为 500μm，采用阻焊层定义，阻焊层开窗直径为 300μm。

（3）C4 凸点制造

C4 焊料凸点的制作通过在积层基板顶部模板印刷 Sn3Ag0.5Cu 焊膏得到，模板为 29μm 厚的不锈钢板。在焊料回流过程中，由于熔融焊料的表面张力，最终可以得到图 5.64 所示的光滑无顶的直径 30μm 的焊球。

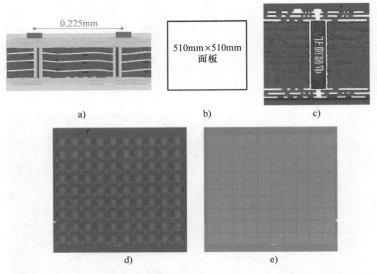

图 5.62 2-2-2 积层封装基板面板：a）示意图；b）面板尺寸；c）截面图；d）制造好的积层基板俯视图；e）制造好的积层基板底视图

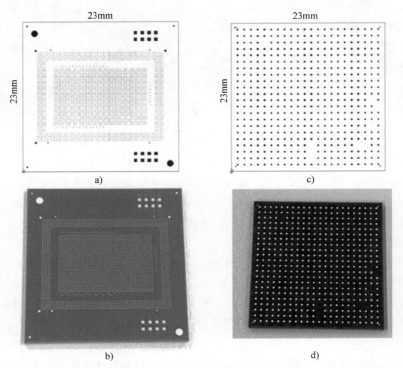

图 5.63 单颗积层封装基板：a）积层封装基板俯视示意图；b）制造好的积层封装基板俯视图；c）积层封装基板底视示意图；d）制造好的积层封装基板底视图

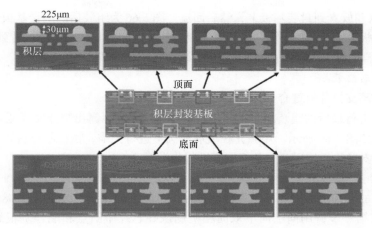

图 5.64 制作完成的带 C4 凸点的 2-2-2 积层封装基板 SEM 图像

(4) 翘曲测量

不同温度下积层封装基板(BU)、玻璃支撑片窄金属线宽/线距 RDL 基板(RDL(G))以及有机支撑片窄金属线宽/线距 RDL 基板(RDL(O))的翘曲利用 TherMoire Platform 采用阴影云纹(shadow Moire)进行测量,测量结果如图 5.65 所示。从图中可以看到 BU 和 RDL(G)的翘曲非常小,相比之下 RDL(O)的翘曲非常大。这是由于玻璃支撑片和 RDL 基板之间的热膨胀系数不匹配小于有机支撑片和 RDL 基板之间的热膨胀系数不匹配。因此,在本研究中,混合基板的制备采用积层封装基板与玻璃支撑片窄金属线宽/线距 RDL 基板的组合。

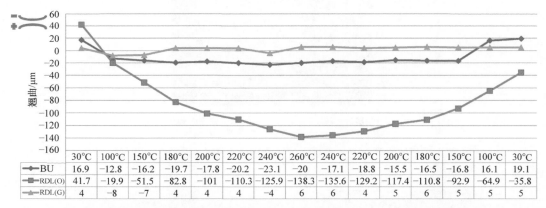

	30℃	100℃	150℃	180℃	200℃	220℃	240℃	260℃	240℃	220℃	200℃	180℃	150℃	100℃	30℃
BU	16.9	-12.8	-16.2	-19.7	-17.8	-20.2	-23.1	-20	-17.1	-18.8	-15.5	-16.5	-16.8	16.1	19.1
RDL(O)	41.7	-19.9	-51.5	-82.8	-101	-110.3	-125.9	-138.3	-135.6	-129.2	-117.4	-110.8	-92.9	-64.9	-35.8
RDL(G)	4	-8	-7	-4	4	5	5	6	5	4	4	5	4	5	5

图 5.65 积层封装基板(BU)、玻璃支撑片窄金属线宽/线距 RDL 基板(RDL(G))、有机支撑片窄金属线宽/线距 RDL 基板(RDL(O))的阴影云纹(shadow Moire)翘曲测量

(5) 混合基板制造

在完成玻璃支撑片的窄线宽/线距 RDL 基板和带有焊料凸点的积层封装基板后,就可如图 5.60 所示制作混合基板。第一步,用向上向下两个相机对准封装基板的焊料凸点和玻璃支撑片 RDL 基板上的焊盘位置;第二步,在焊料凸点和焊盘上都涂覆助焊剂;第三步,将玻璃支撑

片 RDL 基板拾取和放置到封装基板上，然后回流。图 5.66 为组装后的混合基板俯视图，图 5.67 为典型的混合基板截面 SEM 图像，图 5.68 为相应的放大图像。从图中可以看到，窄金属线宽/线距 RDL 基板、焊点、底部填充料以及积层封装基板都已组装正确。

4. 封装体最后组装

（1）临时玻璃支撑片解键合

在对封装体进行最后的组装前，需要将玻璃支撑片去除以露出窄金属线宽/线距 RDL 基板上的键合焊盘，如图 5.60 所示。用激光在玻璃晶圆上进行扫描，牺牲层材料转变为粉末状，很容易地去除玻璃支撑片。然后，对基板表面进行化学清洗。

（2）芯片-混合基板键合

带有微凸点的芯片（见图 5.55）和混合基板（见图 5.66、图 5.67 和图 5.68）准备好后，进行最终的键合。第一步，用向上向下双向相机确认好芯片微凸点和混合基板焊盘的位置；第二步，在微凸点和焊盘上都添加助焊剂；第三步，拾取芯片并放置在混合基板上进行回流。图 5.56b 为 2 颗芯片异质集成封装体组装后的实物图。

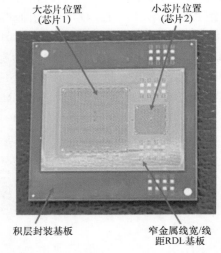

图 5.66　混合基板＝在积层封装基板上焊接含玻璃支撑片的窄金属线宽/线距 RDL 基板

图 5.67　混合基板的截面图

第 5 章 2D、2.1D 和 2.3D IC 集成 241

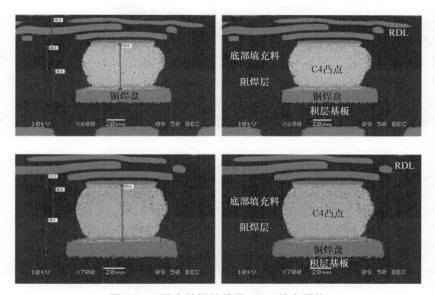

图 5.68 混合基板的截面 SEM 放大照片

（3）底部填充

图 5.69 所示为对 2 颗芯片完成底部填充后的最终组装体截面图。从图中可以看到，芯片、微凸点、底部填充料、RDL 基板、C4 凸点和积层封装基板都组装正确。

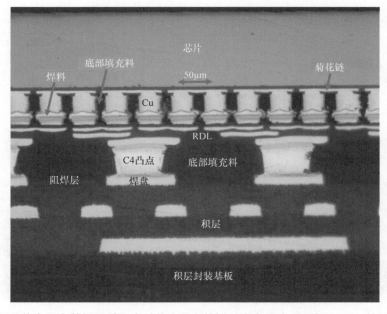

图 5.69 2 颗芯片在混合基板异质集成封装（混合基板＝窄金属线宽/线距 RDL 基板＋积层基板）

242 半导体先进封装技术

5. 热循环仿真

（1）结构和有限元模型

如图 5.70 所示，对于 2 颗芯片混合基板异质集成封装的组装体热循环仿真中，重点关注微凸点焊点和 C4 凸点焊点的可靠性。图 5.71 所示为采用的广义平面应变有限元模型。在连接芯片和 RDL 基板的微凸点焊点（A、B、C、D）及连接 RDL 基板和积层基板的 C4 凸点焊点（E、F）等关键位置进行网格细化。3 层 RDL 基板包括铜导线层和 ABF 介质层，在仿真中采用等效体将其代替。

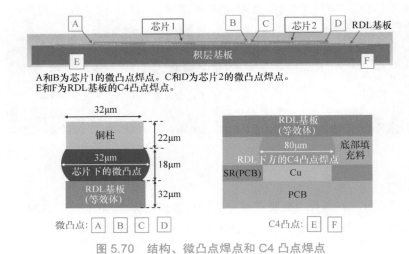

图 5.70　结构、微凸点焊点和 C4 凸点焊点

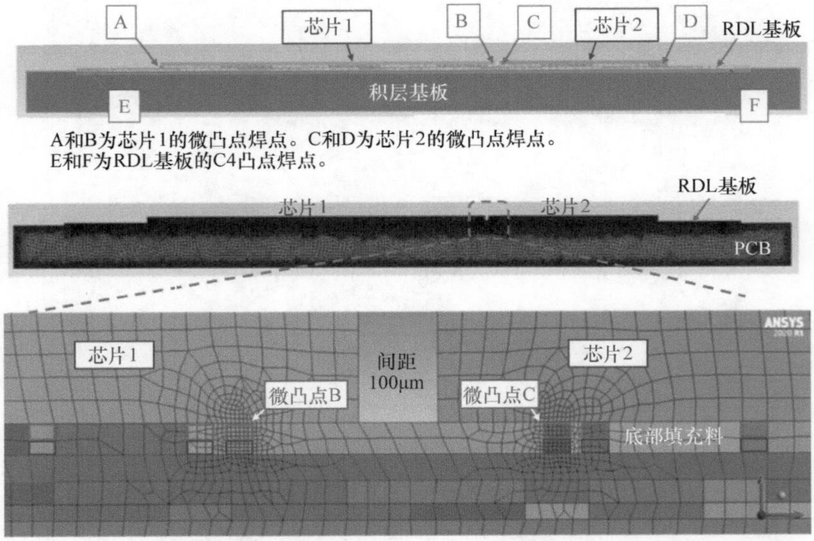

图 5.71　2 颗芯片在混合基板异质集成封装 PCB 组装的有限元模型

(2) 仿真用到的材料特性

表 5.1 给出了模型所用到的材料特性。除了 Sn3Ag0.5Cu 之外，所有材料特性都假设是常量，Sn3Ag0.5Cu 的特性与时间和温度相关，表 5.1 给出了其连续性方程。

(3) 动力学边界条件

图 5.72 所示为仿真的温度边界条件。从图中可以看到温度在 −40℃ ⇌ 85℃ 之间变化，每个循环周期时间为 60min，其中升温、降温、高温保持、低温保持时间都是 15min。

表 5.1 仿真所用的材料特性

材料	热膨胀系数 / (10^{-6}/℃)	杨氏模量 /GPa	泊松比
铜	16.3	121	0.34
PCB	$\alpha_x=\alpha_y=18$ $\alpha_z=70$	$E_x=E_y=22$ $E_z=10$	0.28
阻焊层（RDL）	7.0	7.0	0.3
焊料	$21+0.017T$（℃）	$49-0.07T$（℃）	0.3
EMC（ABF）	7.0	7.0	0.3
硅	2.8	131	0.278
阻焊层（PCB）	39	4.1	0.3
RDL 等效体	14.08	70.53	0.32
底部填充料	50	4.5	0.35

注：焊料连续性方程为 $\dfrac{d\varepsilon}{dt}=500000\left[\sinh(0.01\sigma)\right]^5 \exp\left(-\dfrac{5800}{T}\right)$，其中 T 的单位为 K，σ 的单位为 MPa。

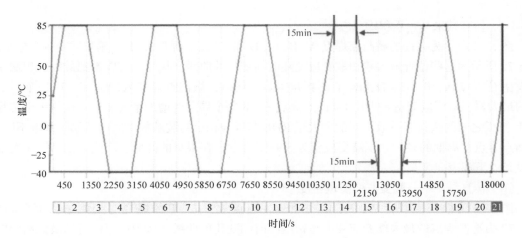

图 5.72 温度边界条件

(4) 仿真结果——变形

图 5.73 为 2 颗芯片混合基板异质集成封装组装体的变形后形状（彩色云图）和变形前形状（黑色实线）。从图 5.73a 可以看到 450s（85℃）时，混合基板膨胀量大于芯片，整个结构呈现

凹面变形（笑脸型）；从图 5.73b 可以看到 2250s（-40℃）时，混合基板的收缩量大于芯片，整个结构呈现凸面变形（哭脸型）。这些结果都与预期相符。

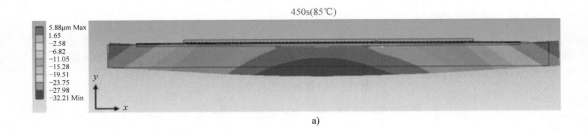

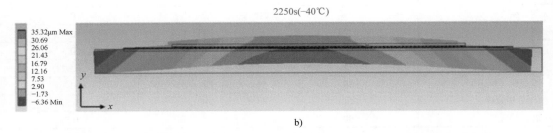

图 5.73　变形后形状（彩色云图）和变形前形状（黑色实线）：a）450s（85℃）时；b）2250s（-40℃）时

（5）仿真结果——累积蠕变应变

图 5.74 所示为关键的微凸点焊点 A、B、C、D 和 C4 凸点焊点 E、F 的累积蠕变应变分布图，图 5.75 所示为这些位置最大累积蠕变应变随时间变化的曲线。从图中可以看到最大累积蠕变应变发生在焊点 A、B、C、D、E、F 的拐角附近。同时，微凸点焊点 A、B、C、D 每个周期增加的最大累积蠕变应变量至少比 C4 凸点焊点 E、F 每个周期的增加量大 4 倍。这是由于芯片和 RDL 基板之间的热膨胀系数不匹配要大于 RDL 基板与积层封装基板之间的热膨胀系数不匹配。微凸点焊点每周期的最大累积蠕变应变为 5.93% 且发生在很小的区域。因此，该结构在大多数服役条件下都表现可靠。

（6）仿真结果——蠕变应变能量密度

图 5.76 为微凸点焊点 A、B、C、D 和 C4 凸点焊点 E、F 的蠕变应变能量密度分布图，图 5.77 为这些焊点的最大蠕变应变能量密度随时间变化的曲线。从图中可以看到最大蠕变应变能量密度发生在焊点 A、B、C、D、E、F 的拐角附近。同时，微凸点焊点 A、B、C、D 每周期的最大蠕变应变能量密度至少比 C4 凸点焊点 E、F 的大 5 倍。如前所述，这同样是因为芯片和 RDL 基板之间的热膨胀系数不匹配要大于 RDL 基板与积层封装基板之间的热膨胀系数不匹配。微凸点焊点每周期最大蠕变应变能量密度仅为 2.63MPa，且只发生在很小的区域。同样，这说明该结构在大多数服役条件下都表现可靠。

第 5 章 2D、2.1D 和 2.3D IC 集成

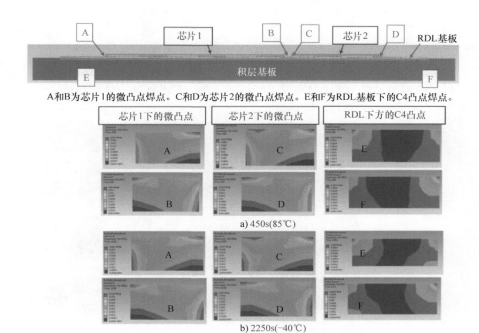

A 和 B 为芯片 1 的微凸点焊点。C 和 D 为芯片 2 的微凸点焊点。E 和 F 为 RDL 基板下的 C4 凸点焊点。

a) 450s(85℃)

b) 2250s(−40℃)

图 5.74 微凸点焊点 A、B、C、D 和 C4 凸点焊点 E、F 的累积蠕变应变分布图

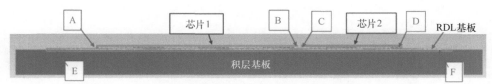

A 和 B 为芯片 1 的微凸点焊点。C 和 D 为芯片 2 的微凸点焊点。E 和 F 为 RDL 基板下的 C4 凸点焊点。

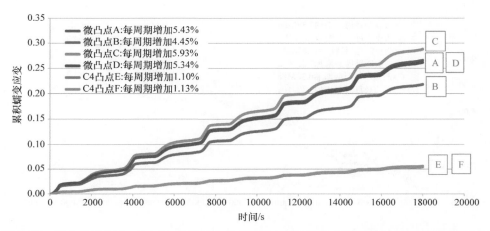

图 5.75 微凸点焊点 A、B、C、D 和 C4 凸点焊点 E、F 的最大累积蠕变应变随时间变化曲线

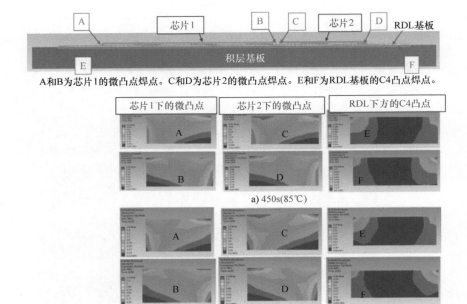

a) 450s(85℃)

b) 2250s(-40℃)

图 5.76 微凸点焊点 A、B、C、D 和 C4 凸点焊点 E、F 的蠕变应变能量密度分布图

A和B为芯片1的微凸点焊点。C和D为芯片2的微凸点焊点。E和F为RDL基板的C4凸点焊点。

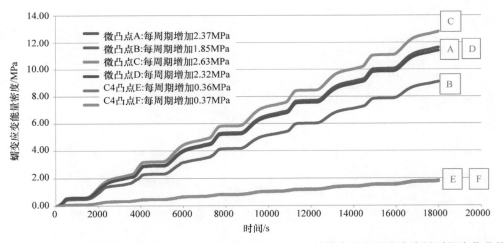

图 5.77 微凸点焊点 A、B、C、D 和 C4 凸点焊点 E、F 的最大累积蠕变应变随时间变化曲线

5.13 总结和建议

一些重要的结论和建议总结如下：

1）基于有机基板的 2D IC 集成已经在诸如 SiP 和同质集成等应用中大规模量产，并将继续广泛使用。

2）由于良率的原因，基于积层封装基板顶部薄膜层的 2.1D IC 集成还没有得到大规模量产。基于 EMIB（英特尔）的 2.1D IC 集成已经小规模量产。台积电的 LSI 仍处于验证阶段，预计在 2021 年年中或 2022 年年初量产。

3）基于 PCB 工艺的无芯板有机转接板 2.3D IC 集成已经小规模量产。日月光计划在 2021 年实现基于扇出型技术（先上晶或后上晶）的无芯板无机/有机 RDL 转接板的 2.3D IC 集成的大规模量产。

参 考 文 献

1. Lau, J. H., P. Tzeng, C. Lee, C. Zhan, M. Li, J. Cline, K. Saito, Y. Hsin, P. Chang, Y. Chang, J. Chen, S. Chen, C. Wu, H. Chang, C. Chien, C. Lin, T. Ku, R. Lo, and M. Kao, "Redistribution Layers (RDLs) for 2.5D/3D IC Integration", *IMAPS Transactions, Journal of Microelectronic Packaging*, Vol. 11, No. 1, First Quarter 2014, pp. 16–24.
2. Lau, J. H., "8 Ways to Make RDLs for FOW/PLP", *Chip Scale Review*, Vol. 22, May/June 2018, pp. 11–19.
3. Lau, J. H., "Redistribution-Layers for Heterogeneous Integrations", *Chip Scale Review*, Vol. 23, January/February 2019, pp. 20–25.
4. Kripesh, V., V. Rao, A. Kumar, G. Sharma, K. Houe, X. Zhang, K. Mong, N. Khan, and J. H. Lau, "Design and Development of a Multi-Die Embedded Micro Wafer Level Package", *IEEE/ECTC Proceedings*, 2008, pp. 1544–1549.
5. Khong, C., A. Kumar, X. Zhang, S. Gaurav, S. Vempati, V. Kripesh, J. H. Lau, and D. Kwong, "A Novel Method to Predict Die Shift During Compression Molding in Embedded Wafer Level Package", *IEEE/ECTC Proceedings*, 2009, pp. 535–541.
6. Sharma, G., S. Vempati, A. Kumar, N. Su, Y. Lim, K. Houe, S. Lim, V. Sekhar, R. Rajoo, V. Kripesh, and J. H. Lau, "Embedded Wafer Level Packages with Laterally Placed and Vertically Stacked Thin Dies", *IEEE/ECTC Proceedings*, 2009, pp. 1537-1543. Also, *IEEE Transactions on CPMT*, Vol. 1, No. 5, May 2011, pp. 52–59.
7. Kumar, A., D. Xia, V. Sekhar, S. Lim, C. Keng, S. Gaurav, S. Vempati, V. Kripesh, J. H. Lau, and D. Kwong, "Wafer Level Embedding Technology for 3D Wafer Level Embedded Package", *IEEE/ECTC Proceedings*, 2009, pp. 1289–1296.
8. Lim, Y., S. Vempati, N. Su, X. Xiao, J. Zhou, A. Kumar, P. Thaw, S. Gaurav, T. Lim, S. Liu, V. Kripesh, and J. H. Lau, "Demonstration of High Quality and Low Loss Millimeter Wave Passives on Embedded Wafer Level Packaging Platform (EMWLP)", *IEEE/ECTC Proceedings*, 2009, pp. 508-515. Also, *IEEE Transactions on Advanced Packaging*, Vol. 33, 2010, pp. 1061–1071.
9. Lau, J. H., "Patent Issues of Fan-Out Wafer/Panel-Level Packaging", *Chip Scale Review*, Vol. 19, November/December 2015, pp. 42–46.
10. Lau, J. H., N. Fan, and M. Li, "Design, Material, Process, and Equipment of Embedded Fan-Out Wafer/Panel-Level Packaging", *Chip Scale Review*, Vol. 20, May/June 2016, pp. 38–44.
11. Lau, J. H., M. Li, M. Li, T. Chen, I. Xu, X. Qing, Z. Cheng, N. Fan, E. Kuah, Z. Li, K. Tan, Y. Cheung, E. Ng, P. Lo, K. Wu, J. Hao, S. Koh, R. Jiang, X. Cao, R. Beica, S. Lim, N. Lee, C. Ko, H. Yang, Y. Chen, M. Tao, J. Lo, and R. Lee, "Fan-Out Wafer-Level Packaging for Heterogeneous Integration", *IEEE Transactions on CPMT*, 2018, September 2018, pp. 1544–1560.
12. Lau, J. H., M. Li, Y. Lei, M. Li, I. Xu, T. Chen, Q. Yong, Z. Cheng, K. Wu, P. Lo, Z. Li, K. Tan, Y. Cheung, N. Fan, E. Kuah, C. Xi, J. Ran, R. Beica, S. Lim, N. Lee, C. Ko, H. Yang, Y. Chen, M. Tao, J. Lo, and R. Lee, "Reliability of Fan-Out Wafer-Level Heterogeneous

Integration", *IMAPS Transactions, Journal of Microelectronics and Electronic Packaging*, Vol. 15, Issue: 4, October 2018, pp. 148–162.
13. Ko, C.T., H. Yang, J. H. Lau, M. Li, M. Li, C. Lin, J. W. Lin, T. Chen, I. Xu, C. Chang, J. Pan, H. Wu, Q. Yong, N. Fan, E. Kuah, Z. Li, K. Tan, Y. Cheung, E. Ng, K. Wu, J. Hao, R. Beica, M. Lin, Y. Chen, Z. Cheng, S. Koh, R. Jiang, X. Cao, S. Lim, N. Lee, M. Tao, J. Lo, and R. Lee, "Chip-First Fan-Out Panel-Level Packaging for Heterogeneous Integration", *IEEE Transactions on CPMT*, September 2018, pp. 1561–1572.
14. Ko, C. T., H. Yang, J. H. Lau, M. Li, M. Li, C. Lin, J. Lin, C. Chang, J. Pan, H. Wu, Y. Chen, T. Chen, I. Xu, P. Lo, N. Fan, E. Kuah, Z. Li, K. Tan, C. Lin, R. Beica, M. Lin, C. Xi, S. Lim, N. Lee, M. Tao, J. Lo, and R. Lee, "Design, Materials, Process, and Fabrication of Fan-Out Panel-Level Heterogeneous Integration", *IMAPS Transactions, Journal of Microelectronics and Electronic Packaging*, Vol. 15, Issue: 4, October 2018, pp. 141–147.
15. Lau, J. H., "Recent Advances and Trends in Fan-Out Wafer/Panel-Level Packaging", *ASME Transactions, Journal of Electronic Packaging*, Vol. 141, December 2019, pp. 1–27.
16. Lau, J. H., "Recent Advances and Trends in Heterogeneous Integrations", *IMAPS Transactions, Journal of Microelectronics and Electronic Packaging*, Vol. 16, April 2019, pp. 45–77.
17. Bu, L., F. Che, M. Ding, S. Chong, and X. Zhang, "Mechanism of Moldable Underfill (MUF) Process for Fan-Out Wafer Level Packaging", *IEEE/EPTC Proceedings*, 2015, pp. 1–7.
18. Che, F., D. Ho, M. Ding, and D. Woo, "Study on Process Induced Wafer Level Warpage of Fan-Out Wafer Level Packaging", *IEEE/ECTC Proceedings*, 2016, pp. 1879–1885.
19. Rao, V., C. Chong, D. Ho, D. Zhi, C. Choong, S. Lim, D. Ismael, and Y. Liang, "Development of High Density Fan Out Wafer Level Package (HD FOWLP) with Multilayer Fine Pitch RDL for Mobile Applications", *IEEE/ECTC Proceedings*, 2016, pp. 1522–1529.
20. Chen, Z., F. Che, M. Ding, D. Ho, T. Chai, V. Rao, "Drop Impact Reliability Test and Failure Analysis for Large Size High Density FOWLP Package on Package", *IEEE/ECTC Proceedings*, 2017, pp. 1196–1203.
21. Lim, T., and D. Ho, "Electrical design for the development of FOWLP for HBM integration", *IEEE/ECTC Proceedings*, 2018, pp. 2136–2142.
22. Ho, S., H. Hsiao, S. Lim, C. Choong, S. Lim, and C. Chong, "High Density RDL build-up on FO-WLP using RDL-first Approach", *IEEE/EPTC Proceedings*, 2019, pp. 23–27.
23. Boon, S., D. Wee, R. Salahuddin, and R. Singh, "Magnetic Inductor Integration in FO-WLP using RDLfirst Approach", *IEEE/EPTC Proceedings*, 2019, pp. 18–22.
24. Hsiao, H., S. Ho, S. S. Lim, W. Ching, C. Choong, S. Lim, H. Hong, and C. Chong, "Ultra-thin FO Packageon-Package for Mobile Application", *IEEE/ECTC Proceedings*, 2019, pp. 21–27.
25. Lin, B., F. Che, V. Rao, and X. Zhang, "Mechanism of Moldable Underfill (MUF) Process for RDL-1st Fan-Out Panel Level Packaging (FOPLP)", *IEEE/ECTC Proceedings*, 2019, pp. 1152–1158.
26. Sekhar, V., V. Rao, F. Che, C. Choong, and K. Yamamoto, "RDL-1st Fan-Out Panel Level Packaging (FOPLP) for Heterogeneous and Economical Packaging", *IEEE/ECTC Proceedings*, 2019, pp. 2126–2133.
27. Huemoeller, R. and C. Zwenger, "Silicon wafer integrated fan-out technology," *Chip Scale Review*, March/April 2015, pp. 34–37.
28. Hiner, D., M. Kelly, R. Huemoeller, and R. Reed, "Silicon interposer-less integrated module - SLIM," *IMAPS/Device Packaging*, March 2015.
29. Hiner, D., M. Kolbehdari, M. Kelly, Y. Kim, W. Do, J. Bae, "SLIM™ advanced fan-out packaging for high performance multi-die solutions," *IEEE/ECTC Proceedings*, May 2017, pp. 575–580.
30. Kim, Y., J. Bae, M. Chang, A. Jo, J. Kim, S. Park, et al., "SLIM™, high density wafer-level fan-out package development with sub-micron RDL," *IEEE/ECTC Proceedings,* May 2017, pp. 18–13.
31. Zwenger, C., G. Scott, B. Baloglu, M. Kelly, W. Do, W. Lee, and J. Yi, "Electrical and Thermal Simulation of SWIFT™ High-density Fan-out PoP Technology", *IEEE/ECTC Proceedings*, May 2017, pp. 1962–1967.
32. Scott, G., J. Bae, K. Yang, W. Ki, N. Whitchurch, M. Kelly, C. Zwenger, J. Jeon, and T. Hwang, "Heterogeneous Integration Using Organic Interposer Technology", *IEEE/ECTC Proceedings*, May 2020, pp. 885–892.

33. Ma, M., S. Chen, P. I. Wu, A. Huang, C. H. Lu, A. Chen, C. Liu, and S. Peng, "The development and the integration of the 5 μm to 1 μm half pitches wafer level Cu redistribution layers", *IEEE/ECTC Proceedings*, May 2016, pp. 1509–1514.
34. Lau, J. H., C. Ko, T. Peng, K. Yang, T. Xia, P. Lin, J. Chen, P. Huang, T. Tseng, E. Lin, L. Chang, C. Lin, and W. Lu, "Chip-Last (RDL-First) Fan-Out Panel-Level Packaging (FOPLP) for Heterogeneous Integration", *IMAPS Transactions, Journal of Microelectronics and Electronic Packaging*, Vol. 17, No. 3, October 2020, pp. 89–98.
35. Lau, J. H., C. Ko, K. Yang, C. Peng, T. Xia, P. Lin, J. Chen, P. Huang, H. Liu, T. Tseng, E. Lin, and L. Chang, "Panel-Level Fan-Out RDL-first Packaging for Heterogeneous Integration", *IEEE Transactions on CPMT*, Vol. 10, No. 7, July 2020, pp. 1125–1137.
36. Shimizu, N., W. Kaneda, H. Arisaka, N. Koizumi, S. Sunohara, A. Rokugawa, and T. Koyama, "Development of Organic Multi Chip Package for High Performance Application", *IMAPS Proceedings of International Symposium on Microelectronics*, October 2013, pp. 414–419.
37. Oi, K., S. Otake, N. Shimizu, S. Watanabe, Y. Kunimoto, T. Kurihara, T. Koyama, M. Tanaka, L. Aryasomayajula, and Z. Kutlu, "Development of New 2.5D Package with Novel Integrated Organic Interposer Substrate with Ultra-fine Wiring and High Density Bumps", *IEEE/ECTC Proceedings*, May 2014, pp. 348–353.
38. Uematsu, Y., N. Ushifusa, and H. Onozeki, "Electrical Transmission Properties of HBM Interface on 2.1-D System in Package using Organic Interposer", *IEEE/ECTC Proceedings*, May 2017, pp. 1943–1949.
39. Chen, W., C. Lee, M. Chung, C. Wang, S. Huang, Y. Liao, H. Kuo, C. Wang, and D. Tarng, "Development of novel fine line 2.1 D package with organic interposer using advanced substrate-based process", *IEEE/ECTC Proceedings*, May 2018, pp. 601–606.
40. Huang, C., Y. Xu, Y. Lu, K. Yu, W. Tsai, C. Lin, C. Chung, "Analysis of Warpage and Stress Behavior in a Fine Pitch Multi-Chip Interconnection with Ultrafine-Line Organic Substrate (2.1D)", *IEEE/ECTC Proceedings*, May 2018, pp. 631–637.
41. Islam, N., S. Yoon, K. Tan, and T. Chen, "High Density Ultra-Thin Organic Substrate for Advanced Flip Chip Packages", *IEEE/ECTC Proceedings*, May 2019, pp. 325–329.
42. Chiu, C., Z. Qian, and M. Manusharow, "Bridge interconnect with air gap in package assembly," *US Patent No. 8,872,349*, 2014.
43. Mahajan, R., R. Sankman, N. Patel, D. Kim, K. Aygun, Z. Qian, et al., "Embedded multi-die interconnect bridge (EMIB) – a high-density, high-bandwidth packaging interconnect," *IEEE/ECTC Proceedings*, May 2016, pp. 557–565.
44. Hsiung, C., and a. Sundarrajan, "Methods and Apparatus for Wafer-Level Die Bridge", US 10,651,126 B2, Filed on December 8, 2017, Granted on May 12, 2020.
45. Li, L., P. Chia, P. Ton, M. Nagar, S. Patil, J. Xue, J. DeLaCruz, M. Voicu, J. Hellings, B. Isaacson, M. Coor, and R. Havens, "3D SiP with Organic Interposer for ASIC and Memory Integration", *IEEE/ECTC Proceedings,* May 2016, pp. 1445–1450.
46. Pendse, R., "Semiconductor Device and Method of Forming Extended Semiconductor Device with Fan-Out Interconnect Structure to Reduce Complexity of Substrate", *US 9,484,319 B2*, Filed: December 23, 2011, Granted: November 1, 2016.
47. Yoon, S., P. Tang, R. Emigh, Y. Lin, P. Marimuthu, and R. Pendse, "Fanout Flipchip eWLB (Embedded Wafer Level Ball Grid Array) Technology as 2.5D Packaging Solutions", *IEEE/ECTC Proceedings*, 2013, pp. 1855–1860.
48. Chen, N., "Flip-Chip Package with Fan-Out WLCSP", *US 7,838,975 B2*, Filed: February 12, 2009, Granted: November 23, 2010.
49. Chen, N. C., T. Hsieh, J. Jinn, P. Chang, F. Huang, J. Xiao, A. Chou, B. Lin, "A Novel System in Package with Fan-out WLP for high speed SERDES application", IEEE/ECTC Proceedings, May 2016, pp. 1496–1501.
50. Lin, Y., W. Lai, C. Kao, J. Lou, P. Yang, C. Wang, and C. Hseih, "Wafer warpage experiments and simulation for fan-out chip on substrate," *IEEE/ECTC Proceedings*, May 2016, pp. 13–18.
51. Kwon, W., S. Ramalingam, X. Wu, L. Madden, C. Huang, H. Chang, et al., "Cost-effective and high-performance 28 nm FPGA with new disruptive silicon-less interconnect technology (SLIT)," *Proc. of Inter. Symp. on Micro.*, October 2014, pp. 599–605.
52. Liang, F., H. Chang, W. Tseng, J. Lai, S. Cheng, M. Ma, et al., "Development of non-TSV interposer (NTI) for high electrical performance package," *IEEE/ECTC Proceedings,* May 2016, pp. 31–36.

53. Suk, K., S. Lee, J. Kim, S. Lee, H. Kim, S. Lee, P. Kim, D. Kim, D. Oh, and J. Byun, "Low Cost Si-less RDL Interposer Package for High Performance Computing Applications", *IEEE/ECTC Proceedings*, May 2018, pp. 64–69.
54. You, S., S. Jeon, D. Oh, K. Kim, J. Kim, S. Cha, G. Kim, "Advanced Fan-Out Package SI/PI/Thermal Performance Analysis of Novel RDL Packages", *IEEE/ECTC Proceedings*, May 2018, pp. 1295–1301.
55. Chang, K., C. Huang, H. Kuo, M. Jhong, T. Hsieh, M. Hung, C. Wang, "Ultra High Density IO Fan-Out Design Optimization with Signal Integrity and Power Integrity", *IEEE/ECTC Proceedings*, May 2019, pp. 41–46.
56. Lai, W., P. Yang, I. Hu, T. Liao, K. Chen, D. Tarng, and C. Hung, "A Comparative Study of 2.5D and Fan-out Chip on Substrate: Chip First and Chip Last", *IEEE/ECTC Proceedings*, May 2020, pp. 354–360.
57. Fang, J., M. Huang, H. Tu, W. Lu, P. Yang, "A Production-worthy Fan-Out Solution – ASE FOCoS Chip Last", *IEEE/ECTC Proceedings*, May 2020, pp. 290–295.
58. Lin, Y., M. Yew, M. Liu, S. Chen, T. Lai, P. Kavle, C. Lin, T. Fang, C. Chen, C. Yu, K. Lee, C. Hsu, P.. Lin, F. Hsu, and S. Jeng, "Multilayer RDL Interposer for Heterogeneous Device and Module Integration", *IEEE/ECTC Proceedings*, May 2019, pp. 931–936.
59. Miki, S., H. Taneda, N. Kobayashi, K. Oi, K. Nagai, T. Koyama, "Development of 2.3D High Density Organic Package using Low Temperature Bonding Process with Sn-Bi Solder", *IEEE/ECTC Proceedings*, May 2019, pp. 1599–1604.
60. Murayama, K., S. Miki, H. Sugahara, and K. Oi, "Electro-migration evaluation between organic interposer and build-up substrate on 2.3D organic package", *IEEE/ECTC Proceedings*, May 2020, pp. 716–722.
61. Lau, J. H., G. Chen, R. Chou, C. Yang, and T. Tseng, "Fan-Out (RDL-First) Panel-Level Hybrid Substrate for Heterogeneous Integration", *IEEE/ECTC Proceedings*, May 2021.

第 6 章

2.5D IC 集成

6.1 引言

半导体产业将 2.5D IC 集成技术定义为先采用无源硅通孔（through-silicon via，TSV）转接板来支撑芯片，随后再将其连接至封装基板上[1-111]。有源 TSV 转接板（将在第 7 章叙述）集成有互补金属氧化物半导体（complementary metal oxide semiconductor，CMOS）器件，但是无源 TSV 转接板仅包含 TSV 及再布线层（redistribution layer，RDL），是一片无功能的硅。现阶段 2.5D IC 集成技术已被台积电（为赛灵思和英伟达代工）、联电（为 AMD 代工）等代工厂大规模量产（high volume manufacturing，HVM），相关内容将在本章讲述，2.5D IC 集成技术的近期进展也会被简要提及。首先简要介绍一下 2.5D IC 集成技术的起源。

6.2 Leti 的 SoW 技术（2.5D IC 集成技术的起源）

2.5D IC 集成技术的早期应用之一是法国电子信息技术研究所（Leti）开发的晶上系统（system-on-wafer，SoW）技术，如图 6.1 所示。可以看到带有 TSV 及 RDL 的硅晶圆上集成有各类系统芯片，如专用集成电路（application specific IC，ASIC）、存储芯片、电源管理集成电路（power management IC，PMIC）、微机电系统（micro-electro-mechanical system，MEMS）。划片后每个独立单元会成为一个系统或子系统，可以安装到有机封装基板上或者自成一体。

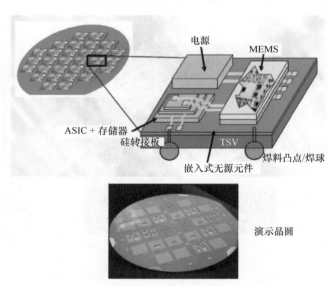

图 6.1 Leti 的 SoW 技术（2.5D IC 集成技术的起源）[1]

6.3 IME 的 2.5D IC 集成技术

在 IEEE/ECTC 2008 会议上,新加坡微电子研究所 (institute of microelectronics,IME) 就 2.5D IC 集成技术发表了三篇文章,下面简要对其进行描述。

6.3.1 2.5D IC 集成的三维非线性局部及全局分析

参考文献 [3,6] 研究了 2.5D IC 集成局部及全局热机械效应的非线性仿真。图 6.2 为 2007 年所开发的当时最大的无源 TSV 转接板[3,6],用于支持输入/输出端口(Input/Output,I/O)数达 11000 的大尺寸芯片(21mm×21mm)。对于局部效应,仿真结果[3,6]表明,由于 Si(2.5×10^{-6}/℃)和 Cu(17.5×10^{-6}/℃)在局部区域竖直方向上显著的热膨胀系数(thermal expansion coefficient,TEC)差异,在加热过程中,TSV 内部的 Cu 可能从周围 Si 基体中挤出,如图 6.2 所示。如果 TSV Cu 部分地被 SiO_2 覆盖,那么在加热过程中因局部热膨胀失配,Cu 可能(被挤出)造成 SiO_2 破裂,如图 6.2 所示,这被称为铜挤出(Cu pumping)现象。参考文献 [3,6] 表明,当深宽比(厚度/直径)大于 5,应力及应变对 TSV 深宽比的依赖性不大。

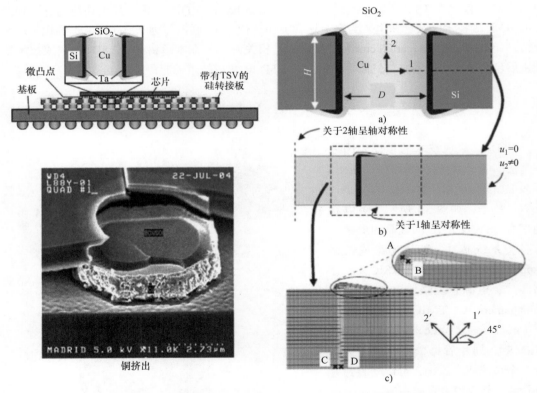

图 6.2 包含 TSV 转接板的封装体:a)TSV 的简化示意图; b)施加边界条件后的四分之一 TSV 模型;c)关键区域的网格划分

A、B 两点的对角线应变曲线（见图 6.2）绘制于图 6.3a、b（通孔已填充的情况）和图 6.3c、d（通孔未填充的情况）。在通孔已填充的情况中，如图所示对角线应变和通孔直径之间存在线性关系。与预期相符，铜中的应变低于二氧化硅中的。在含 RDL 但通孔未填充的情况中，对角线应变随着直径增大而减小，这是因为二氧化硅厚度/直径比降低，导致铜中膨胀相对较小。

当硅转接板含有填铜 TSV 时，其有效 TEC 会增大，这导致在温度循环过程中转接板膨胀和收缩的程度会超过芯片，在焊点处形成疲劳载荷。这种全局失配对焊点可靠性产生的影响将在下面的仿真中加以研究。仿真中的全局模型包含芯片组装在含 TSV 的硅转接板上，经历从 -40 ~ 125℃ 的三次温度循环。为了提高计算效率，考虑该模型的对角线切片，图 6.4 展示了模型的示意图、施加的边界条件及关键区域的网格划分情况。分析结果表明，随着转接板高度（H）的增加，焊料中的蠕变应变能量密度增加，如图 6.5a 所示。对于高度为 50 ~ 200μm 的转接板，最大应变能量密度点发生在焊点的顶部。

另一方面，当转接板高度为 300 ~ 500μm 时，最大应变能量会出现在焊点的底部表面。当转接板的膨胀开始对焊料造成剪切载荷时，最大应变能量位置就会转移到底部。这一效应可以在图 6.5b 中看到，H 最小的转接板上焊点在其底面上具有低得多的应变能量密度。使用底部填充料可将非弹性应变能量密度降低 4 倍。如图 6.5c 所示，增大节距（P）会导致蠕变应变能量密度下降。这是由于铜含量降低减小了硅芯片和含 TSV 的硅转接板之间的不匹配度。

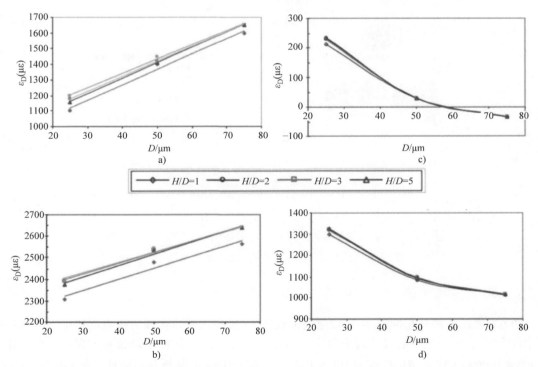

图 6.3　带有 RDL 的已填充通孔中关键位置的对角线应变值（ε_D）：a）铜（A 点）；b）二氧化硅（B 点）。带有 RDL 的未填充通孔中关键位置的对角线应变值（ε_D）：c）铜（A 点）；d）二氧化硅（B 点）

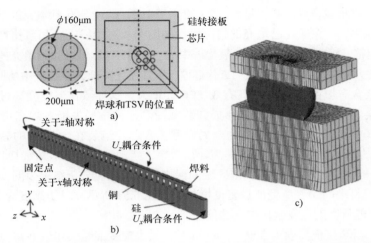

图 6.4 a）全局模型示意图；b）施加边界条件后的对角线切片模型；c）关键区域的网格划分

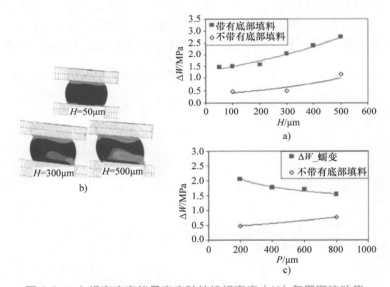

图 6.5 a）蠕变应变能量密度随转接板高度（H）每周期波动值；
b）三种转接板高度下的蠕变应变能量密度；c）蠕变应变能量密度随通孔节距（P）每周期波动值

6.3.2 用于电气和流体互连的 2.5D IC 集成技术

图 6.6 展示了两片相同的 TSV 转接板（载体），带有用于热管理的微通道和用于电气连接的 TSV[4, 7-10]。整个系统的尺寸约为拳头的四分之一大小，每个 TSV 转接板支撑着一个尺寸为 10mm × 10mm、（热）功率耗散为 100W 的芯片。这一 TSV 转接板是由圆片-圆片堆叠（wafer-on-wafer，WoW）技术将两块硅晶圆键合制备而成的，利用深反应离子刻蚀（deep reactive ion etching，DRIE）技术制作的优化后的液态制冷通道结构嵌入在转接板中。如图 6.6 及图 6.7 所

示,通过一个迷你泵将冷水从热交换器中通过微通道泵入转接板入口端,流过微通道,从出口端流出后再回到热交换器。选择硅作为转接板材料是因为其是一种合适的材料,采用 DRIE 技术在同一个基体上可以完成电学(TSV)及流体结构(微通道)的集成。这两片转接板的不同在于底部硅转接板没有出口。在 WoW 键合完成后,贯穿载体的电互连由侧壁金属化的 TSV 实现(这时铜填充并非必要的)。

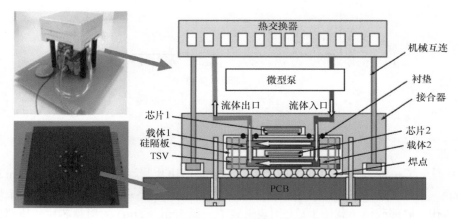

图 6.6　TSV 转接板堆叠结构实现的集成式液冷系统

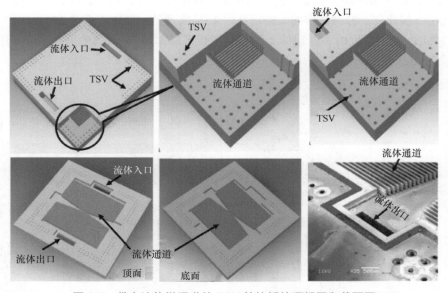

图 6.7　带有流体微通道的 TSV 转接板的顶视图和截面图

图 6.8 展示了压力下降测量值,可以看到带有两个入口和两个出口的载体在散热上的表现优于只含一个入口、一个出口的载体,更多的细节在参考文献 [4,7-10] 中阐释。

256 半导体先进封装技术

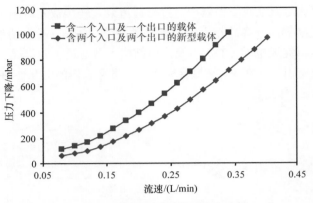

图 6.8 压力下降测量值 - 流速曲线

6.3.3 双面堆叠无源 TSV 转接板

图 6.9 展示了一种包含三种不同芯片构成的堆叠模块 [5, 11]。模块的尺寸为 12mm × 12mm，厚为 1.3mm。顶部 TSV 转接板（载体 2）尺寸为 12mm × 12mm × 0.2mm，含有 168 个周边排布已填充的通孔。底部载体（载体 1）上倒装组装了一颗 5mm × 5mm 的射频芯片。顶部载体（载体 2）上组装了一颗 5mm × 5mm 的逻辑芯片和两个堆叠起来的 3mm × 6mm 的引线键合芯片。载体 2 进行了模塑（采用递模成型工艺）以保护引线键合芯片。TSV 转接板有两层金属，采用 SiO_2 作为介质 / 钝化层，穿过载体的电连接是由 TSV 实现的。载体 1 通过 250μm 的 SAC305 焊球安装在 FR4 PCB 上。同时，对载体 1 的组装进行了底部填充（165℃固化 3h）。更多的技术信息可见参考文献 [5，11]。

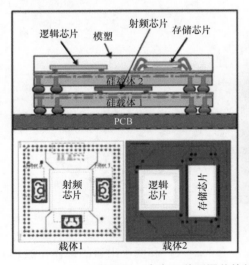
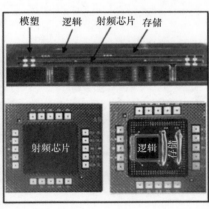

图 6.9 含有三种不同芯片的双面堆叠 TSV 转接板

6.3.4 作为应力(可靠性)缓冲的 TSV 转接板

在 ECTC2009 会议中,IME 提出使用 TSV 转接板作为应力缓冲[12, 13]。图 6.10 展示了一个粘接在含有填铜 TSV 的转接板上的大尺寸芯片(21mm×21mm),随后这一转接板被粘接在一个有机封装基板上。参考文献 [12,13] 中显示使用填铜 TSV 转接板作为应力释放(可靠性)缓冲层,减小了大尺寸芯片铜低介电常数焊盘处的应力(从 250MPa 减到 125MPa)。这对于小特征尺寸器件更加重要,因为其在铜低介电常数焊盘处所能允许的应力更小。如果在芯片和转接板之间加入一种特殊的底部填充料(具有非常小的填料尺寸),则芯片铜低介电常数焊盘处的应力将会如图 6.10 所示进一步下降到 42MPa。

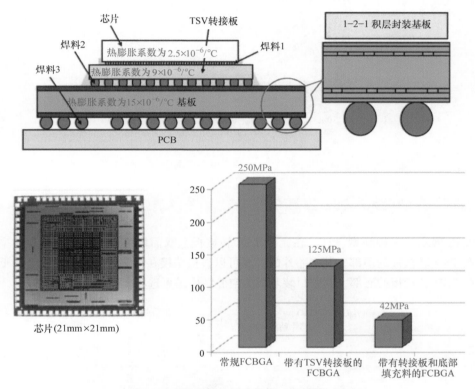

图 6.10 作为应力缓冲层的 TSV 转接板

图 6.11a 展示了 TSV 转接板的尺寸对于封装热性能的影响,可以看到当 TSV 转接板尺寸从 21mm×21mm 增加到 45mm×45mm 时,热阻(R_{ja})降低了大约 14%。这表明封装热性能可以通过增大 TSV 转接板尺寸的方式提升。

图 6.11b 展示了 TSV 转接板的厚度对于封装热性能的影响,可以看到 TSV 转接板越厚,热阻越小。这是因为扩展效应会随着转接板厚度的增大而强化。然而,对于小尺寸的转接板,这一效应可以忽略不计。

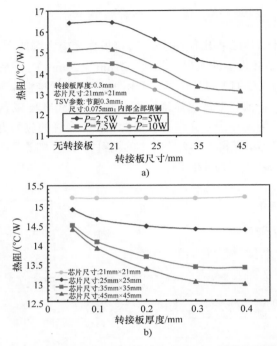

图 6.11 a) 转接板尺寸和 b) 转接板厚度对于封装热性能的影响

6.4 中国香港科技大学双面集成芯片的 TSV 转接板技术

图 6.12 展示了一种新型设计,在无源 TSV 转接板上双面均集成芯片[14-16]。高功率倒装芯片粘接在 TSV 转接板的顶部,其中芯片的底面可以直接粘接在散热片甚至是热沉上。低功率芯片粘接在 TSV 转接板的底部,其中封装基板的空腔结构是可选项。

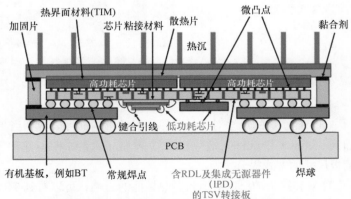

在TSV转接板与高功率和低功率倒装芯片之间填充特殊底部填充料。
在TSV转接板及有机基板之间填充常规底部填充料。
对引线键合存储器堆叠进行包封。

图 6.12 双面集成芯片的 TSV 转接板

6.5 中国台湾"工业技术研究院"的 2.5D IC 集成

自从 2011 年起，中国台湾"工业技术研究院"（China Taiwan Industrial Technology Research Institute，ITRI）已在 2.5D IC 集成[17-41]方面发表很多文章，部分会在本节中简要描述。

6.5.1 双面集成芯片 TSV 转接板的热管理

图 6.13 所示为双面集成芯片的 TSV 转接板，这是图 6.12 所示的新设计的一种旧版本。该结构的尺寸、材料特性、热边界条件在参考文献 [17-21] 中有具体阐述。芯片 1 的尺寸为 12mm×12mm，芯片 2 的尺寸为 20mm×20mm，TSV 转接板的尺寸为 30mm×30mm×100μm，在芯片 1 的背面有散热片。

环境温度假定为 20℃，PCB 顶部和底部的边界条件为 h（热传递系数）=10W/（m²·K），以模仿自然对流条件。对于散热片，我们使用热阻 R_{ca} 代替制冷模块以简化模型。R_{ca} 设定为 0.3℃/W，以模拟在带风扇或液体制冷模块高效热沉的作用下散热片的表现。芯片 1 的功率为 75W，芯片 2 为 15W，SiP 总的功率消耗为 90W。使用的仿真工具为 Icepak，通过有限体积法解决热传导问题。

SiP 的边界条件为顶部散热片 R_{ca}=0.3℃/W，PCB 两面则为 h=10W/（m²·K）。图 6.13 展示了所分析 SiP 温度分布的截面图。如图 6.13 所示，在环境温度为 20℃的情况下，芯片 1 和芯片 2 的平均温度分别为 109℃和 121℃。值得注意的是，虽然芯片 2 的发热较低，其温度依然高于芯片 1。这是因为在芯片 1 背面有散热片，因此为芯片 2 提供一个降低温度的方案是必要的。

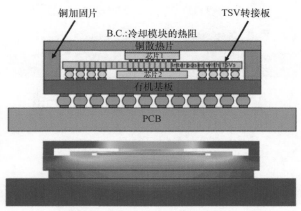

图 6.13 双面集成芯片的 TSV 转接板的热学分析

图 6.14 为图 6.12 中底部芯片通过热界面材料加上铜热塞的示意图，图 6.15 所示为分析的边界条件。除了 PCB 上有一个空腔之外，芯片和转接板的尺寸以及边界条件均与图 6.13 中相同。

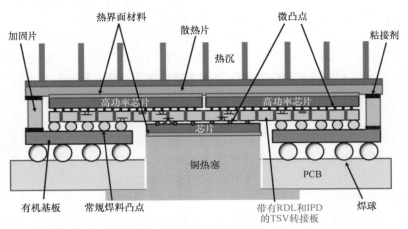

图 6.14 双面集成芯片的 TSV 转接板,底部芯片背面粘接有铜热塞

图 6.15 所示为仿真结果。可以看到芯片 2 的温度从原来的 121℃降低到 98℃。同时,芯片 1 的温度从原来的 109℃降低到 102℃,这一结果表明转接板底部的芯片可以通过使用图 6.14 和图 6.15 中简单的热塞结构获得有效冷却。

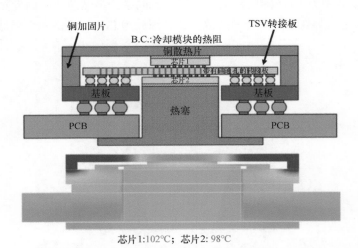

图 6.15 面向双面集成芯片的 TSV 转接板及底部芯片背面粘接有铜热塞结构的热性能分析

6.5.2 应用于 LED 含嵌入式流体微通道的 TSV 转接板

图 6.16 为含嵌入式流体微通道(见图 6.7)TSV 转接板在发光二极管(light-emitting diode,LED)热管理的应用[22]。它包括(见图 6.17)100 颗均匀分布在 TSV 转接板顶部完全相同的 LED 和均匀分布于底部完全相同的 4 颗逻辑芯片(如 ASIC)。LED 的尺

寸为 1mm×1mm×300μm，IC 芯片尺寸为 6mm×6mm×300μm，TSV 转接板尺寸为 25mm×25mm×1.4mm。微通道高度为 700μm，散热鳍片宽度为 0.1mm、节距为 1.25mm。流体入口/出口开口为 20mm×1.5mm，材料特性和边界条件见参考文献 [22]。

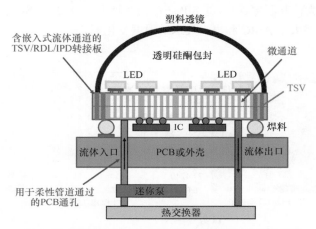

图 6.16　应用于 LED 热管理的含微通道 TSV 转接板

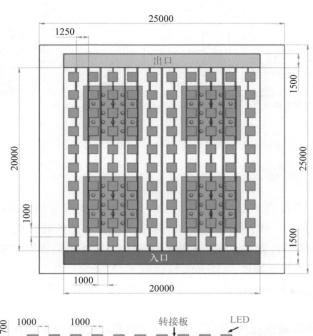

图 6.17　分析的 LED、ASIC 结构尺寸

图 6.18 所示为仿真结果：①接近入口的 LED、芯片温度是最低的，接近出口的温度最高（水带走了热量）；②邻近转接板中心的 LED、芯片温度高于邻近转接板边缘（平行于流体方向）的芯片；③LED 温度通常高于 ASIC。参考文献 [22] 还表明：①LED 功率越大，其温度越高；②ASIC 功率越大，其温度越高；③SiP 芯片功率越大，LED、ASIC、出口处水温度越高；④LED 平均温度变化率大于 ASIC 和出口处水的温度变化率。

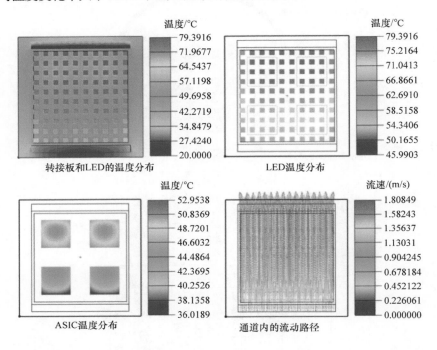

图 6.18 在 LED 功耗 2W、ASIC 功耗 10W、流速 1.26L/min 下的仿真结果

6.5.3 集成有片上系统和存储立方的 TSV 转接板

图 6.19 所示为中国台湾"工业技术研究院"的 3D 2.5D IC 集成测试结构的截面图[23-26]，包含支撑着四颗存储芯片、一颗热测试芯片和一颗机械测试芯片的转接板。为了便于拾取、放置、保护芯片不受恶劣环境条件影响，该结构进行了模塑。转接板的顶面和底面有 RDL。应力传感器（机械测试芯片）放置在厚度为 100μm 转接板（12.3mm × 12.2mm）顶面，转接板上制作有集成无源元件（integrated passive device，IPD）。图 6.20 为样品照片、截面 SEM 图像、测试装置的 X 射线图[23-26]。

这一测试结构可以被演化用于如下情形：①在既不存在存储芯片堆叠，机械测试及热测试芯片上也没有 TSV，转接板可能是逻辑芯片、微处理器或 SoC 的情况下，面向宽 I/O 存储应用；②在不存在机械测试/热测试芯片、转接板为逻辑芯片的情况下面向宽 I/O DRAM 应用；③在不存在存储芯片堆叠且机械测试及热测试芯片上无 TSV 的情况下应用于宽 I/O 界面。

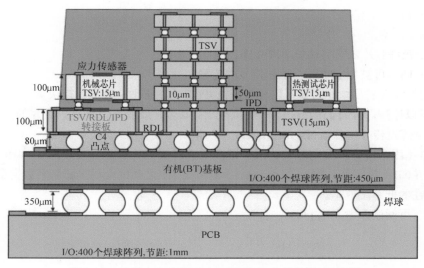

图 6.19 支撑一个存储立方和两颗 SoC 的 TSV 转接板

图 6.20 组装后的测试结构，典型截面的 SEM 图像和结构的 X 射线图像

6.5.4 半嵌入式 TSV 转接板

图 6.21 所示为半嵌入式 TSV 转接板（带有应力释放空隙），其顶面可集成多个芯片。这一设计的好处是

1）小外形。

2）方便使用任何无 TSV 的芯片。

3）短设计周期。

4）低制造成本。

5）RDL 允许芯片之间实现短距离互连。

6）大量 TSV 可被用于电源、地和信号的传输。

7）可返修（在底部填充料前可以检修基板 /PCB 上的芯片 - 转接板模块）。

8）热可以通过芯片背面的散热片 / 热沉散出，以及 / 或者通过带有散热片的热塞结构（图中未展示）从焊点传导至基板 /PCB 的背面。

9）可靠（因为应力释放间隙结构减少了嵌入式转接板（热膨胀系数为 $(6\sim8)\times10^{-6}/℃$）和有机基板 /PCB（热膨胀系数为 $(15\sim18.5)\times10^{-6}/℃$）之间的全局热膨胀系数失配程度）。

10）潜在的低系统成本。

更多细节可阅读参考文献 [30]。

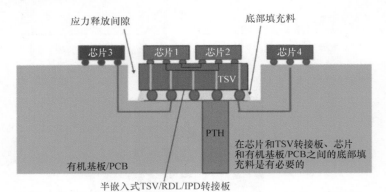

图 6.21 带有应力释放间隙半嵌入式有机基板的 TSV 转接板

6.5.5 双面粘接芯片的 TSV 转接板

图 6.22 为双面粘接芯片的 TSV 转接板[32, 33]，可以看到转接板在其顶部粘接一颗 CPU 芯片，底部粘接两颗存储芯片。转接板尺寸为 28mm×28mm×150μm，CPU/ASIC 尺寸为 22mm×18mm×400μm，DRAM 为 10mm×10mm×100μm。封装基板为 40mm×40mm×950μm，PCB 为 114.3mm×101.6mm×1600μm。所有芯片和转接板之间的微焊点直径为 25μm、节距为 250μm。常规焊点直径为 150μm、节距为 250μm。焊球直径为 600μm、节距为 1000μm。

在 -40~125℃ 的温度循环载荷下，转接板和有机（BT）基板之间位于拐角处的常规焊点上，最大蠕变应变能量密度随时间变化的曲线如图 6.22 所示，同时提供了 von Mises 应力云图。可以看到每个周期的蠕变应变能量密度是 0.018MPa，远低于 0.1MPa，因此在底部填充料的保护下，在大部分环境条件下芯片都是可靠的。底部填充非常有效。

组装体截面的 SEM 图像如图 6.23 所示，可以看到无源转接板顶部有一颗较大较厚的芯片、底部芯片较小较薄。图 6.23 展示了顶部芯片和转接板之间的微焊点放大图，可以观察到：①转接板上的 UBM 为 Cu 和 Ni；②焊料转变为金属间化合物（intermetallic compound，IMC）

Cu_6Sn_5;③大芯片上的 UBM 为 Cu。类似地,带有 TSV 的转接板和小芯片之间的微焊点放大也展示于图 6.23 中了。

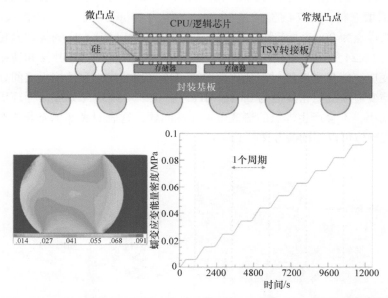

图 6.22 双面粘接芯片的 TSV 转接板,及拐角处焊点最大蠕变应变能量密度随时间变化曲线

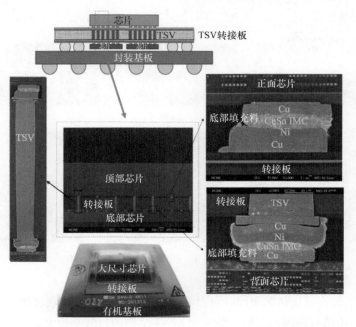

图 6.23 各组件的 SEM 图像,如 TSV、微凸点、芯片和样品

6.5.6 双面集成芯片的 TSV 转接板

图 6.24 所示为双面集成芯片的 TSV 转接板的示意图[34-37]，可以看到顶部有两颗芯片，底部有一颗芯片。封装基板的尺寸为 35mm×35mm×970μm，芯片和转接板之间、转接板和封装基板之间有底部填充料。TSV 的直径为 10μm、节距为 150μm；芯片、转接板之间和转接板、封装基板之间的焊点为 90μm、节距为 125μm；封装基板和 PCB 之间的焊球直径为 600μm、节距为 1000μm。图 6.24 所示为非线性分析的有限元模型。图 6.25 为芯片 2 和转接板之间位于拐角处焊点的蠕变应变能量密度随时间变化曲线，可以看到芯片 2 和转接板之间处于拐角处的焊点每个周期的蠕变应变能量密度值为 0.0107MPa。在 -25 ~ 125℃、60min 循环周期的环境条件下，这一量级太小，不会产生焊点的热疲劳可靠性问题，这表明底部填充非常有效。对于其他焊点位置的蠕变响应可以阅读参考文献 [34-37]。

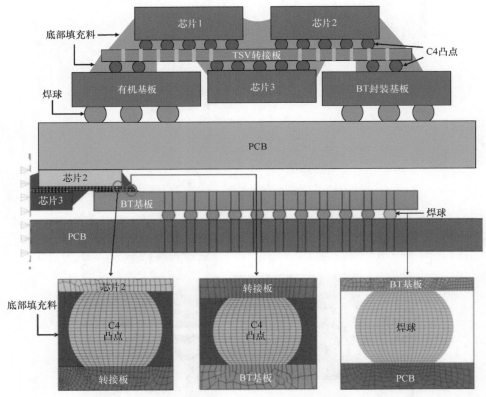

图 6.24　双面集成芯片的 TSV 转接板，及仿真的有限元模型

图 6.26 为完整组装模块的截面图。可以看到转接板共集成了三颗芯片，并填充了底部填充料，转接板被焊接（填有底部填充料）到一个 4-2-4 结构的封装基板上。

第 6 章 2.5D IC 集成 267

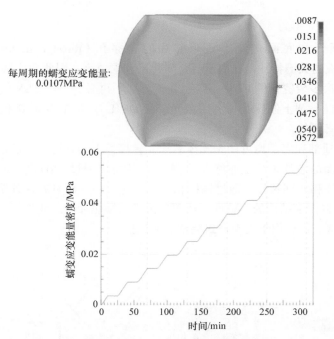

图 6.25 蠕变应变能量密度云图和随时间变化曲线

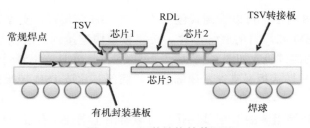

图 6.26 组装结构的截面图

6.5.7 TSH 转接板

图 6.27 为在顶面和底面支撑了数个芯片的硅穿孔（through silicon hole，TSH）转接板[38,39]。TSH 转接板的关键特征是孔内没有金属化层和铜填充，因此介质层、阻挡层、种子层、通孔填充、用于移除多余覆铜的 CMP、露铜等工艺均不再必要。与 TSV 转接板相比，TSH 转接板仅需要在一片硅晶圆上制作孔（利用激光或者 DRIE），与 TSV 转接板相同的是，RDL 在 TSH 转接板中也是必要的。

TSH 转接板的顶部和底部均可用于支撑芯片，转接板上的孔可以让底面的芯片信号通过铜柱和焊料传输到顶面的芯片（反之亦然）。同侧的芯片可以通过 TSH 转接板的 RDL 相互通信。在物理上，顶部芯片和底部芯片是通过铜柱和微焊点连接的。所有芯片外围的引脚也均可焊接在 TSH 转接板上，以保证结构的完整性，从而抵御机械和热冲击。此外，TSH 转接板底部外围分布有常规焊点，这些焊点用于实现与封装基板的连接。

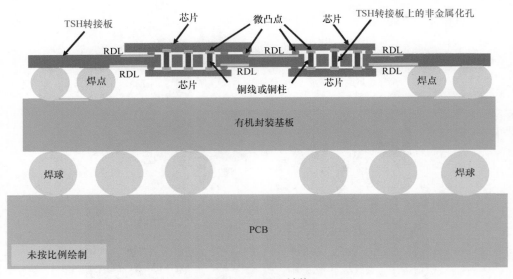

图 6.27 TSH 封装

测试结构如图 6.28 所示，可以看到其包含了一片 TSH 转接板，支撑着一颗带有铜柱的顶部芯片与一颗带有 UBM 和焊料的底部芯片。转接板模块连接到封装基板上，然后再连接到 PCB 上。最终组装好的测试装置如图 6.29 所示。可以看出，PCB 支撑着封装基板，封装基板支撑着 TSH 转接板，TSH 转接板又支撑着顶部芯片。底部芯片被 TSH 转接板挡住了，所以无法看到。

图 6.29 还显示了最终组装的 X 射线图像。可以看出①铜柱没有接触到 TSH 的侧壁，②铜柱基本处于 TSH 中心位置。图 6.28 还显示了封装体横截面的 SEM 图像，包括了该结构的所

有关键元素，如顶部芯片、TSH 转接板、底部芯片、封装基板、PCB、微凸点、焊点、焊球、TSH 和铜柱。通过 X 射线和 SEM 图像可以看出，该结构各个关键部件的制造准确无误。

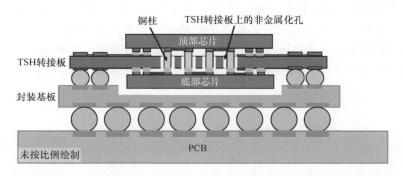

图 6.28　TSH 测试结构及其截面图

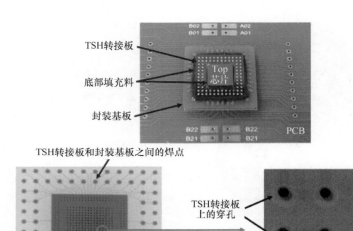

图 6.29　组装结构，铜柱和硅孔的 X 射线图

6.6 台积电的 CoWoS 技术

在台积电（TSMC）2011 年第三季度的投资者会议上，张忠谋博士（Morris Chang，台积电创始人）在没有任何预警的情况下，宣布公司将进军封装测试领域，震惊了所有人。第一款产品是基板上晶圆上芯片（chip-on-wafer-on-substrate，CoWoS），即在硅转接板上集成了逻辑计算芯片和存储芯片，随后将其安装在封装基板上，今天工业界将 CoWoS 认定为 2.5D IC 集成。

6.7 赛灵思/台积电的 2.5D IC 集成

自 2011 年起，赛灵思（Xilinx）就已在 2.5D IC 集成领域发表了多篇文章[46-60]。如图 6.30 所示，为了实现更高的器件良率（以节约成本），一颗非常大的片上系统（system-on-chip，SoC）被划分为四颗更小的、由台积电 28nm 工艺所制作的现场可编程门阵列（field programmable gate array，FPGA）芯片。FPGA 芯片之间超过 10000 条的横向互连主要由 TSV 转接板上 0.4μm（最小值）节距的 RDL 实现，RDL 的金属层和介质层的最小厚度约为 1μm。如图 6.31 和图 6.32 所示，每颗 FPGA 包含超过 50000 个的节距为 45μm 的微凸点（在 TSV 转接板上有超过 200000 个微凸点）。这样无源 TSV/RDL 转接板可用于极窄节距、高 I/O 数、高性能和高密度的半导体 IC 应用。2013 年 10 月 20 日，赛灵思和台积电[61]联合宣布采用 28nm 工艺代工的 Virtex-7 HT 系列芯片正式投产，声称是工业界首个量产的 2.5D IC 集成产品。赛灵思 Virtex-7 HT FPGA 搭载了多达 16 个 28.05Gbit/s 和 72 个 13.1Gbit/s 的收发器。

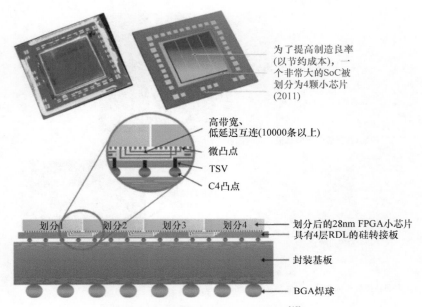

图 6.30 赛灵思的划分 FPGA[46]

第 6 章 2.5D IC 集成

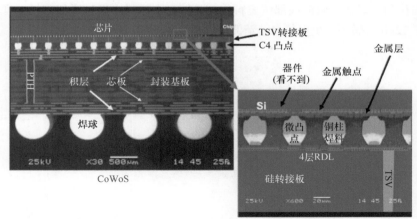

> RDL：0.4μm 节距线宽及线距
> 在 45μm 节距下每个 FPGA 有超过 50000 个微凸点
> 转接板支撑了超过 200000 个微凸点

图 6.31　赛灵思／台积电为 FPGA 开发的 CoWoS 技术[52]

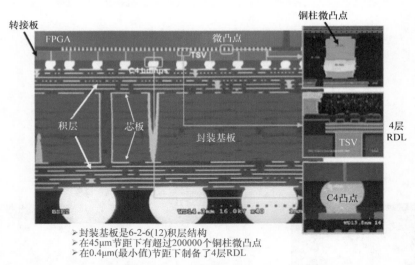

> 封装基板是 6-2-6(12) 积层结构
> 在 45μm 节距下有超过 200000 个铜柱微凸点
> 在 0.4μm（最小值）节距下制备了 4 层 RDL

图 6.32　赛灵思／台积电的 CoWoS（6-2-6 积层封装基板）[52]

现阶段，赛灵思和台积电的工作已经远远超过上述范畴。图 6.33 为一个测试结构，它包含了一个 31.5mm×41.7mm×100μm 的 TSV 转接板，采用台积电的 CoWoS XLTM 65nm 后道工艺制造，其上包含三颗 FPGA 芯片和两个 HBM，封装基板尺寸为 55mm×55mm×1.9mm。第一批热循环测试结果在规定的 1200 次循环完成之前便产生了一些失效。图 6.34 所示为截面扫描电子显微镜（scanning electron microscopy，SEM）失效分析结果，SEM 照片显示 C4 底部填充料中有一条裂纹，从转接板边缘延伸到了 C4 凸点区域。裂纹主要是沿着转接板边缘分布，偶尔会沿着 C4 的铜柱出现。导致应力失效的主要原因是基板和芯片 - 转接板之间的热膨胀系数不匹配。由于固化和热老化导致的底部填充料收缩是第二个原因。通过增加基板厚度，热循环测试次数通过了 1200 次循环。更多信息，请阅读参考文献 [59]。

图 6.33 赛灵思/台积电的 VIRTEX 产品[60]

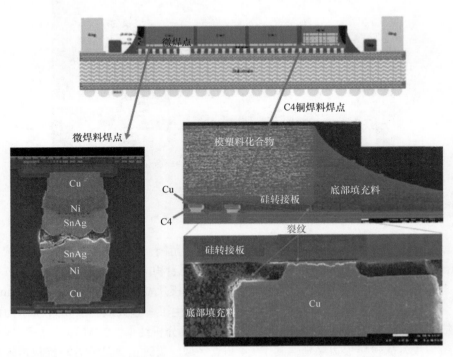

图 6.34 赛灵思/台积电 VIRTEX 产品的 SEM 图像[60]

6.8 Altera/台积电的 2.5D IC 集成

图 6.35 为阿尔特拉（Altera）2.5D IC 集成的截面图[62, 63]，可以看到 TSV 转接板通过铜柱和焊料帽微凸点支撑着芯片，随后 TSV 转接板由 C4 凸点键合到 6-2-6 封装基板上，其中 TSV 转接板由台积电 CoWoS 技术制作。遗憾的是，这一产品未应用至大规模量产中。

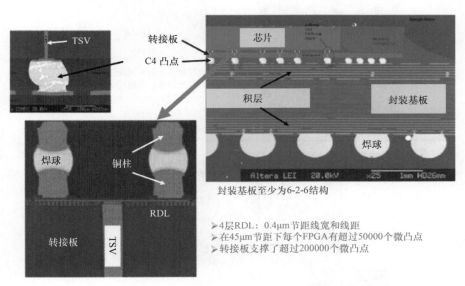

图 6.35 Altera/台积电的 CoWoS[63]

6.9 AMD/联电的 2.5D IC 集成

图 6.36 所示为超微半导体（AMD）在 2015 年下半年出货的 Radeon R9 Fury X 图形处理单元（Graphics Processing Unit，GPU）。该 GPU 采用台积电 28nm 工艺技术，配套有海力士（Hynix）制造的四个 HBM（高带宽存储器，将在第 7 章讨论）立方。每个 HBM 由四个带有铜柱+焊料帽凸点的 DRAM 和逻辑底座组成，它们包含穿透的 TSV。每个 DRAM 芯片有 1000 个以上的 TSV。GPU 和 HBM 位于 TSV 转接板（28mm×35mm）上面，该转接板由联电（UMC）采用 65nm 工艺制造。带有可控塌陷芯片连接（controlled collapse chip connection，C4）凸点的 TSV 转接板，与在 4-2-4 结构的有机封装基板（由 Ibiden 制造）的最终组装由日月光（ASE）完成。

截面的一些 SEM 图像如图 6.37 所示，可以看出 GPU 和 HBM 是由带有微凸点（铜柱+焊料帽）的 TSV 转接板支撑，TSV 转接板采用 C4 凸点由 4-2-4 积层封装基板支撑。

274 半导体先进封装技术

图 6.36　AMD/ 联电的 GPU（Fiji）

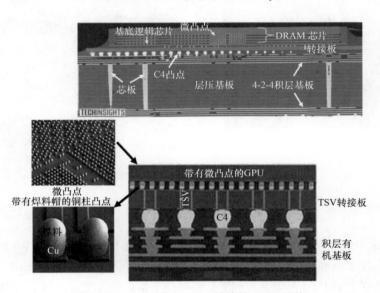

图 6.37　AMD/ 联电 GPU 模块的 SEM 图像

6.10　英伟达 / 台积电的 2.5D IC 集成

图 6.38 和图 6.39 为英伟达（NVidia）Pascal 100 GPU，它在 2016 年下半年出货。该 GPU 采用台积电的 16nm 工艺技术，配套有三星制造的四个 HBM2（16GB）[43]。每个 HBM2 由四颗带有铜柱 + 焊料帽凸点的 DRAM 和一颗基础逻辑芯片组成，它们带有 TSV。每颗 DRAM 芯片都有超过 1000 个 TSV。GPU 和 HBM2 用微凸点连接在 TSV 转接板（1200mm^2）上，该转接板

采用台积电 64nm 工艺制造。TSV 转接板则通过 C4 凸点连接到 5-2-5 有机封装基板上。

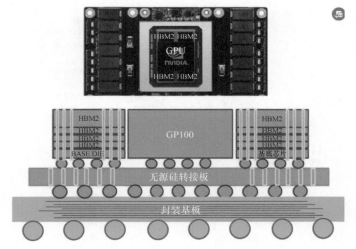

图 6.38　英伟达 / 台积电的 P100 [43]

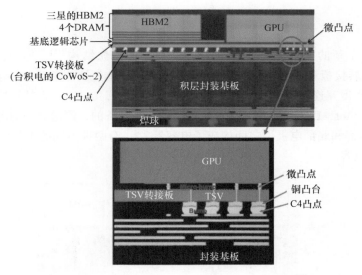

图 6.39　英伟达 / 台积电 P100 的 SEM 图像 [43]

6.11　台积电 CoWoS 路线图

台积电关于 CoWoS 的路线图如图 6.40 所示 [64]。可以看到，2011 年的第一代版本（1.0 倍最大光罩尺寸大约 33mm × 26mm = 858mm²）是为 2013 年的赛灵思芯片准备的。多年来，CoWoS 技术发展的重点是为不断增加的硅转接板尺寸提供支持，台积电计划把转接板尺寸扩大到3倍最大光罩（2021年）和4倍最大光罩（2023年），从而支撑封装中的处理器和HBM堆栈。

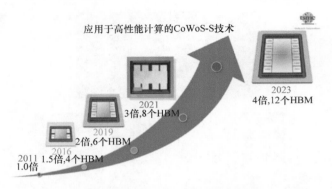

图 6.40　台积电的 CoWoS 路线图

6.12　2.5D IC 集成的近期进展

本节中简要描述一些关于 2.5D IC 集成的近期进展。

6.12.1　台积电的集成有深槽电容 CoWoS

图 6.41 为一个新的 CoWoS 平台上的高性能计算的概念结构[65]，它包括一颗逻辑芯片、HBM2E、一片硅转接板和一片基板。逻辑芯片和 HBM2E 首先被并排键合在硅转接板上，形成具有窄节距和高密度互连布线的晶圆上芯片（chip-on-wafer，CoW）。在硅转接板中，深槽电容（deep trench capacitor，DTC）是用高深宽比的硅蚀刻法制备的，深槽电容的高 k 介质在深宽比超过 10 的硅槽顶部和底部电极层之间构成了电容器[65]。有两种不同的工艺顺序可用于在硅转接板中实现深槽电容。

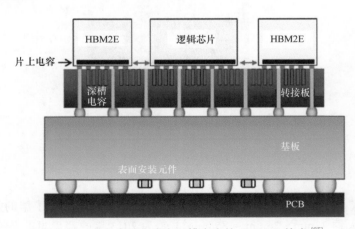

图 6.41　台积电的集成有深槽电容的 CoWoS 技术[65]

图 6.42a 所示为深槽电容的名义电容密度与电压的关系,它被定义在深槽电容结构的等效全表面积上。在外加电压为零、100kHz 条件下,采用三用表(电感、电容、电阻测量计)测量出的高介电常数介电膜的电容密度约为 $300nF/mm^2$。它提供的电容密度比金属 - 绝缘体 - 金属电容高一个数量级。图 6.42b 所示为两条名义的 I-V 曲线,分别是高介电常数介电膜在 25℃和 100℃下的测量值。可以看出,即使在 10℃的测试温度下,在 ±1.35V 偏压下测得的漏电流仍低于 $1fA/\mu m^2$,这种优异的特性避免了深槽电容中的额外功率浪费[65]。

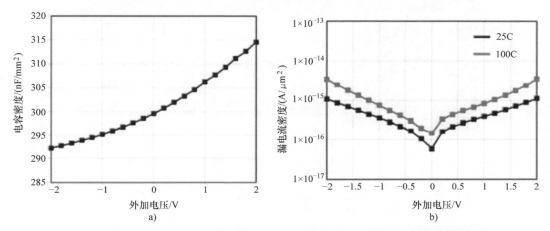

图 6.42 a)电容密度与外加电压的关系;b)漏电流密度与外加电压的关系[65]

6.12.2 IME 2.5D IC 集成的非破坏性失效定位方法

在参考文献 [66] 中,IME 开发了一种物理失效分析(physical failure analysis,PFA)的方法,以准确定位 2.5D IC 集成中的失效。图 6.43 所示为 2.5D IC 结构中开路缺陷的位置。PFA 证实了该缺陷的存在,并揭示了该开路失效是靠近基板和转接板界面的通孔后侧的金属层中所出现的裂纹导致的。这种缺陷可能是应力从硅转接板的边缘向基板方向蔓延所致[66]。

6.12.3 Fraunhofer 的光电转接板

图 6.44 所示为单模路由器的概念性示意图。该转接板装配在基于玻璃的光印刷电路板(optical printed circuit board,OPCB)上,其中 OPCB 的光层和硅转接板之间的互连是通过一个耦合镜在垂直方向上完成的,如图 6.44 所示[67]。对于传输操作,从 OPCB 引出的 12 个光通道均被馈送到一个单独的光电二极管(photodiode,PD)和其各自的电子跨阻放大器上(transimpedance amplifier,TIA)。然后 TIA 可以对传入的信号进行光电转换,而接收到的电信号在驱动垂直腔表面发射激光器(vertical-cavity surface-emitting laser,VCSEL)的调制操作之前则被传送到一个电子驱动放大器。然后,每个 VCSEL 通过电流注入被调谐到不同的波长,以匹配硅层上阵列波导光栅(arrays waveguide grating,AWG)复用器的信道间隔。从图中可以观察到使用底层金属堆叠的晶圆正面和背面用于电气连接的常规 TSV(左),以及 TSV 底部没有金属层的

所谓光学 TSV（右）。

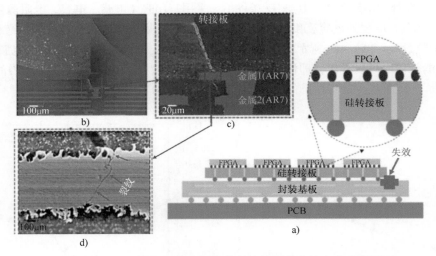

图 6.43 IME 在 TSV 转接板中非破坏性失效定位方法示意图

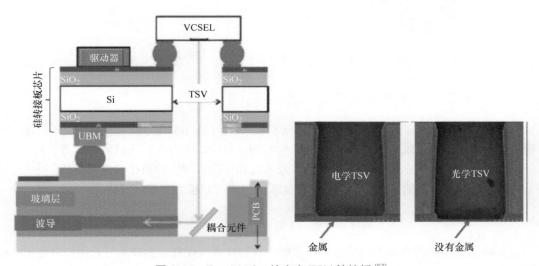

图 6.44 Fraunhofer 的光电 TSV 转接板[67]

6.12.4 Dai Nippon/AGC 的玻璃转接板

图 6.45 为大日本印刷株式会社（Dai Nippon）/艾杰旭（AGC）用于高频和高速应用的玻璃转接板，特别是用于封装天线（antenna-in-package，AiP）领域[68]。它们的基本结构包括在石英衬底上的共面波导（coplanar waveguide，CPW），以及用于连接顶部和底部电路的石英通孔（through quartz via，TQV）。图 6.46 为工艺流程，同时展示了典型的玻璃通孔（through-glass via，TGV）和铜线。转接板的厚度为400μm，TGV的顶部直径约为80μm，底部直径为50μm。铜线的线宽和线距为2μm。

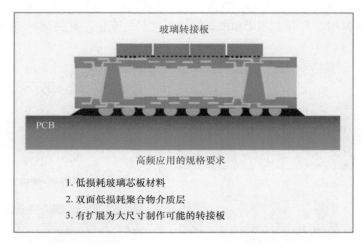

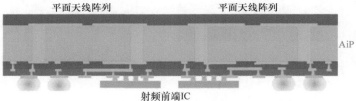

图6.45 Dai Nippon/AGC 用于封装天线的玻璃转接板[68]

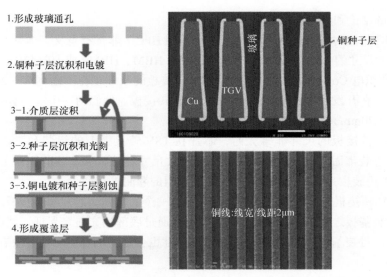

图6.46 工艺流程：玻璃转接板上的TGV及RDL布线[68]

6.12.5 富士通的多层玻璃转接板

图 6.47 所示为富士通（Fujitsu）的多层玻璃转接板。TGV 是通过激光诱导深蚀刻（laser induced deep etching，LIDE）制作的。TGV 通过丝网印刷导电浆料进行填充。节距为 40μm 的微凸点（铜柱 + 焊料帽）实现了芯片和玻璃转接板之间的互连。有关该封装的更多信息，请阅读参考文献 [69]。

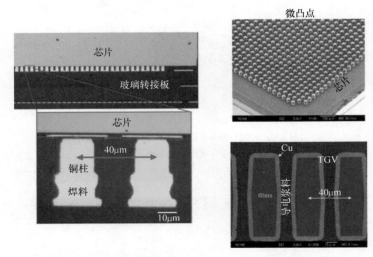

图 6.47 富士通填充有导电膏的 TGV 玻璃转接板[69]

6.13 总结和建议

一些重要的结论和建议总结如下。

1）2.5D IC 集成对于高性能和高密度应用（如 HPC）非常有意义。通常无源 TSV 转接板用于支撑非常大尺寸的 SoC，如 CPU、GPU 和一些 HBM、HBM2 或 HBM2E。

2）根据台积电 CoWoS 路线图，无源 TSV 转接板的尺寸正在不断增大。2020 年 9 月，台积电宣称他们正在开发尺寸约为 2400mm^2 的 TSV 转接板。他们的一个潜在客户正在尝试使用积层封装基板（70mm×78mm）来支撑 TSV 转接板。

3）当 TSV 转接板的尺寸非常大时，芯片在 TSV 转接板上的组装和 TSV 转接板在积层封装基板上的组装都面临巨大的挑战（原因在于大尺寸转接板和基板的翘曲）。另外，模组在 PCB 上的 SMT 组装也是一个很大的挑战（由于模组的翘曲）。

4）当 TSV 转接板的尺寸非常大时，可靠性是一个大问题。芯片和转接板之间的微凸点以及转接板和封装基板之间的 C4 凸点的可靠性可以通过底部填充料来保证。然而，由于封装基板的尺寸问题，封装基板和 PCB 之间的焊点可靠性也会成为问题，可能需要在 PCB 上进行底部填充来解决。

5）关于 2.5D IC 集成的更多信息，参见参考文献 [92-96]。

参 考 文 献

1. Souriau, J., O. Lignier, M. Charrier, and G. Poupon, "Wafer Level Processing Of 3D System in Package for RF and Data Applications", *IEEE/ECTC Proceedings*, 2005, pp. 356–361.
2. Henry, D., D. Belhachemi, J-C. Souriau, C. Brunet-Manquat, C. Puget, G. Ponthenier, J. Vallejo, C. Lecouvey, and N. Sillon, "Low Electrical Resistance Silicon Through Vias: Technology and Characterization", *IEEE/ECTC Proceedings*, 2006, pp. 1360–1366.
3. Selvanayagam, C., J. H. Lau, X. Zhang, S. Seah, K. Vaidyanathan, and T. Chai, "Nonlinear Thermal Stress/Strain Analysis of Copper Filled TSV (Through Silicon Via) and Their Flip-Chip Microbumps", *IEEE/ECTC Proceedings*, May 2008, pp. 1073–1081.
4. Yu, A., N. Khan, G. Archit, D. Pinjalal, K. Toh, V. Kripesh, S. Yoon, and J. H. Lau, "Fabrication of Silicon Carriers with TSV Electrical Interconnection and Embedded Thermal Solutions for High Power 3-D Package", *IEEE/ECTC Proceedings*, May 2008, pp. 24–28.
5. Khan, N., V. Rao, S. Lim, H. We, V. Lee, X. Zhang, E. Liao, R. Nagarajan, T. C. Chai, V. Kripesh, and J. H. Lau, "Development of 3-D Silicon Module With TSV for System in Packaging", *IEEE/ECTC Proceedings*, May 2008, pp. 550–555.
6. Selvanayagam, C., J. H. Lau, X. Zhang, S. Seah, K. Vaidyanathan, and T. Chai, "Nonlinear Thermal Stress/Strain Analyses of Copper Filled TSV (Through Silicon Via) and Their Flip-Chip Microbumps", *IEEE Transactions on Advanced Packaging*, Vol. 32, No. 4, November 2009, pp. 720–728.
7. Khan, N., L. Yu, P. Tan, S. Ho, N. Su, H. Wai, K. Vaidyanathan, D. Pinjala, J. H. Lau, T. Chuan, "3D Packaging with Through Silicon Via (TSV) for Electrical and Fluidic Interconnections", *IEEE/ECTC* Proceedings, May, 2009, pp. 1153–1158.
8. Yu, A., N. Khan, G. Archit, D. Pinjala, K. Toh, V. Kripesh, S. Yoon, and J. H. Lau, "Fabrication of Silicon Carriers With TSV Electrical Interconnections and Embedded Thermal Solutions for High Power 3-D Packages", *IEEE Transactions on CPMT*, Vol. 32, No. 3, September 2009, pp. 566–571.
9. Tang, G. Y., S. Tan, N. Khan, D. Pinjala, J. H. Lau, A. Yu, V. Kripesh, and K. Toh, "Integrated Liquid Cooling Systems for 3-D Stacked TSV Modules", *IEEE Transactions on CPMT*, Vol. 33, No. 1, March 2010, pp. 184–195.
10. Khan, N., H. Li, S. Tan, S. Ho, V. Kripesh, D. Pinjala, J. H. Lau, and T. Chuan, "3-D Packaging With Through-Silicon Via (TSV) for Electrical and Fluidic Interconnections", *IEEE Transactions on CPMT*, Vol. 3, No. 2, February 2013, pp. 221–228.
11. Khan, N., V. Rao, S. Lim, H. We, V. Lee, X. Zhang, E. Liao, R. Nagarajan, T. C. Chai, V. Kripesh, and J. H. Lau, "Development of 3-D Silicon Module With TSV for System in Packaging", *IEEE Transactions on CPMT*, Vol. 33, No. 1, March 2010, pp. 3–9.
12. Zhang, X., T. Chai, J. H. Lau, C. Selvanayagam, K. Biswas, S. Liu, D. Pinjala, et al., "Development of Through Silicon Via (TSV) Interposer Technology for Large Die (21x21mm) Fine-pitch Cu/low-k FCBGA Package", *IEEE/ECTC Proceedings*, May 2009, pp. 305–312.
13. Chai, T. C., X. Zhang, J. H. Lau, C. S. Selvanayagam, D. Pinjala, et al., "Development of Large Die Fine-Pitch Cu/low-*k* FCBGA Package with through Silicon via (TSV) Interposer", *IEEE Transactions on CPMT*, Vol. 1, No. 5, May 2011, pp. 660–672.
14. Lau, J. H., S. Lee, M. Yuen, J. Wu, C. Lo, H. Fan, and H. Chen, "Apparatus having thermal-enhanced and cost-effective 3D IC integration structure with through silicon via interposer". US Patent No: 8,604,603, Filed Date: February 19, 2010, Date of Patent: December 10, 2013.
15. Lau, J. H., Y. S. Chan, and R. S. W. Lee, "3D IC Integration with TSV Interposers for High-Performance Applications", *Chip Scale Review*, Vol. 14, No. 5, September/October, 2010, pp. 26–29.
16. Lau, J. H., M. S. Zhang, and S. W. R. Lee, "Embedded 3D Hybrid IC Integration System-in-Package (SiP) for Opto-Electronic Interconnects in Organic Substrates", *ASME Transactions, Journal of Electronic Packaging*, Vol. 133, September 2011, pp. 1–7.
17. Chien, J., Y. Chao, J. H. Lau, M. Dai, R. Tain, M. Dai, P. Tzeng, C. Lin, Y. Hsin, S. Chen, J. Chen, C. Chen, C. Ho, R. Lo, T. Ku, and M. Kao, "A Thermal Performance Measurement Method for Blind Through Silicon Vias (TSVs) in a 300 mm Wafer", *IEEE/ECTC Proceedings*, June 2011, pp. 1204–1210.

18. Chien, H. C., J. H. Lau, Y. Chao, R. Tain, M. Dai, S. T. Wu, W. Lo, and M. J. Kao, "Thermal Performance of 3D IC Integration with Through-Silicon Via (TSV)", *Proceedings of IMAPS International Conference*, Long Beach, CA, October 2011, pp. 25–32.
19. Chien, H. C., J. H. Lau, Y. Chao, R. Tain, M. Dai, S. T. Wu, W. Lo, and M. J. Kao, "Thermal Performance of 3D IC Integration with Through-Silicon Via (TSV)", *IMAPS Transactions, Journal of Microelectronic Packaging,* Vol. 9, 2012, pp. 97–103.
20. Chien, J., J. H. Lau, Chao, Y., M. Dai, R. Tain, L. Li, P. Su, J. Xue, M. Brillhart, "Thermal Evaluation and Analyses of 3D IC Integration SiP with TSVs for Network System Applications", *IEEE/ECTC Proceedings*, May 2012, pp. 1866–1873.
21. Chien, H., J. H. Lau, T. Chao, M. Dai, and R. Tain, "Thermal Management of Moore's Law Chips on Both sides of an Interposer for 3D IC integration SiP," *IEEE/ICEP Proceedings*, Japan, April 2012, pp. 38–44.
22. Lau, J. H., H. C. Chien, and R. Tain, "TSV Interposers with Embedded Microchannels for 3D IC and LED Integration", *ASME Paper no. InterPACK2011–52204*, Portland, OR, July 2011.
23. Lau, J. H., C-J Zhan, P-J Tzeng, C-K Lee, M-J Dai, H-C Chien, Y-L Chao, et al., "Feasibility Study of a 3D IC Integration System-in-Packaging (SiP) from a 300 mm Multi-Project Wafer (MPW)", *IMAPS International Symposium on Microelectronics*, October 2011, pp. 446–454.
24. Lau, J. H., C-J Zhan, P-J Tzeng, C-K Lee, M-J Dai, H-C Chien, Y-L Chao, et al., "Feasibility Study of a 3D IC Integration System-in-Packaging (SiP) from a 300 mm Multi-Project Wafer (MPW)", *IMAPS Transactions, Journal of Microelectronic Packaging*, Vol. 8, No. 4, Fourth Quarter 2011, pp. 171–178.
25. Zhan, C., P. Tzeng, J. H. Lau, M. Dai, H. Chien, C. Lee, S. Wu, et al., "Assembly Process and Reliability Assessment of TSV/RDL/IPD Interposer with Multi-Chip-Stacking for 3D IC Integration SiP", *IEEE/ECTC Proceedings*, May 2012, pp. 548–554.
26. Tzeng, P., J. H. Lau, M. Dai, S. Wu, H. Chien, Y. Chao, C. Chen, S. Chen, C. Wu, C. Lee, C. Zhan, J. Chen, Y. Hsu, T. Ku, and M. Kao, "Design, Fabrication, and Calibration of Stress Sensors Embedded in a TSV Interposer in a 300 mm Wafer", *IEEE/ECTC Proceedings*, San Diego, CA, May 2012, pp. 1731–1737.
27. Sheu, S., Z. Lin, J. Hung,, J. H. Lau, P. Chen, S. Wu, K. Su, C. Lin, S. Lai, T. Ku, W. Lo, M. Kao, "An Electrical Testing Method for Blind Through Silicon Vias (TSVs) for 3D IC Integration", *IMAPS Transactions, Journal of Microelectronic Packaging*, Vol. 8, No. 4, Fourth Quarter 2011, pp. 140–145.
28. Chen, J. C., J. H. Lau, P. J. Tzeng, S. Chen, C. Wu, C. Chen, H. Yu, Y. Hsu, S. Shen, S. Liao, C. Ho, C. Lin, T. K. Ku, and M. J. Kao, "Effects of Slurry in Cu Chemical Mechanical Polishing (CMP) of TSVs for 3-D IC Integration", *IEEE Transactions on CPMT*, Vol. 2, No. 6, June 2012, pp. 956–963.
29. Lau, J. H., and G. Y. Tang, "Effects of TSVs (through-silicon vias) on thermal performances of 3D IC integration system-in-package (SiP)", *Journal of Microelectronics Reliability*, Vo. 52, Issue 11, November 2012, pp. 2660–2669.
30. Lau, J. H., S. T. Wu, and H. C. Chien, "Nonlinear Analyses of Semi-Embedded Through-Silicon Via (TSV) Interposer with Stress Relief Gap Under Thermal Operating and Environmental Conditions", *IEEE EuroSime Proceedings, Chapter 11: Thermo-Mechanical Issues in Microelectronics*, Lisbon, Portugal, April 2012, pp. 1/6–6/6.
31. Wu, C., S. Chen, P. Tzeng, J. H. Lau, Y. Hsu, J. Chen, Y. Hsin, C. Chen, S. Shen, C. Lin, T. Ku, and M. Kao, "Oxide Liner, Barrier and Seed Layers, and Cu-Plating of Blind Through Silicon Vias (TSVs) on 300 mm Wafers for 3D IC Integration", *IMAPS Transactions, Journal of Microelectronic Packaging*, Vol. 9, No. 1, First Quarter 2012, pp. 31–36.
32. Li, L., P. Su, J. Xue, M. Brillhart, J. H. Lau, P. Tzeng, C. Lee, C. Zhan, M. Dai, H. Chien, and S. Wu, "Addressing Bandwidth Challenges in Next Generation High Performance Network Systems with 3D IC Integration," *IEEE ECTC Proceedings*, San Diego, CA, May 2012, pp. 1040–1046.
33. Lau, J. H., P. Tzeng, C. Zhan, C. Lee, M. Dai, J. Chen, Y. Hsin, S. Chen, C. Wu, L. Li, P. Su, J. Xue, and M. Brillhart, " Large Size Silicon Interposer and 3D IC Integration for System-in-Packaging (SiP)", *Proceedings of the 45th IMAPS International Symposium on Microelectronics,* September 2012, pp. 1209–1214.

34. Lau, J. H., P. Tzeng, C. Lee, C. Zhan, M. Li, J. Cline, K. Saito, Y. Hsin, P. Chang, Y. Chang, J. Chen, S. Chen, C. Wu, H. Chang, C. Chien, C. Lin, T. Ku, R. Lo, and M. Kao, (Redistribution Layers (RDLs) for 2.5D/3D IC Integration", *Proceedings of the 46th IMAPS International Symposium on Microelectronics,* October 2013, pp. 434–441.
35. Wu, S. T., H. Chien, J. H. Lau, M. Li, J. Cline, and M. Ji, "Thermal and Mechanical Design and Analysis of 3D IC Interposer with Double-Sided Active Chips", *IEEE/ECTC Proceedings,* May 2013, pp. 1471–1479.
36. P. J., Tzeng, J. H. Lau, C. Zhan, Y. Hsin, P. Chang, Y. Chang, J. Chen, S. Chen, C. Wu, C. Lee, H. Chang, C. Chien, C. Lin, T. Ku, M. Kao, M. Li, J. Cline, K. Saito, and M. Ji, "Process Integration of 3D Si Interposer with Double-Sided Active Chip Attachments", *IEEE/ECTC Proceedings,* May 2013, pp. 86–93.
37. Lau, J. H., P. Tzeng, C. Lee, C. Zhan, M. Li, J. Cline, K. Saito, Y. Hsin, P. Chang, Y. Chang, J. Chen, S. Chen, C. Wu, H. Chang, C. Chien, C. Lin, T. Ku, R. Lo, and M. Kao, "Redistribution Layers (RDLs) for 2.5D/3D IC Integration", *IMAPS Transactions, Journal of Microelectronic Packaging,* Vol. 11, No. 1, First Quarter 2014, pp. 16–24.
38. Lau, J. H., C. Lee, C. Zhan, S. Wu, Y. Chao, M. Dai, R. Tain, H. Chien, C. Chien, R. Cheng, Y. Huang, Y. Lee, Z. Hsiao, W. Tsai, P. Chang, H. Fu, Y. Cheng, L. Liao, W. Lo, and M. Kao, "Low-Cost TSH (Through-Silicon Hole) Interposers for 3D IC Integration", *Proceedings* of *IEEE/ECTC,* May 2014, pp. 290–296.
39. Lau, J. H., C. Lee, C. Zhan, S. Wu, Y. Chao, M. Dai, R. Tain, H. Chien, J. Hung, C. Chien, R. Cheng, Y. Huang, Y. Lee, Z. Hsiao, W. Tsai, P. Chang, H. Fu, Y. Cheng, L. Liao, W. Lo, and M. Kao, "Low-Cost Through-Silicon Hole Interposers for 3D IC Integration", *IEEE Transactions on CPMT,* Vol. 4, No. 9, September 2014, pp. 1407–1419.
40. Hsieh, M. C., S. T. Wu, C. J. Wu, and J. H. Lau, "Energy Release Rate Estimation for Through Silicon Vias in 3-D Integration", *IEEE Transactions on CPMT,* Vol. 4, No. 1, January 2014, pp. 57–65.
41. Lee, C. C., C. S. Wu,, K. S. Kao, C. W. Fang, C. J. Zhan, J. H. Lau, and T. H. Chen, "Impact of high density TSVs on the assembly of 3D-ICs packaging", *Microelectronic Engineering,* Vol. 107, July 2013, pp. 101–106.
42. Che, F., M. Kawano, M. Ding, Y. Han, and S. Bhattacharya, "Co-design for Low Warpage and High Reliability in Advanced Package with TSV-Free Interposer (TFI)", *Proceedings of IEEE/ECTC,* May 2017, pp. 853–861.
43. Hou, S., W. Chen, C. Hu, C. Chiu, K. Ting, T. Lin, W. Wei, W. Chiou, V. Lin, V. Chang, C. Wang, C. Wu, and D. Yu, "Wafer-Level Integration of an Advanced Logic-Memory System Through the Second-Generation CoWoS Technology", *IEEE Transactions on Electron Devices,* October 2017, pp. 4071–4077.
44. Lau, J. H., and G. Tang, "Thermal Management of 3D IC Integration with TSV (Through Silicon Via)", *IEEE/ECTC Proceedings,* May 2009, pp. 635–640.
45. Lau, J. H., "TSV Manufacturing Yield and Hidden Costs for 3D IC Integration", *IEEE/ECTC Proceedings, May* 2010, pp. 1031–1041.
46. Banijamali, B., S. Ramalingam, K. Nagarajan, and R. Chaware, "Advanced Reliability Study of TSV Interposers and Interconnects for the 28 nm Technology FPGA", *Proceedings of IEEE/ECTC,* May 2011, pp. 285–290.
47. Kim, N., D. Wu, D. Kim, A. Rahman, and P. Wu, "Interposer Design Optimization for High Frequency Signal Transmission in Passive and Active Interposer using Through Silicon Via (TSV)", *IEEE/ECTC Proceedings,* May 2011, pp. 1160–1167.
48. Banijamali, B., S. Ramalingam, N. Kim, C. Wyland, N. Kim, D. Wu, J. Carrel, J. Kim, and Paul Wu, "Ceramics versus low-CTE Organic packaging of TSV Silicon Interposers", *IEEE/ECTC Proceedings,* May 2011, pp. 573–576.
49. Chaware, R., K. Nagarajan, and S. Ramalingam, "Assembly and Reliability Challenges in 3D Integration of 28 nm FPGA Die on a Large High Density 65 nm Passive Interposer", *Proceedings of IEEE/ECTC,* May 2012, San Diego, CA, pp. 279–283.
50. Banijamali, B., S. Ramalingam, H. Liu and M. Kim, "Outstanding and Innovative Reliability Study of 3D TSV Interposer and Fine Pitch Solder Micro-bumps", *Proceedings of IEEE/ECTC,* San Diego, CA, May 2012, pp. 309–314.

51. Kim, N., D. Wu, J. Carrel, J. Kim, and P. Wu, "Channel Design Methodology for 28 Gb/s SerDes FPGA Applications with Stacked Silicon Interconnect Technology", *IEEE/ECTC Proceedings,* May 2012, pp. 1786–1793.
52. Banijamali, B., C. Chiu, C. Hsieh, T. Lin, C. Hu, S. Hou, et al., "Reliability evaluation of a CoWoS-enabled 3D IC package," *IEEE/ECTC Proceedings*, May 2013, pp. 35–40.
53. Hariharan, G., R. Chaware, L. Yip, I. Singh, K. Ng, S. Pai, M. Kim, H. Liu, and S. Ramalingam, "Assembly Process Qualification and Reliability Evaluations for Heterogeneous 2.5D FPGA with HiCTE Ceramic", *IEEE/ECTC Proceedings,* May 2013, pp. 904–908.
54. Kwon, W., M. Kim, J. Chang, S. Ramalingam, L. Madden, G. Tsai, S. Tseng, J. Lai, T. Lu, and S. Chin, "Enabling a Manufacturable 3D Technologies and Ecosystem using 28 nm FPGA with Stack Silicon Interconnect Technology", *IMAPS Proceedings of International Symposium on Microelectronics*, Orlando, FL, October 2013, pp. 217–222.
55. Banijamali, B., T. Lee, H. Liu, S. Ramalingam, I. Barber, J. Chang and M. Kim, and L. Yip, "Reliability Evaluation of an Extreme TSV Interposer and Interconnects for the 20 nm Technology CoWoS IC Package", *IEEE/ECTC Proceedings,* May 2015, pp. 276–280.
56. Hariharan, G., R. Chaware, I. Singh, J. Lin, L. Yip, K. Ng, and S. Pai, "A Comprehensive Reliability Study on a CoWoS 3D IC Package", *IEEE/ECTC Proceedings,* May 2015, pp. 573–577.
57. Chaware, R., G. Hariharan, J. Lin, I. Singh, G. O'Rourke, K. Ng, S. Pai, C. Li, Z. Huang, and S. Cheng, "Assembly Challenges in Developing 3D IC Package with Ultra High Yield and High Reliability", *IEEE/ECTC Proceedings,* May 2015, pp. 1447–1451.
58. Xu, J., Y. Niu, S. Cain, S. McCann, H. Lee, G. Ahmed, and S. Park, "The Experimental and Numerical Study of Electromigration in 2.5D Packaging", *IEEE/ECTC Proceedings,* May 2018, pp. 483–489.
59. McCann, S., H. Lee, G. Ahmed, T. Lee, S. Ramalingam, "Warpage and Reliability Challenges for Stacked Silicon Interconnect Technology in Large Packages", *IEEE/ECTC Proceedings,* May 2018, pp. 2339–2344.
60. Wang, H., J. Wang, J. Xu, V. Pham, K. Pan, S. Park, H. Lee, and G. Ahmed, "Product Level Design Optimization for 2.5D Package Pad Cratering Reliability during Drop Impact", *IEEE/ECTC Proceedings,* May 2019, pp. 2343–2348.
61. http://press.xilinx.com/2013–10-20-Xilinx-and-TSMCReach-Volume-Production-on-all-28nm-CoWoS-based-All-Programmable-3D-IC-Families.
62. Xie, J., H. Shi, Y. Li, Z. Li, A. Rahman, K. Chandrasekar, D. Ratakonda, M. Deo, K. Chanda, V. Hool, M. Lee, N. Vodrahalli, D. Ibbotson, and T. Verma, "Enabling the 2.5D Integration", *Proceedings of IMAPS International Symposium on Microelectronics*, September 2012, San Diego, CA, pp. 254–267.
63. Li, Z., H. Shi, J. Xie, and A. Rahman, "Development of an Optimized Power Delivery System for 3D IC Integration with TSV Silicon Interposer", *Proceedings of IEEE/ECTC*, May 2012, pp. 678–682.
64. https://semiwiki.com/semiconductor-manufacturers/tsmc/290560-highlights-of-the-tsmc-technology-symposium-part-2/.
65. Chen, W., C. Lin, C. Tsai, H. Hsia, K. Ting, S. Hou, C. Wang, and D. Yu, "Design and Analysis of Logic-HBM2E Power Delivery System on CoWoS® Platform with Deep Trench Capacitor", *IEEE/ECTC Proceedings,* May 2020, pp. 380–385.
66. Bhuvanendran, S., N. Gourikutty, K. Chua, J. Alton, J. Chinq, R. Umralkar, V. Chidambaram1, and S. Bhattacharya, "Non-destructive fault isolation in through-silicon interposer based system in package", *IEEE/EPTC Proceedings*, December 2020, pp. 281–285.
67. Sirbu, B., Y. Eichhammer, H. Oppermann, T. Tekin, J. Kraft, V. Sidorov, X. Yin, J. Bauwelinck, C. Neumeyr, and F. Soares, "3D Silicon Photonics Interposer for Tb/s Optical Interconnects in Data Centers with double-side assembled active components and integrated optical and electrical Through Silicon Via on SOI", *IEEE/ECTC Proceedings*, May 2020, pp. 1052–1059.
68. Tanaka, M., S. Kuramochi, T. Dai, Y. Sato, and N. Kidera, "High Frequency Characteristics of Glass Interposer", *IEEE/ECTC Proceedings*, May 2020, pp. 601–610.
69. Iwai, T., T. Sakai, D. Mizutani, S. Sakuyama, K. Iida, T. Inaba, H. Fujisaki, A. Tamura, and Y. Miyazawa, "Multilayer Glass Substrate with High Density Via Structure for All Inorganic Multi-chip Module", *IEEE/ECTC Proceedings*, May 2020, pp. 1952–1957.

70. Ding, Q., H. Liu, Y. Huan, and J. Jiang, "High Bandwidth Low Power 2.5D Interconnect Modeling and Design", *IEEE/ECTC Proceedings*, May 2020, pp. 1832–1837.
71. Kim, M., H. Liu, D. Klokotov, A. Wong, T. To, and J. Chang, "Performance Improvement for FPGA due to Interposer Metal Insulator Metal Decoupling Capacitors (MIMCAP)", *IEEE/ECTC Proceedings*, May 2020, pp. 386–392.
72. Bhuvanendran, S., N. Gourikutty, Y. Chow, J. Alton, R. Umralkar, H. Bai, K. Chua, and S. Bhattacharya, "Defect Localization in Through-Si-Interposer Based 2.5DICs", *IEEE/ECTC Proceedings*, May 2020, pp. 1180–1185.
73. Hsiao, Y., C. Hsu, Y. Lin, and C. Chien, "Reliability and Benchmark of 2.5D Non-molding and Molding Technologies", *IEEE/ECTC Proceedings*, May 2019, pp. 461–466.
74. Pares, G., J. Michel, E. Deschaseaux, P. Ferris, A. Serhan, and A. Giry, "Highly Compact RF Transceiver Module using High Resistive Silicon Interposer with Embedded Inductors and Heterogeneous Dies Integration", *IEEE/ECTC Proceedings*, May 2019, pp. 1279–1286.
75. Okamoto, D., Y. Shibasaki, D. Shibata, and T. Hanada, F. Liu, M. Kathaperumal, and R. Tummala, "Fabrication and Reliability Demonstration of 3 μm Diameter Photo Vias at 15 μm Pitch in Thin Photosensitive Dielectric Dry Film for 2.5 D Glass Interposer Applications", *IEEE/ECTC Proceedings*, May 2019, pp. 2112–2116.
76. Ravichandran, S., S. Yamada, G. Park, H. Chen, T. Shi, C. Buch, F. Liu, V. Smet, V. Sundaram, and R. Tummala, "2.5D Glass Panel Embedded (GPE) Packages with Better I/O Density, Performance, Cost and Reliability than Current Silicon Interposers and High-Density Fan-Out Packages", *IEEE/ECTC Proceedings*, May 2018, pp. 625–630.
77. Wang, J., Y. Niu, S. Park, A. Yatskov, "Modeling and design of 2.5D package with mitigated warpage and enhanced thermo-mechanical reliability", *IEEE/ECTC Proceedings*, May 2018, pp. 2471–2477.
78. Okamoto, D., Y. Shibasaki, D. Shibata, T. Hanada, F. Liu, V. Sundaram, R. Tummala, "An Advanced Photosensitive Dielectric Material for High-Density RDL with Ultra-Small Photo-Vias and Ultra-Fine Line/Space in 2.5D Interposers and Fan-Out Packages.
79. Cai1, H., S. Ma, J. Zhang, W. Xiang, W. Wang, Y. Jin, J. Chen, L. Hu, and S. He, "Thermal and Electrical characterization of TSV interposer embedded with Microchannel for 2.5D integration of GaN RF devices", *IEEE/ECTC Proceedings*, May 2018, pp. 2150–2156.
80. Hong, J., K. Choi, D. Oh, S. Shao, H. Wang, Y. Niu, and V. Pham, "Design Guideline of 2.5D Package with Emphasis on Warpage Control and Thermal Management", *IEEE/ECTC Proceedings*, May 2018, pp. 682–692.
81. Nair, C., B. DeProspo, H. Hichri, M. Arendt, F. Liu, V. Sundaram, and R. Tummala, "Reliability Studies of Excimer Laser-Ablated Microvias Below 5 Micron Diameter in Dry Film Polymer Dielectrics for Next Generation, Panel-Scale 2.5D Interposer RDL", *IEEE/ECTC Proceedings*, May 2018, pp. 1005–1009.
82. Lai, C., H. Li, S. Peng, T. Lu, and S. Chen, "Warpage Study of Large 2.5D IC Chip Module", *IEEE/ECTC Proceedings*, May 2017, pp. 1263–1268.
83. Shih, M., C. Hsu, Y. Chang, K. Chen, I. Hu, T. Lee, D. Tarng, and C. Hung, "Warpage Characterization of Glass Interposer Package Development", *IEEE/ECTC Proceedings*, May 2017, pp. 1392–1397.
84. Agrawal, A., S. Huang, G. Gao, L. Wang, J. DeLaCruz, and L. Mirkarimi, "Thermal and Electrical Performance of Direct Bond Interconnect Technology for 2.5D and 3D integrated Circuits", *IEEE/ECTC Proceedings*, May 2017, pp. 989–998.
85. Choi, S., J. Park, D. Jung, J. Kim, H. Kim, K. Kim, "Signal Integrity Analysis of Silicon/Glass/Organic Interposers for 2.5D/3D Interconnects", *IEEE/ECTC Proceedings*, May 2017, pp. 2139–2144.
86. Wang, X., Q. Ren, and M. Kawano, "Yield Improvement of Silicon Trench Isolation for One-Step TSV", *IEEE/EPTC Proceedings*, December 2020, pp. 22–26.
87. Ren, Q., W. Loh, S. Neo, and K. Chui, "Temporary Bonding and De-bonding Process for 2.5D/3D Applications", *IEEE/EPTC Proceedings*, December 2020, pp. 27–31.
88. Chuan, P. and S. Tan, "Glass Substrate Interposer for TSV-integrated Surface Electrode Ion Trap", pp. 262–265. *IEEE/EPTC Proceedings*, December 2020, pp. 262–265.
89. Loh, W. and K. Chui, "Wafer Warpage Evaluation of Through Si Interposer (TSI) with Different Temporary Bonding Materials", *IEEE/EPTC Proceedings*, December 2020, pp. 268–272.

90. Lau, J. H., *Heterogeneous Integrations*, Springer, New York, 2019.
91. Lau, J. H., *Fan-Out Wafer-Level Packaging*, Springer, New York, 2018.
92. Lau, J. H., *3D IC Integration and Packaging*, McGraw-Hill, New York, 2016.
93. Lau, J. H., *Through-Silicon Via (TSV) for 3D Integration*, McGraw-Hill, New York, 2013.
94. Lau, J. H., *Reliability of RoHS compliant 2D & 3D IC Interconnects*, McGraw-Hill, New York, 2011.
95. Lau, J. H., "Overview and Outlook of 3D IC Packaging, 3D IC Integration, and 3D Si Integration", *ASME Transactions, Journal of Electronic Packaging*, December 2014, Vol. 136, Issue 4, pp. 1–15.
96. Lau, J. H., "Overview and Outlook of TSV and 3D Integrations", *Journal of Microelectronics International*, Vol. 28, No. 2, 2011, pp. 8–22.
97. Lau, J. H., "Critical Issues of 3D IC Integrations", *IMAPS Transactions, Journal of Microelectronics and Electronic Packaging, First Quarter Issue*, 2010, pp. 35–43.
98. Lau, J. H., "Design and Process of 3D MEMS Packaging", *IMAPS Transactions, Journal of Microelectronics and Electronic Packaging, First Quarter Issue*, 2010, pp. 10–15.
99. Lau, J. H., Lee, R., Yuen, M., and Chan, P., "3D LED and IC Wafer Level Packaging", *Journal of Microelectronics International,* Vol. 27, Issue 2, 2010, pp. 98–105.
100. Lau, J. H., "3D IC Integration with a Passive Interposer", *Proceedings of SMTA International Conference*, Chicago, IL, September 2014, pp. 11–19.
101. Lau, J. H., "The Role and Future of 2.5D IC Integration", *IPC APEX EXPO Proceedings*, Las Vegas, NE, March 2014, pp. 1–14.
102. Chen, J., J. H. Lau, T. Hsu, C. Chen, P. Tzeng, P. Chang, C. Chien, Y. Chang, S. Chen, Y. Hsin, S. Liao, C. Lin, T. Ku, and M. Kao, "Challenges of Cu CMP of TSVs and RDLs Fabricated from the Backside of a Thin Wafer", *IEEE International 3D Systems Integration Conference*, San Francisco, CA, October 2013, pp. 1–5.
103. Lau, J. H., H. C. Chien, S. T. Wu, Y. L. Chao, W. C. Lo, and M. J. Kao, "Thin-Wafer Handling with a Heat-Spreader Wafer for 2.5D/3D IC Integration", *Proceedings of the 46th IMAPS International Symposium on Microelectronics,* Orlando, FL, October 2013, pp. 389–396.
104. Hung, J. F., J. H. Lau, P. Chen, S. Wu, S. Hung, S. Lai, M. Li, S. Sheu, Z. Lin, C. Lin, W. Lo, and M. Kao, "Electrical Performance of Through-Silicon Vias (TSVs) for High-Frequency 3D IC Integration Applications", *Proceedings of the 45th IMAPS International Symposium on Microelectronics,* September 2012, pp. 1221–1228.
105. Lau, J. H., "Supply Chains for 3D IC Integration Manufacturing", *Proceedings of IEEE Electronic Materials and Packaging Conference*, December 2012, pp. 72–78.
106. Lau, J. H., S. T. Wu, and H. C. Chien, "Thermal-Mechanical Responses of 3D IC Integration with a Passive TSV Interposer", *IEEE EuroSime Proceedings, Chapter 5: Reliability Modeling*, Lisbon, Portugal, April 2012, pp. 1/8 – 8/8.
107. Lau, J. H., "The Most Cost-Effective Integrator (TSV Interposer) for 3D IC Integration System-in-Package (SiP)", *ASME Paper no. InterPACK2011–52189*, Portland, OR, July 2011.
108. Lau, J. H., and X. Zhang, "Effects of TSV Interposer on the Reliability of 3D IC Integration SiP", *ASME Paper no. InterPACK2011-52205*, Portland, OR, July 2011.
109. Lau, J. H., "State-of-the-art and Trends in Through-Silicon Via (TSV) and 3D Integrations, *ASME Paper no. IMECE2010-37783*.
110. Lau, J. H., Y. S. Chan, S. W. R. Lee, "Thermal-Enhanced and Cost-Effective 3D IC Integration with TSV Interposers for High-Performnnance Applications", *ASME Paper no. IMECE2010-40975*.
111. Lau, J. H., "Evolution and Outlook of TSV and 3D IC/Si Integration" *IEEE/EPTC Proceedings*, Singapore, December 2010, pp. 560–570.

第 7 章

3D IC 集成和 3D IC 封装

7.1 引言

3D 集成[1-70]至少包括 3D IC 集成和 3D IC 封装两个概念。首先,顾名思义,3D IC 集成和 3D IC 封装都是在垂直方向堆叠芯片。在本书中,3D IC 集成和 3D IC 封装之间的主要区别在于 3D IC 集成使用了硅通孔(through silicon via,TSV)[36-38],但 3D IC 封装却没有。

7.2 3D IC 封装

有许多不同种类的 3D IC 封装。图 7.1 和图 7.2 仅示意性地展示了一些。图 7.1a 展示了使用引线键合技术的存储芯片堆叠。图 7.1b 为两个芯片面对面通过焊料凸点倒装键合在一起,然后再用引线键合实现下一层互连。图 7.1c 展示了两个背对背键合的芯片;底部芯片通过焊料凸点倒装键合到基板上,顶部芯片通过引线键合连接到基板上。图 7.1d 的两个芯片面对面通过焊料凸点连接的倒装芯片,顶部芯片再通过焊球连接到基板上。图 7.2a 展示了应用处理器芯片组(应用处理器 + 存储器)的封装堆叠(package-on-package,PoP)。可以看到,在底部封装中,应用处理器通过焊料凸点倒装键合到积层封装基板并完成底部填充;顶部封装则用于封装存储器,通常采用交叉堆叠和引线键合的方式连接到无芯板有机基板上。图 7.2b 显示了应用处理器芯片组的另一种 PoP 结构。在底部封装中,应用处理器通过再布线层(redistribution-layer,RDL)扇出,然后通过焊球连接在印制电路板(printed circuit board,PCB)上。用于倒装芯片的晶圆凸点成型工序、积层封装基板和底部填充均被省略。上层封装保持不变,仍用于封装存储芯片。

7.2.1 3D IC 封装——引线键合式存储芯片堆叠

图 7.3、图 7.4 和图 7.5 为不同种类的采用引线键合的 3D 存储芯片堆叠。目前,超过 50% 的键合引线已经从 Au 换成 Cu 甚至部分 Ag。如图 7.3、图 7.4 和图 7.5 所示,在引线键合技术中,所有引线都沿芯片的周边(一排或者两排)进行键合。图 7.6 显示了三星为苹果的 iPhone 制造的八颗闪存芯片堆叠的截面。包括基板在内的封装厚度为 0.93mm,芯片堆叠高度为 0.67mm。芯片厚度从 55μm 到 70μm 不等,最厚的芯片位于底部。图 7.7 和图 7.8 所示为三星的固态硬盘

(solid state drives，SSD)。可以看到，16颗48层V-NAND 3D闪存芯片采用引线键合技术堆叠，每个芯片的厚度只有40μm。堆叠芯片采用无芯板有机封装基板封装，通过面阵列焊球连接到PCB上。

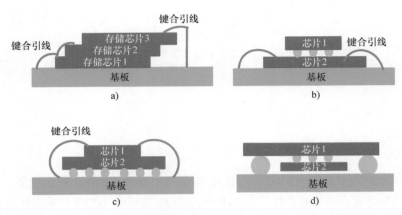

图7.1 3D IC封装：a）使用引线键合的堆叠存储芯片；b）使用引线键合到基板的面对面键合芯片；c）使用引线键合到基板的背对背键合芯片；d）使用焊料凸点/焊球与基板连接的面对面键合芯片

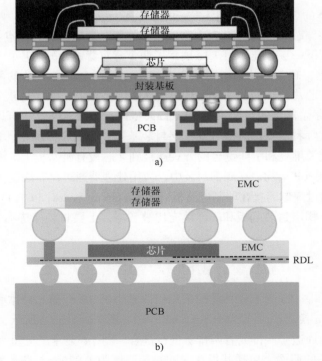

图7.2 3D IC封装：a）采用倒装芯片技术的PoP；b）采用扇出型封装技术的PoP

图 7.3 利用引线键合的三颗芯片堆叠

图 7.4 通过引线键合技术实现的两颗芯片堆叠

图 7.5 两排引线键合的芯片堆叠

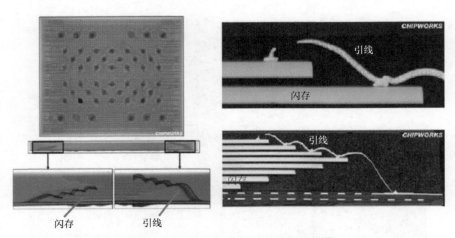

图 7.6 八颗闪存芯片通过引线键合堆叠

第 7 章　3D IC 集成和 3D IC 封装　291

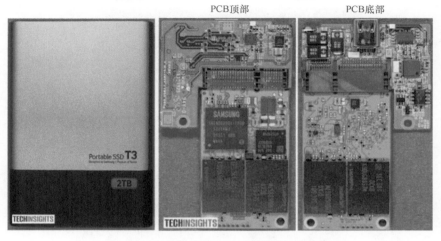

图 7.7　三星的 SSD：PCB 的顶部和底部

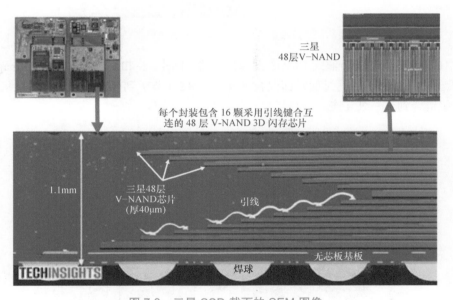

图 7.8　三星 SSD 截面的 SEM 图像

7.2.2　3D IC 封装——面对面键合后引线键合到基板

图 7.9 所示为索尼的 PlayStation CXD53135GG。可以看出，它是一个 5 颗芯片堆叠，同时具有引线键合和焊料凸点倒装互连。双倍数据速率类型 2（double data rate type-2，DDR2）同步动态随机存取存储器（synchronous dynamic random access memory，SDRAM）和隔离芯片通过引线键合进行堆叠。而 1GB 宽 I/O SDRAM 和处理器芯片通过面对面的焊料凸点进行倒装互连，然后从处理器芯片通过引线键合到下一级互连。

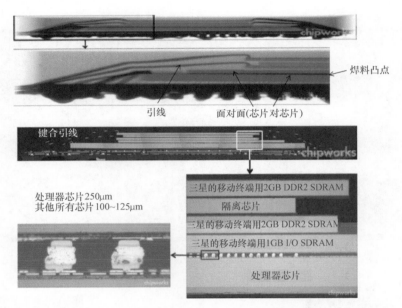

图 7.9 索尼的 5 颗芯片堆叠（面对面键合后通过引线键合到下一级互连）

图 7.10 所示为索尼的背照式 CMOS 图像传感器（backside illuminated CMOS image sensor，BI-CIS）[1, 2]。BI-CIS 芯片通过面对面混合键合（见 8.5.3 节）到处理器芯片，然后从处理器芯片通过引线键合到下一级互连。

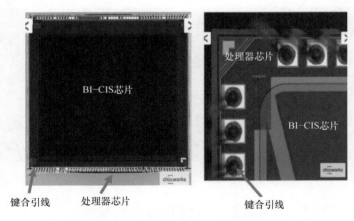

图 7.10 索尼带有键合引线面向下一级互连的面对面键合 [1]

7.2.3 3D IC 封装——背对背键合后引线键合到基板

图 7.11 所示为英特尔为 iPhone XR 提供的第二重要芯片组，调制解调器芯片组。可以看出，基带应用处理器（application processor，AP）通过焊料凸点倒装键合到 3 层嵌入式布线基板

(embedded trace substrate, ETS)上。DRAM 芯片贴装在 AP 的背面，并通过引线键合到 ETS 上。这是一个使用背对背键合后引线键合到基板上的例子。

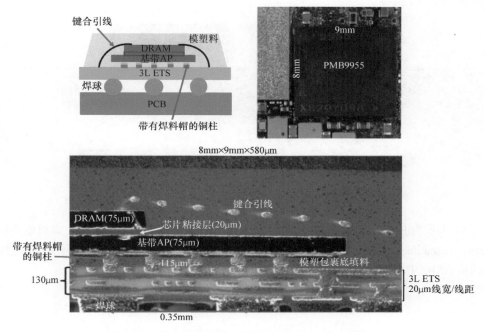

图 7.11 英特尔带有键合引线至 ETS 的背对背键合

7.2.4 3D IC 封装——面对面键合后通过凸点/焊球到基板上

图 7.12 所示为面对面键合后通过焊球到下一级互连。可以看出，存储器通过逻辑芯片上的 C2 凸点（带有 AuSn 焊料帽的铜柱）进行键合。然后，逻辑芯片通过 SnAgCu 焊球连接到基板上[3]。

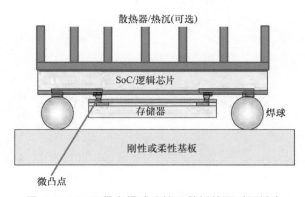

图 7.12 IME 带有焊球连接至基板的面对面键合

图 7.13 所示为佐治亚理工学院的 3D IC 封装[4]。可以看出，芯片与布线芯片通过芯片上焊料凸点和布线芯片的扇出 RDL 进行面对面键合。然后，信号通过玻璃通孔（through-glass via，TGV）连接到焊点/焊球。

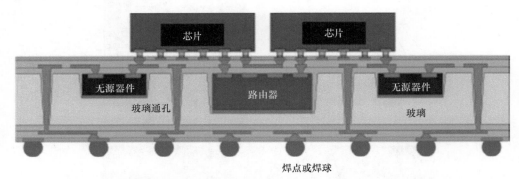

图 7.13 佐治亚理工学院使用 TMV 及焊球至下一互连层的面对面键合[4]

图 7.14 所示为 Fraunhofer[5] 的芯片对芯片、面对面键合技术。可以看出，微机电系统（micro-electro-mechanical system，MEMS）芯片面对面键合至 ASIC 芯片上，信号可以通过焊球或键合引线从 ASIC 中引出。

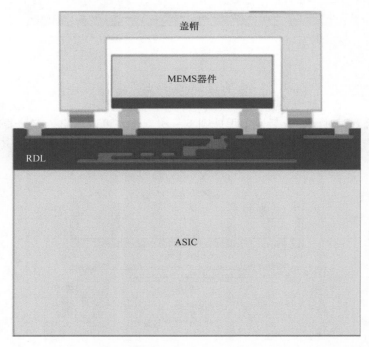

图 7.14 IZM 使用键合引线或焊球从 ASIC 至下一互连层的面对面键合[5]

图 7.15 所示为通过 C4 凸点到基板的面对面键合。可以看出，ASIC 芯片面对面（通过微凸点 + 焊料帽）键合到 FPGA 上[6]。然后，将 FPGA 通过 C4 焊点键合到封装基板上。

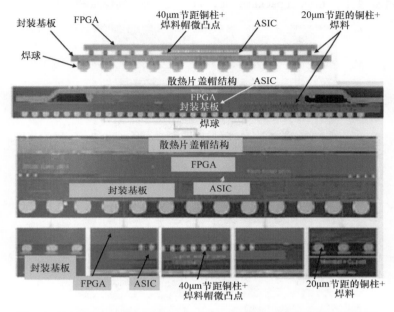

图 7.15　ASIC 与 FPGA 利用 C2 凸点面对面键合后连接至封装基板

图 7.16 所示为 Amkor 的 Double-POSSUM 封装[7]。可以看出，封装实际上是由两级嵌套芯片定义的。三个子芯片倒装互连到较大的母芯片上，然后再连接到最大的祖母芯片上，再将祖母芯片倒装互连到封装基板上。子、母芯片之间的凸点是 C2 凸点，母、祖母芯片之间以及祖母芯片与封装基板之间采用的是 C4 凸点。

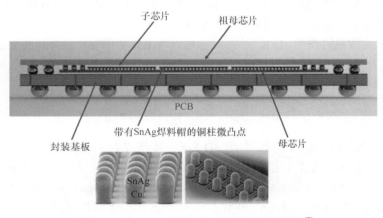

图 7.16　Amkor 的 Double-POSSUM 封装[7]

图 7.17 所示为 3D 嵌入式 MEMS 器件及其配对控制芯片的俯视图和截面图[8]。功能性 MEMS 和 ASIC 器件的厚度约为 600μm。此外，还包括一个近似面对面的芯片到芯片组装，其表面仅由焊点（来自倒装芯片 B）和单层 RDL（来自芯片 A 的扇出）隔开。互连路径是从 PCB、焊球、封装通孔（through package via，TPV）、RDL 到焊点和芯片。

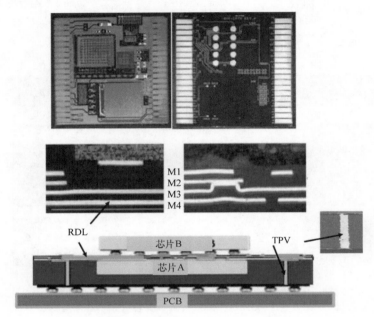

图 7.17 面对面键合通过 TPV 至焊球连接至 PCB

7.2.5 3D IC 封装——面对背

图 7.18 所示为一个 3D 扇出晶圆级封装[9, 10]，对芯片 1 和芯片 2 进行模塑和 RDL 制作。芯片 3 通过焊料凸点面朝下倒装到芯片 1 和芯片 2 扇出封装的背面。互连路径是从 PCB、焊球到下层 RDL、芯片 1 和芯片 2、TMV 以及上层 RDL 和芯片 3。

图 7.19 所示为矽品科技的 3D IC 封装[11]。可以看出，顶部芯片键合到底部芯片的背面。顶部芯片和底部芯片之间的互连是通过扇出 RDL、微凸点（铜柱 + 焊料帽）和 TMV 实现的。芯片和封装基板之间通过 C4 焊点实现互连。

7.2.6 3D IC 封装——SiP 中的埋入式芯片（面对面）

图 7.20 所示为带有埋入式芯片的 3D SiP。从图 7.20a 可以看出，传统的 SiP 基板（3.5mm×3.5mm）上并排有 5 个芯片（IC 和传感器）（U1~U5）。图 7.20b 所示为采用 Fujikura 的晶圆和板级嵌入式封装（wafer and board level embedded package，WABE）技术[12]将 U1 芯片埋入到柔性基板（2.5mm×2.5mm）中，其他 4 个芯片位于基板顶部表面（传感器芯片和嵌入式 IC 是面对面放置的），封装尺寸减少了 50%。

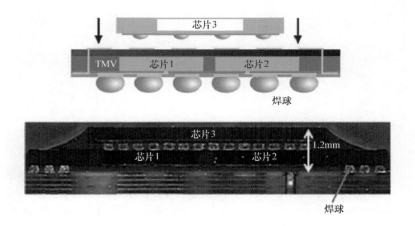

图 7.18 面对背键合后通过 TMV 连接到 PCB 上的焊球[9]

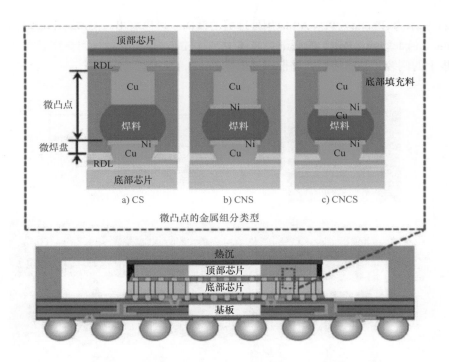

图 7.19 面对背键合后通过扇出型 RDL、微凸点和 TMV 连接下一互连层[11]

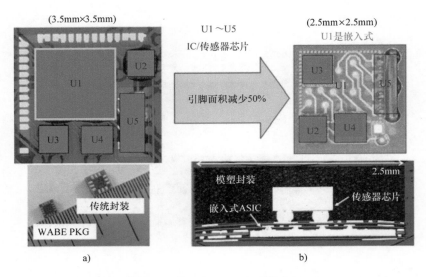

图 7.20　SiP 中埋入式芯片（面对面型）

7.2.7　3D IC 封装——采用倒装芯片技术的 PoP

图 7.21 所示为 iPhone 6 Plus A9 应用处理器的截面。可以看出，A9 应用处理器是在封装堆叠（package-on-package，PoP）的底部封装体中，通过 C4 焊点的批量回流完成在 2-2-2 有机积层封装基板上的倒装芯片连接，并进行底部填充。

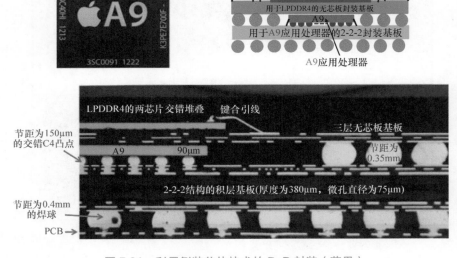

图 7.21　利用倒装芯片技术的 PoP 封装（苹果）

如图 7.22 所示，Amkor[13]首先研究了基板上通过非导电浆料（TC-nonconductive paste，TC-NCP）底部填充料与带有 C2 凸点芯片的大压力热压键合（thermocompression bonding，TCB），已用于高通骁龙应用处理器的组装，该处理器用于三星 Galaxy 智能手机。封装基板，也就是模塑芯板嵌入式封装（molded core embedded package，MCeP）由 Shinko 提供。NCP 底部填充料可以通过旋涂、针头滴涂或真空辅助等方式实施。

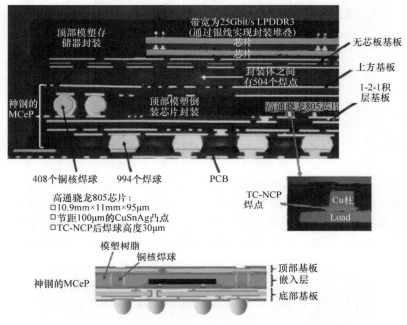

图 7.22　带有倒装芯片技术的 PoP 封装（高通）[13]

图 7.23 所示为恩智浦的 3D IC 封装[14]。可以看出，底部封装体在 4 层的 ETS 上封装了一颗处理器，顶部封装体封装存储芯片。顶部和底部封装体之间的互连通过双层堆叠焊球和无 TSV 转接板实现。

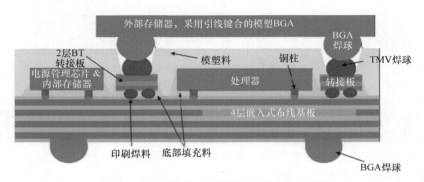

图 7.23　带有倒装芯片技术的 PoP 封装（恩智浦）

图 7.24 所示为 Shinko 的 3D IC 封装[15]。可以看出，底部封装体封装存储芯片，顶部封装体封装 ASIC 芯片。顶部和底部封装体之间的互连通过 Shinko 的铜芯焊球实现。

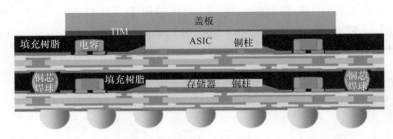

图 7.24　带有倒装芯片技术的 PoP 封装（Shinko）[15]

7.2.8　3D IC 封装——采用扇出技术的 PoP

为了获得比图 7.21、图 7.22、图 7.23 和图 7.24 所示 PoP 更薄的封装体，图 7.25 所示为星科金朋提出的 3D 扇出型 PoP[16] 的截面扫描电子显微镜（scanning electron microscope，SEM）图像。它包含扇出型封装的底部封装体（即取代图 7.21、图 7.22、图 7.23、图 7.24 中所示的焊点倒装芯片封装）和存储器封装的顶部封装体。可以看到：①扇出型封装厚度只有 450μm，②扇出型封装中是应用处理器，③顶部封装体厚度为 520μm，封装了采用引线键合的存储芯片，④互连的路径是从 PCB、焊球、RDL 到处理器，以及从焊球、RDL 到存储芯片。

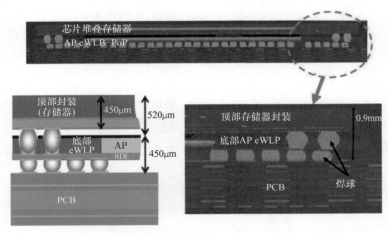

图 7.25　带有扇出型技术的 PoP 封装（星科金朋）[16]

图 7.26 所示为 iPhone 手机中 AP 芯片组的 PoP 示意图和实物。AP（A12）和移动 DRAM 的 PoP 采用台积电的 InFO-WLP 技术制造[17, 18]。基本上，用于所有 AP（A10、A11、A12、A13 和 A14）的 PoP 工艺平台非常相似。不过，由于台积电采用 7nm 工艺制造 A12，即使其中包含更多功能（例如人工智能等应用），其芯片尺寸也比 A11 略小。为了获得更好的电性能，少量的集成无源器件（integrated passive device，IPD）通过焊点倒装至图 7.26 所示的扇出型封

装底部。扇出型封装共有三层 RDL，最小金属线宽和线距为 8μm；封装的焊球节距为 0.35mm。近期，台积电的 5nm 工艺已用于 A14 处理器，图 7.27 所示为台积电的另一种 PoP[19]。可以看到底部封装体采用扇出型封装技术来封装芯粒（chiplet）。在底部封装体和顶部封装体（用于存储芯片）之间，有一个层压的无 TSV 转接板。

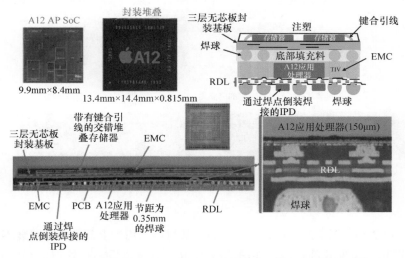

图 7.26 带有扇出型技术的 PoP 封装（苹果/台积电）[17]

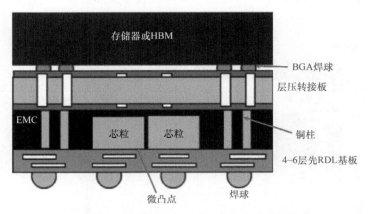

图 7.27 带有扇出型技术的 PoP 封装（台积电）[19]

图 7.28 所示为三星智能手表中的 PoP。上层封装体包含了存储器嵌入式封装堆叠（embedded package-on-package，ePoP），由 2 颗 DRAM 芯片、2 颗 NAND 闪存和 1 颗 NAND 控制芯片组成。这些存储芯片通过引线键合至 3 层的无芯板封装基板上，如图 7.28 所示，上封装体尺寸为 8mm×9.5mm×1mm。底部封装体采用三星的扇出型板级封装技术并排封装 AP 和电源管理芯片（power management IC，PMIC）。AP 芯片尺寸为 5mm×3mm，PMIC 芯片尺寸为 3mm×3mm。关键工艺步骤[20] 是首先在 PCB 上制作空腔，然后将芯片放置在空腔上并层压 EMC，随后将其粘接到支撑片上，制备 RDL 并植球。

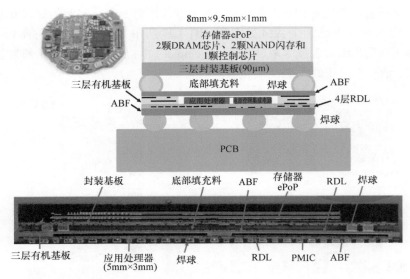

图 7.28 带有扇出型技术的 PoP 封装（三星）[20]

与图 7.25、图 7.26、图 7.27 和图 7.28 中的底部封装采用先上晶扇出型封装不同，IME 采用后上晶（先 RDL）实现底部封装，如图 7.29 所示[21]。可以看到，处理器和先 RDL 基板之间有微凸点。此外，与图 7.27 和图 7.28 不同，底部封装和顶部封装之间没有焊球（用于存储芯片）。

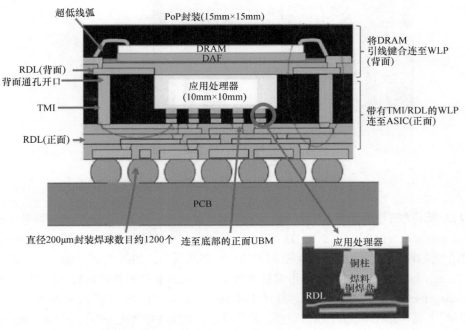

图 7.29 带有扇出型技术的 PoP 封装（IME）[21]

7.2.9 总结和建议

一些重要的结论和建议总结如下:

1) 3D IC 封装(无 TSV)是无晶圆厂设计公司的重要竞争方向,主要由外包半导体组装和测试公司(outsourced semiconductor assembly and test,OSAT)制造。如英特尔、IBM、三星、海力士、美光、恩智浦、英飞凌等公司也制造其自己的 3D IC 封装产品。最近,甚至台积电、三星等晶圆代工厂也开始为其先进工艺节点进行 3D IC 封装的研发。

2) 3D IC 封装种类非常繁多(尽情发挥您的想象力),一切皆有可能。

7.3 3D IC 集成

如前所述,3D IC 集成是将芯片通过 TSV 进行三维堆叠,如图 7.30 所示。从图 7.30a 可以看出,DRAM 和基本逻辑芯片是通过 TSV、微凸点和底部填充料堆叠的。图 7.30b 为一个高带宽存储器(通过微凸点)连接在带有 TSV 的逻辑芯片上。图 7.30c 所示为一个无凸点芯片混合键合在另一个带有 TSV 的无凸点芯片上。

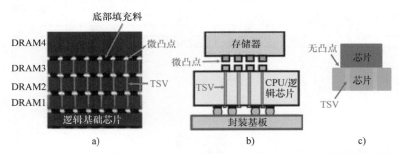

图 7.30 3D IC 集成:a) 带有微凸点和 TSV 的存储器堆叠;b) 存储器堆叠在带有 TSV 的逻辑芯片上;c) 带有 TSV 的芯片对芯片的无凸点混合键合

7.3.1 3D IC 集成——HBM 标准

图 7.31 所示为高带宽存储器(high bandwidth memory,HBM)、HBM2 和 HBM2E。它们与 SoC 一起使用,并且是由宽带蜂窝网络的第五代技术标准(5th generation technology standard for broadband cellular networks,5G)和人工智能(artificial intelligence,AI)驱动的高性能计算(high-performance computing,HPC)应用的必备组成部分[22],如图 7.32 所示。目前全球只有三星和海力士在大规模量产 HBM 芯片/模组。最近美光也在尝试研发。

HBM 比第四代双倍数据速率同步动态随机存储器(double data rate 4,DDR4)或第五代图形用双倍速率同步动态随机存储器(graphics double data rate 5,GDDR5)的功耗更低,但带宽更高,芯片更小,因此对显卡供应商而言很有吸引力。HBM 技术的工作原理是将存储芯片垂直堆叠在一起,存储芯片通过 TSV 和微凸点互连。此外,每个芯片有两个 128 位通道,HBM 的总线比其他类型的 DRAM 更宽。第一颗 HBM 立方于 2013 年由海力士生产。

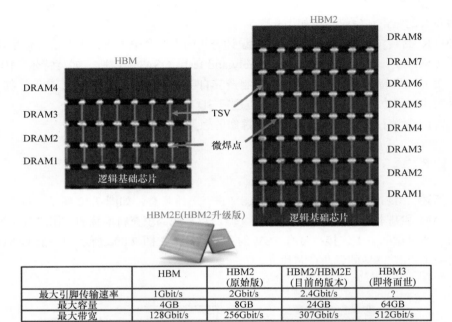

图 7.31 HBM、HBM2、HBM2E、HBM3

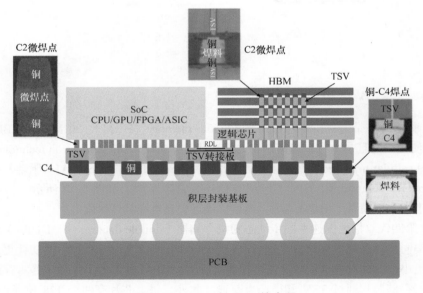

图 7.32 HPC 中 HBM 的应用

HBM2 于 2016 年首次亮相,2018 年 12 月 JEDEC 更新了 HBM2 标准。更新后的标准通常称为 HBM2 和 HBM2E(以表示与初始 HBM2 标准的差别)。HBM2 标准允许每个堆栈最多容纳 12 颗裸片,最大容量为 24 GB。该标准还将内存带宽固定为 307Gbit/s,通过 1024 位内存接口交付,每个堆栈由 8 个独立的通道分隔。最初,HBM2 标准要求堆栈中最多有 8 颗裸片(与 HBM 一样),总带宽为 256Gbit/s。

虽然尚未推出,但 HBM3 标准目前正在讨论中。根据 Ars Technica 的报告,HBM3 有望支持高达 64GB 的容量和高达 512Gbit/s 的速度。HBM3 将在每个堆栈中提供更多的裸片,并在相似的功率预算下将每颗裸片的密度提高 2 倍以上。

7.3.2 3D IC 集成——HBM 组装

如图 7.33 所示,三星和海力士都采用带 C2 凸点(铜柱 + 焊料帽)与 DRAM 基于非导电膜(NCF)(带 C2 凸点的晶圆经 NCF 层压后切单)的高键合力 TCB 工艺制造如图 7.34 所示的 3D IC 集成堆栈。这个 3D 存储立方一次堆叠一颗芯片,每颗芯片需要约 10s 的时间使底部填充膜凝胶化、焊料熔化和固化以及膜固化,产率是一个问题。

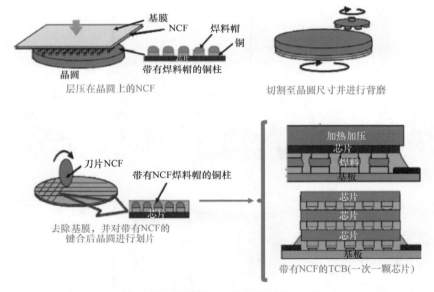

图 7.33 采用 TCB-NCF 堆叠的 HBM

为了解决这个问题,Toray[23, 24] 提出了一种集中式键合方法,如图 7.35 所示。可以看出,带有 NCF 的 C2 芯片在温度为 80℃ 的热台上进行了预键合(键合力为 30N,温度为 150℃,时间小于 1s)。后键合工艺在 80℃ 温度的热台上进行 [第一步(3s):键合力为 50N 和温度为 220~260℃,第二步(7s):键合力为 70N 和温度为 280℃]。传统方法堆叠四颗芯片需要 40s,而集中式 TCB 方法只需不到 14s。集中式键合的一些截面图像如图 7.35 所示,通过优化实现了符合预期的连接。

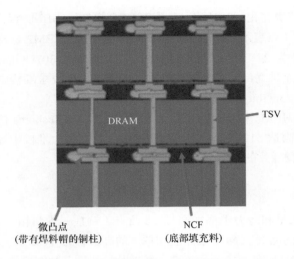

图 7.34 采用 TCB-NCF 进行堆叠的 HBM 图像

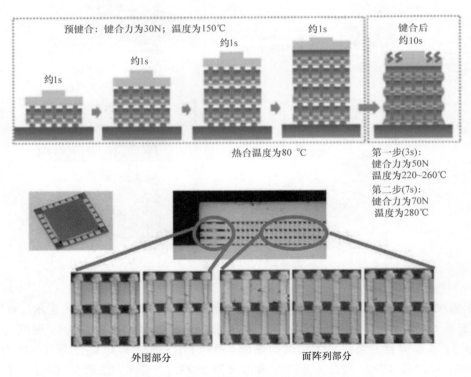

图 7.35 Toray 的 HBM 集中式键合方法[23]

7.3.3 3D IC 集成——采用 TSV 的芯片堆叠

图 7.36 所示为索尼的 ISX014 堆叠式相机传感器[25]。可以看到，背照式 CMOS 图像传感器（backside illuminated CMOS image sensor，BI-CIS）芯片位于处理器芯片的顶部，芯片之间的互连通过 BI-CIS 芯片边缘的 TSV 实现。如图 7.36 所示，信号从处理器芯片的边缘通过引线键合的方式引出。

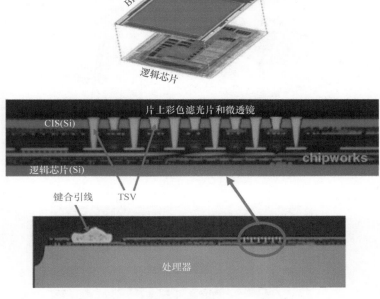

图 7.36 带有 TSV 的芯片对芯片键合（索尼的 CIS）[25]

在美国国防部高级研究计划局（Defense Advanced Research Projects Agency，DARPA）通用异质集成和知识产权复用策略（Common Heterogeneous Integration and Intellectual Property Reuse Strategies，CHIPS）项目的支持下，UCSB 和 AMD[26] 提出了一个应用于未来的超高性能系统，如图 7.37 所示。该系统包括一个中央处理器（central processing unit，CPU）芯粒和几个图形处理器（graphic processing unit，GPU）芯粒，以及 HBM，以上芯粒放置于有源 TSV 转接板（即具有 CMOS 器件的 TSV 转接板）上。

在 2020 年 IEEE 第 32 期高性能芯片研讨会（Hot Chips Symposium，HCS）期间，三星宣布了他们用于高性能应用场景的 3D IC 集成技术 eXtended-Cube（X-Cube），如图 7.38 所示。利用三星的 TSV 技术，X-Cube 实现了速度和功效的显著飞跃，以帮助满足下一代应用的严格性能需求，包括由 5G 和 AI 驱动的 HPC，以及移动和可穿戴设备。从图 7.38 可以看出，顶部芯片（通常是 HBM）位于具有 TSV 的逻辑芯片之上（或称为有源 TSV 转接板）。这两个芯片之间的互连是通过 C2 微凸点实现的。

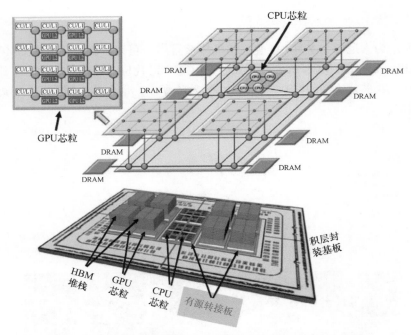

图 7.37 芯片与有源 TSV 转接板键合（AMD/UCSB）[26]

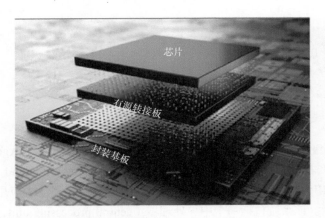

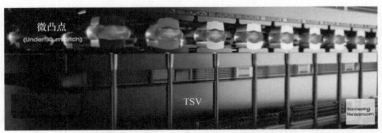

图 7.38 芯片与有源 TSV 转接板键合（三星 X-Cube）

图 7.39 所示为一些有源 TSV 转接板，如英特尔所提出来的。图 7.39a 是 2018 年 12 月发布的 FOVEROS 技术，含有 CMOS 器件的 TSV 转接板（有源转接板）就像一颗芯片一样，与芯粒或 SoC 进行了面对面的热压键合。图 7.39b、c 是英特尔于 2019 年 7 月发布的另一项新技术，称为全方位互连（omni-directional interconnect，ODI）技术。图 7.39b 所示为 ODI 类型 1，有源 TSV 转接板（芯片）位于大芯片（如 SoC）下方，并与 SoC 面对面热压键合。图 7.39c 所示为 ODI 类型 2，有源 TSV 转接板（桥，即芯片）位于下方并与芯粒 /SoC 片间互连。ODI 类型 3 是类型 2 的一种特殊情况，其中有源转接板（或基本逻辑芯片）直接连接了 SoC/ 芯粒。英特尔还发布了用于芯片对芯片互连界面的管理数据输入输出（management data input/output，MDIO），用于替代当前的高级接口总线（advanced interface bus，AIB）。所有这些 TSV 转接板上的异质集成技术都是为了实现极高性能，而有源 TSV 转接板则是为了获得较之更高的性能。2020 年 7 月，英特尔发布了采用 FOVEROS 技术的"Lakefield"处理器产品，如图 7.40[27-29] 所示。

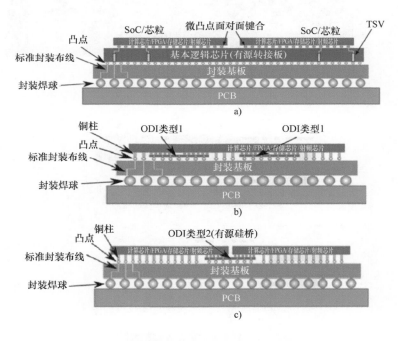

图 7.39　芯片与有源 TSV 转接板键合（英特尔）[27]

图 7.41 和图 7.42 所示为 IME 的带有 TSV 的存储芯片和逻辑芯片通过微凸点键合。测试结构的设计、材料、工艺和制备已在参考文献 [30] 中描述。该结构的 SEM 图像，尤其是 TSV 部分，如图 7.41 所示。互连结构的微凸点（铜柱 + 锡料帽）和凸点下金属化（under bump metallization，UBM）（化学镀镍浸金）如图 7.42 所示。

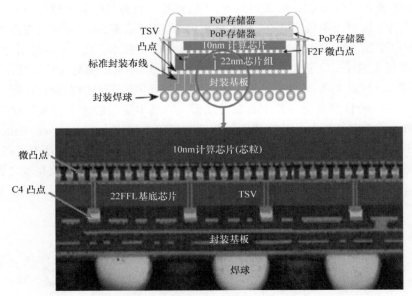

图 7.40 芯片与有源 TSV 转接板（英特尔 FOVEROS）[27]

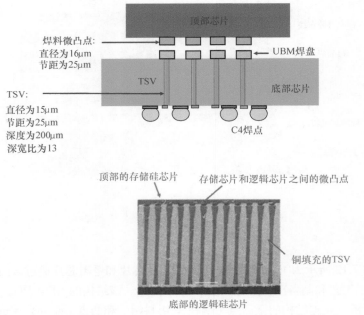

图 7.41 带有 TSV 的芯片对芯片键合（IME）

第 7 章 3D IC 集成和 3D IC 封装 311

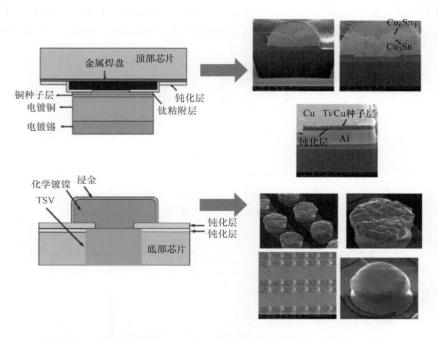

图 7.42 带有 TSV 互连的芯片对芯片键合（IME）

7.3.4 3D IC 集成——采用 TSV 的无凸点混合键合芯片堆叠

在英特尔架构日（2020 年 8 月 13 日），英特尔宣布将 Cu-Cu 混合键合用于其 FOVEROS 技术，如图 7.43 所示。更多细节将在 9.8 节中描述。

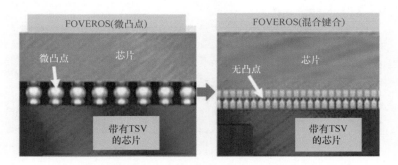

图 7.43 无凸点混合键合（英特尔）

图 7.44 所示为台积电的部分 3DFabric 技术[31-34]，它由前端集成芯片系统（system on integrated chips，SoIC）技术和后端 SoIC + InFO 技术组成。对于前端 SoIC，如图 7.44a 所示，台积电使用无凸点混合键合将芯片键合到另一个带有 TSV 的芯片的顶部。对于后端 SoIC + InFO，

如图 7.44b 所示,台积电使用集成先上晶和面朝上扇出型(InFO)技术来封装 SoIC。更多细节将在 9.9 节中阐述。如图 7.45[34] 所示,台积电还应用无凸点混合键合不同数量的 DRAM 以集成 HBM。可以看出,由于该技术是无凸点的,因此封装比传统的微凸点倒装芯片技术更薄。

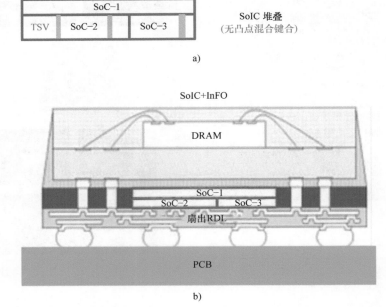

图 7.44 无凸点混合键合(台积电):a)前端 SoIC;b)后端 SoIC+InFO[33]

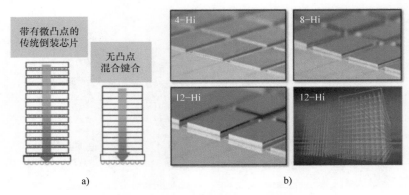

图 7.45 无凸点混合键合(台积电):
a)厚度对比:带有微凸点的传统倒装芯片与无凸点混合键合对比;b)利用无凸点混合键合实现的不同叠层[34]

图 7.46 所示为 ARM 带有 TSV 芯片的无凸点混合键合示意图。图 7.46a 是无凸点面对面键合,而图 7.46b 是无凸点面对背键合[35]。图 7.46 还绘制了所有 3D 架构的最高稳态温度(相较

于 2D 基准）。可以看出，对于 3D 逻辑和内存分区 CPU，由于具有更高的功率密度，逻辑层的温度更高。对于 CPU 堆叠的情况，底部芯片 CPU 温度更高，因为它与散热器距离较远。在不同的 3D 架构中，与 2D 基准相比，内存上逻辑芯片 3D 设计的温升约为 5℃，而逻辑上内存芯片 3D 设计的温升为 9℃。

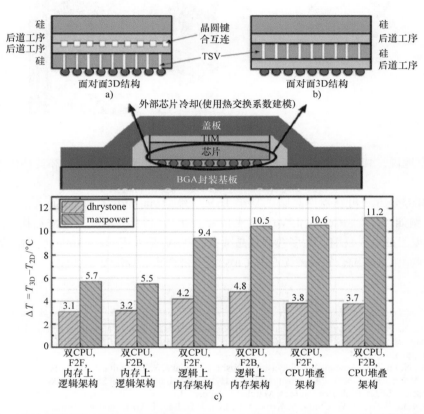

图 7.46　ARM 带有 TSV 的无凸点混合键合芯片：a）面对面混合键合；b）面对背混合键合[35]

7.3.5　3D IC 集成——无 TSV 的无凸点混合键合芯片堆叠

图 7.47 为索尼的 IMX260 BI-CIS[1, 2]。可以看出，BI-CIS 芯片（无 TSV）通过无凸点混合键合与处理器芯片互连。自 2016 年以来，索尼一直在大批量销售这款产品。更多细节详见 8.5.3 节。

7.3.6　总结和建议

一些重要的结论和建议总结如下：

1）3D IC 封装是设计公司和 OSAT 的重要竞争领域。然而近期，一些系统制造商和晶圆代工厂也开始利用自己的先进节点半导体器件开发 PoP 等 3D IC 封装。

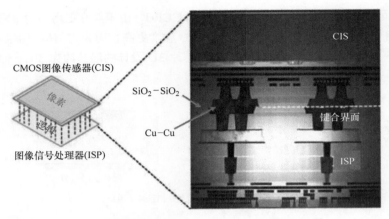

图 7.47 索尼无 TSV 的无凸点混合键合芯片 [1]

2）坦率地说，3D IC 集成（带有 TSV）通常由台积电等晶圆代工厂和英特尔等 IDM 进行制造。

3）只有部分方法可在器件晶圆上制作 TSV，例如 via-middle 或 via-last[36-38]。同时，只有部分方法（例如，CoC、CoW 和 WoW）支持带有 TSV 芯片的堆叠，互连结构/材料也不多，例如微凸点或无凸点。目前，大多数器件晶圆中的 TSV 是采用 via-middle 方式制造的，组装过程通过微凸点由 CoW 键合完成。无凸点 CoW 混合键合也在逐渐受到关注。

参 考 文 献

1. Kagawa, Y., N. Fujii, K. Aoyagi, Y. Kobayashi, S. Nishi, N. Todaka, et al., "Novel stacked CMOS image sensor with advanced Cu2Cu hybrid bonding," *Proceedings of IEEE/IEDM*, Dec. 2016, pp. 8.4.1–8.4.4.
2. Kagawa, Y., N. Fujii, K. Aoyagi, Y. Kobayashi. S. Nishi, N. Todaka, S. Takeshita, J. Taura, H. Takahashi, Y. Nishimura, et al., "An Advanced CuCu Hybrid Bonding for Novel Stack CMOS Image Sensor", *IEEE/EDTM Proceedings*, March 2018, pp. 1–3.
3. Lim, S., V. Rao, W. Hnin, W. Ching, V. Kripesh, C. Lee, J. H. Lau, J. Milla, and A. Fenner, "Process Development and Reliability of Microbumps", *IEEE Transactions on CPMT*, Vol. 33, No. 4, December 2010, pp. 747–753.
4. Ravichandran, S., M. Kathaperumal, M. Swaminathan, and R. Tummala, "Large-body-sized Glass-based Active Interposer for High-Performance Computing", *IEEE/ECTC Proceedings*, May 2020, pp. 879–884.
5. Zoschke, K., P. Mackowiak, K. Kröhnert, H. Oppermann, N. Jürgensen, M. Wietstruck, A. Göritz, S. Wipf, M. Kaynak, and K. Lang, "Cap Fabrication and Transfer Bonding Technology for Hermetic and Quasi Hermetic Wafer Level MEMS Packaging", *IEEE/ECTC Proceedings*, May 2020, pp. 432–438.
6. Xie, J., and D. Patterson, "Realizing 3D IC Integration with Face-to-Face Stacking", *Chip Scale Review*, May-June Issue, 2013, pp. 16–19.
7. Sutanto, J., 2012, "POSSUMTM Die Design as a Low Cost 3D Packaging Alternative," *3D Package*, 25, pp. 16–18.
8. Hayes, S., N. Chhabra, T. Duong, Z. Gong, D. Mitchell, and J. Wright, "System-in-Package Opportunities with the Redistributed Chip Package (RCP)", *Proceedings of IWLPC*, November 2011, pp. 10.1–10.7.
9. Yoon, S., J. Caparas, Y. Lin, and P. Marimuthu, "Advanced Low Profile PoP Solution with Embedded Wafer Level PoP (eWLB-PoP) Technology", *Proceedings of IEEE/ECTC*, May 2012, pp. 1250–1254.

10. Jin, Y., J. Teysseyrex, X. Baraton, S. Yoon, Y. Lin, P. Marimuthu, "Development of Next General eWLB (Embedded Wafer Level BGA) Technology", *Proceedings of IWLPC*, November 2011, pp. 7.1–7.7.
11. Chang, N., C. Chung, Y. Wang, C. Lin, P. Su, T. Shih, N. Kao, J. Hung, "3D Micro Bump Interface Enabling Top Die Interconnect to True Circuit Through Silicon Via Wafer", *IEEE/ECTC Proceedings*, May 2020, pp. 1888–1893.
12. Itoi, K., M. Okamoto, Y. Sano, N. Ueta, S. Okude and O. Nakao, T. Tessier, S. Sivaswamy, and G. Stout, "Laminate Based Fan-Out Embedded Die Packaging Using Polyimide Multilayer Wiring Boards", *Proceedings of IWLPC*, November 2011, pp. 7.8–7.14.
13. Lee, M., Yoo, M., Cho, J., Lee, S., Kim, J., Lee, C., Kang, D., Zwenger, C., and Lanzone, R., "Study of Interconnection Process for Fine Pitch Flip Chip," *IEEE/ECTC Proceedings*, May 2009, pp. 720–723.
14. Singh, A., K. Sullivan, G. Leal, and T. Gong, "Assembly challenges with Flip Chip multi-die and interposer-based SiP Modules", *IMAPS Proceedings*, October 2019, pp. T1–T18.
15. Furukawa, T., T. Kasuga, M. Umehara, and Y. Tamadate, "Prospects for automotive SiP modules applyinc IC assembly and packaging technology", *IMAPS Proceedings*, October 2019, pp. M5–M10.
16. Yoon, S., P. Tang, R. Emigh, Y. Lin, P. C. Marimuthu, and R. Pendse, "Fanout Flipchip eWLB (embedded Wafer Level Ball Grid Array) Technology as 2.5D Packaging Solutions", *Proceedings of IEEE/ECTC*, May 2013, pp. 1855–1860.
17. C.-F. Tseng, C.-S. Liu, C.-H. Wu, and D. Yu, "InFO (wafer level integrated fan-out) technology," *Proceedings of IEEE/ECTC*, May 2016, pp. 1–6.
18. C.-C. Hsieh, C.-H. Wu, and D. Yu, "Analysis and comparison of thermal performance of advanced packaging technologies for state-of-the-art mobile applications," *Proceedings of IEEE/ECTC*, May 2016, pp. 1430–1438.
19. Chuang, P., M. Lin, S. Hung, Y. Wu, D. Wong, M. Yew, C. Hsu, L. Liao, P. Lai, P. Tsai, S. Chen, S. Cheng, and S. Jeng, "Hybrid Fan-out Package for Vertical Heterogeneous Integration", *IEEE/ECTC Proceedings*, May 2020, pp. 333–338.
20. Kim, J., I. Choi, J. Park, J. Lee, T. Jeong, J. Byun, Y. Ko, K. Hur, D. Kim, and K. Oh, "Fan-out Panel Level Package with Fine Pitch Pattern", *IEEE/ECTC Proceedings*, May 2018, pp. 52–57.
21. Chong, S., E. Ching, S. Siang, S. Boon, T. Chai, "Demonstration of Vertically Integrated POP using FOWLP Approach", *IEEE/ECTC Proceedings*, May 2020, pp. 873–878.
22. Hou, S., W. Chen, C. Hu, C. Chiu, K. Ting, T. Lin, W. Wei, W. Chiou, V. Lin, V. Chang, C. Wang, C. Wu, and D. Yu, "Wafer-Level Integration of an Advanced Logic-Memory System Through the Second-Generation CoWoS Technology", *IEEE Transactions on Electron Devices*, Vol. 64, No. 10, October 2017, pp. 4071–4077.
23. Matsumura, K., Tomikawa, M., Sakabe, Y., and Shiba, Y., "New Non Conductive Film for high productivity process", *IEEE Proceedings of CPMT Symposium Japan*, 2015, pp. 19–20.
24. Asahi, N., Miyamoto, Y., Nimura, M., Mizutani, Y., and Arai, Y., "High Productivity Thermal Compression Bonding for 3D-IC", *Proceedings of IEEE International 3D Systems Integration Conference*, 2015, pp. TS7.3.1–TS7.3.5.
25. Sukegawa, S., T. Umebayashi, T. Nakajima, H. Kawanobe, K. Koseki, I. Hirota, T. Haruta, et al., "A 1/4-inch 8Mpixel Back-Illuminated Stacked CMOS Image Sensor," *Proceedings of IEEE/ISSCC*, San Francisco, CA, February 2013, pp. 484–486.
26. Stow, D., Y. Xie, T. Siddiqua, and G. H. Loh, "Cost-effective design of scalable high-performance systems using active and passive interposers," *Proc. of IEEE/ACM International Conf. on Computer-Aided Design*, Nov. 2017, pp. 728–735.
27. Ingerly, D., S. Amin, L. Aryasomayajula, A. Balankutty, D. Borst, A. Chandra, K. Cheemalapati, C. Cook, R. Criss, K. Enamul1, W. Gomes, D. Jones, K. Kolluru, A. Kandas, G.. Kim, H. Ma, D. Pantuso, C. Petersburg, M. Phen-givoni, A. Pillai, A. Sairam, P. Shekhar, P. Sinha, P. Stover, A. Telang, and Z. Zell, "Foveros: 3D Integration and the use of Face-to-Face Chip Stacking for Logic Devices", *IEEE/IEDM Proceedings*, December 2019, pp. 19.6.1–19.6.4.
28. Gomes, W., S. Khushu, D. Ingerly, P. Stover, N. Chowdhury, F. O'Mahony, etc., "Lakefield and Mobility Computer: A 3D Stacked 10 nm and 2FFL Hybrid Processor System in 12 × 12 mm^2, 1 mm Package-on-Package", *IEEE/ISSCC Proceedings*, February 2020, pp. 40–41.
29. WikiChip, "A Look at Intel Lakefield: A 3D-Stacked Single-ISA Heterogeneous Penta-Core SoC", https://en.wikichip.org/wiki/chiplet, May 27, 2020.

30. Yu, A. B., J. H. Lau, S. Ho, A. Kumar, W. Hnin, W. Lee, M. Jong, et al., "Fabrication of High Aspect Ratio TSV and Assembly with Fine-Pitch Low-Cost Solder Microbump for Si Interposer Technology with High-Density Interconnects", *IEEE Transactions on CPMT*, Vol. 1, No. 9, September 2011, pp. 1336–1344.
31. Chen, M. F., C. S. Lin, E. B. Liao, W. C. Chiou, C. C. Kuo, C. C. Hu, C. H. Tsai, C. T. Wang and D. Yu, "SoIC for Low-Temperature, Multi-Layer 3D Memory Integration", *IEEE/ECTC Proceedings,* May 2020, pp. 855–860.
32. Chen, Y. H., C. A. Yang, C. C. Kuo, M. F. Chen, C. H. Tung, W. C. Chiou, and D. Yu, "Ultra High Density SoIC with Sub-micron Bond Pitch", *IEEE/ECTC Proceedings,* May 2020, pp. 576–581.
33. Chen, F., M. Chen, W. Chiou, D. Yu, "System on Integrated Chips (SoIC^{TM}) for 3D Heterogeneous Integration", *IEEE/ECTC Proceedings,* May 2019, pp. 594–599.
34. Chen, M., C. Lin, E. Liao, W. Chiou, C. Kuo, C. Hu, C. Tsai, C. Wang and D. Yu, "SoIC for Low-Temperature, Multi-Layer 3D Memory Integration", *IEEE/ECTC Proceedings*, May 2020, pp. 855–860.
35. Mathur, R., C. Chao, R. Liu, N. Tadepalli, P. Chandupatla, S. Hung, X. Xu, S. Sinha, and J. Kulkarni, "Thermal Analysis of a 3D Stacked High-Performance Commercial Microprocessor using Face-to-Face Wafer Bonding Technology", *IEEE/ECTC Proceedings*, May 2020, pp. 541–547.
36. Lau, J. H., *3D IC Integration and Packaging*, McGraw-Hill, New York, 2016.
37. Lau, J. H., *Through-Silicon Via (TSV) for 3D Integration*, McGraw-Hill, New York, 2013.
38. Lau, J. H., *Reliability of RoHS compliant 2D & 3D IC Interconnects*, McGraw-Hill, New York, 2011.
39. Kumahara, K., R. Liang, S. Lee, Y. Miwa, M. Murugesan, H. Kino, T. Fukushima, and T. Tanaka, "Low-temperature multichip-to-wafer 3D integration based on via-last TSV with OER-TEOS-CVD and microbump bonding without solder extrusion", *IEEE/ECTC Proceedings*, May 2020, pp. 1199–1204.
40. Ma, S., Y. Liu, F. Zheng, F. Li, D. Yu, A. Xiao, and X. Yang, "Development and Reliability study of 3D WLCSP for automotive CMOS image sensor using TSV technology", *IEEE/ECTC Proceedings,* May 2020, pp. 461–466.
41. Camara, J., S. Soroushiani, D. Wilding, S. Sayeed, M. Monshi, J. Volakis, S. Bhardwaj, and P. Raj, "Remateable and Deformable Area-Array Interconnects in 3D Smart Wireless Sensor Packages", *IEEE/ECTC Proceedings*, May 2020, pp. 671–676.
42. Kawano, M., X. Wang, and Q. Ren, "Trench Isolation Technology for Cost-effective Wafer-level 3D Integration with One-step TSV", *IEEE/ECTC Proceedings,* May 2020, pp. 1161–1166.
43. Miwa, Y., K. Kumahara, S. Lee, R. Liang, H. Kino, T. Fukushima, and T. Tanaka, "7-μm-thick NCF technology with low-height solder microbump bonding for 3D integration", *IEEE/ECTC Proceedings,* May 2020, pp. 1453–1458.
44. Kaul, A., S. Rajan, M. Hossen, G. May, and M. Bakir, "BEOL-Embedded 3D Polylithic Integration: Thermal and Interconnection Considerations", *IEEE/ECTC Proceedings,* May 2020, pp. 1459–1467.
45. Seo, H., S. Kim, H. Park, G. Kim, and Y. Park, "Effects of two-step plasma treatment on Cu and SiO_2 surfaces for 3D bonding applications", *IEEE/ECTC Proceedings,* May 2020, pp. 1677–1683.
46. Jourdain, A., J. Vos, E. Chery, G. Beyer, G. Plas, E., Walsby K. Roberts, H. Ashraf, D. Thomas, and E. Beyne, "Extreme Wafer Thinning and nano-TSV processing for 3D Heterogeneous Integration", *IEEE/ECTC Proceedings,* May 2020, pp. 42–48.
47. Chen, Y., C. A. Yang, C. C. Kuo, M. F. Chen, C. H. Tung, W. C. Chiou, and D. Yu, "Ultra High Density SoIC with Sub-micron Bond Pitch", *IEEE/ECTC Proceedings,* May 2020, pp. 576–581.
48. Liu, D., P. Chen, and K. Chen*, "A Novel Low-Temperature Cu-Cu Direct Bonding with Cr Wetting Layer and Au Passivation Layer", *IEEE/ECTC Proceedings,* May 2020, pp. 1322–1327.
49. Rahimi, A., P. Somarajan, and Q. Yu, "Modeling and Characterization of Through-Silicon Vias (TSVs) in Radio Frequency Regime in an Active Interposer Technology", *IEEE/ECTC Proceedings,* May 2020, pp. 1383–1389.

50. Lim, T., D. Ho, C. Chong, and S. Bhattacharya, "3D FOWLP Integration", *IEEE/ECTC Proceedings,* May 2020, pp. 1728–1735.
51. Kim, J., K. Yoon, H. Oh, E. Ahn, Y. Shin, and Y. Kim, "Study on Advanced Substrate for Double-side Package to Reduce Module Size", *IEEE/ECTC Proceedings,* May 2020, pp. 1904–1909.
52. Singh, S., and T. Kukal, "LTCC PoP Technology-Based Novel Approach for mm-Wave 5G System for Next Generation Communication System", *IEEE/ECTC Proceedings,* May 2020, pp. 1973–1978.
53. Tsai, M., R. Chiu, D. Huang, F. Kao, E. He, J. Chen, S. Chen, J. Tsai, and Y. Wang, "Innovative Packaging Solutions of 3D Double Side Molding with System in Package for IoT and 5G Application", *IEEE/ECTC Proceedings,* May 2019, pp. 700–706.
54. Mori, K., S. Yamashita, T. Fukuda, M. Sekiguchi, H. Ezawa, and S. Akejima, "3D Fan-Out Package Technology with Photosensitive Through Mold Interconnects", *IEEE/ECTC Proceedings,* May 2019, pp. 1140–1145.
55. Ravichandran, S., S. Yamada, F. Liu, V. Smet, M. Kathaperumal, and R. Tummala, "Low-Cost Non-TSV based 3D Packaging using Glass Panel Embedding (GPE) for Power-efficient, High-Bandwidth Heterogeneous Integration", *IEEE/ECTC Proceedings,* May 2019, pp. 1796–1802.
56. England, L., D.l Fisher, K. Rivera, B. Guthrie, P. Kuo, C. Lee, C. Hsu, F. Min, K. Kang, and C. Weng, "Die-to-Wafer (D2W) Processing and Reliability for 3D Packaging of Advanced Node Logic", *IEEE/ECTC Proceedings,* May 2019, pp. 600–606.
57. Yu, D., Y. Zou, X.Xu, A. Shi, X. Yang, and Z. Xiao, "Development of 3D WLCSP with Black Shielding for Optical Finger Print Sensor for the Application of Full Screen Smart Phone", *IEEE/ECTC Proceedings,* May 2019, pp. 884–889.
58. Kawano, M., X. Wang, and Q. Ren, "New Cost-effective Via-last Approach by "One-step TSV" after Wafer Stacking for 3D Memory Applications", *IEEE/ECTC Proceedings,* May 2019, pp. 1996–2002.
59. Panigrahy, A., S. Bonam, T. Ghosh, S. Rama, K. Vanjari, and S. Singh, "Diffusion enhanced drive sub 100 °C wafer level fine-pitch Cu-Cu thermocompression bonding for 3D IC integration", *IEEE/ECTC Proceedings,* May 2019, pp. 2156–2161.
60. Lee, J., S. Park, Y. Kim, J. Lee, S. Lee, C. Lee. Y. Kwon, C. Lee. J. Kim, N. Kim, Y. Sung, "Three-Dimensional Integrated Circuit (3D-IC) Package Using Fan-out Technology", *IEEE/ECTC Proceedings,* May 2019, pp. 35–40.
61. Jung, J., H. Lee, J. Kim, Y. Park, J. Yu, Y. Park, J. Lim, H. Choi, S. Cho, D. Kim, and S. An, "A Study of 3D Packaging Interconnection Performance affected by Thermal Diffusivity and Pressure Transmission", *IEEE/ECTC Proceedings,* May 2019, pp. 204–209.
62. Coudrain, P., J. Charbonnier, A. Garnier, P. Vivet, R. Vélard, A. Vinci, F. Ponthenier, A. Farcy, R. Segaud, P. Chausse, L. Arnaud, D. Lattard, E. Guthmuller, G. Romano, A. Gueugnot, F. Berger, J. Beltritti, T. Mourier, M. Gottardi, S. Minoret, C. Ribière, G. Romero, P. Philip, Y. Exbrayat, D. Scevola, D. Campos, M. Argoud, N. Allouti, R. Eleouet, C. Tortolero, C. Aumont, D. Dutoit, C. Legalland, J. Michailos, S. Chéramy, and G. Simon, "Active interposer technology for chiplet-based advanced 3D system architectures", *IEEE/ECTC Proceedings,* May 2019, pp. 569–578.
63. Su, A., T. Ku, C. Tsai, K. Yee, and D. Yu, "3D-MiM (MUST-in-MUST) Technology for Advanced System Integration", *IEEE/ECTC Proceedings,* May 2019, pp. 1–6.
64. Watanabe, A., Y. Wang, N. Ogura, P. Raj, V. Smet, M. Tentzeris, and R. Tummala., "Low-Loss Additively-Deposited Ultra-Short Copper-Paste Interconnections in 3D Antenna-Integrated Packages for 5G and IoT Applications", *IEEE/ECTC Proceedings,* May 2019, pp. 972–976.
65. Ahmed, O., G. Jalilvand, H. Fernandez, P. Su, T. Lee, and T. Jiang, "Long-Term Reliability of Solder Joints in 3D ICs under Near-Application Conditions", *IEEE/ECTC Proceedings,* May 2019, pp. 1106–1112.
66. Jouve, A., L. Sanchez, C. Castan, M. Laugier, E. Rolland, B. Montmayeul, R. Franiatte, F. Fournel, and S. Cheramy, "Self-Assembly process for 3D Die-to-Wafer using direct bonding: A step forward toward process automatisation", *IEEE/ECTC Proceedings,* May 2019, pp. 225–234.
67. Hwang, T., D. Oh, J. Kim, E. Song, T. Kim, K. Kim, J. Lee, and T. Kim, "The Thermal Dissipation Characteristics of The Novel System-In-Package Technology (ICE-SiP) for Mobile and 3D High-end Packages", *IEEE/ECTC Proceedings,* May 2019, pp. 614–619.

68. Sirbu, B., Y. Eichhammer, H. Oppermann, T. Tekin, J. Kraft, V. Sidorov, X. Yin, J. Bauwelinck, C. Neumeyr, and F. Soares, "3D Silicon Photonics Interposer for Tb/s Optical Interconnects in Data Centers with double-side assembled active components and integrated optical and electrical Through Silicon Via on SOI", *IEEE/ECTC Proceedings,* May 2019, pp. 1052–1059.
69. Sun, X., N. Pantano, S. Kim, G. Van der Plas and E. Beyne, "Inductive links for 3D stacked chip-to-chip communication", *IEEE/ECTC Proceedings,* May 2019, pp. 1215–1220.
70. Jani, I., D. Lattard, P. Vivet, L. Arnaud, S. Cheramy, E. Beigné, A. Farcy, J. Jourdon, Y. Henrion, E. Deloffre, and H. Bilgen, "Characterization of fine pitch Hybrid Bonding pads using electrical misalignment test vehicle", *IEEE/ECTC Proceedings,* May 2019, pp. 1926–1932.

第 8 章

混合键合

8.1 引言

首先，本章的重点是硅到硅的倒装芯片键合，而不是如第 2、3、5 章所示的硅到有机基板的倒装芯片键合。Cu-Cu 之间至少有两种不同的键合方式，即 Cu-Cu 热压键合（thermal compression bonding，TCB）[1-11] 和低温直接键合互连（混合键合）[12-37]。大多数 Cu-Cu TCB 在高温（通常为 350～400℃）和大压力条件下进行，以驱动 Cu 原子扩散穿过界面来形成一个整体。混合键合（将介质键合与金属键合结合以形成互连）与 Cu-Cu TCB 非常不同，混合键合也被业界广泛称为低温直接键合互连（direct bond interconnect，DBI），首先在室温下进行，然后在 150～300℃下退火。

DBI 是由美国三角研究所（Research Triangle Institute，RTI）发明的，并以 ZiBond（一种直接的氧化物 - 氧化物键合，包括低温下为获得初始高键合强度的晶圆至晶圆键合工艺）为名获得专利。2000 年至 2001 年期间，Fountain、Enguist、Tong 和其他几位同事成立了从 RTI 分拆出来的公司 Ziptronix。在 2004 年和 2005 年之间，基于 ZiBond 技术，Ziptronix 将介质层键合与嵌入式金属键合组合起来，在低温下同时通过晶圆键合形成互连（所谓的 DBI）[36, 37]。Ziptronix 于 2015 年 8 月 28 日被 Tessera 收购，Tessera 于 2017 年 2 月 23 日更名为 Xperi。

Ziptronix DBI 技术的突破是在 2015 年春天，当时索尼已经使用其"ZiBond"氧化物至氧化物键合技术，并将其技术许可扩展至 DBI。现在，DBI 已被广泛应用于智能手机和其他基于图像的设备中的 CMOS 图像传感器市场。

本章将首先简要介绍 Cu-Cu TCB、SiO_2-SiO_2 TCB 和室温（room-temperature，RT）Cu-Cu TCB，然后讨论 DBI 技术，最后将简要介绍混合键合近期的一些进展。

8.2 Cu-Cu TCB

8.2.1 Cu-Cu TCB 的一些基本原理

Cu-Cu 键合互连的优点是相较于其他连接可以提供更低的电阻率、更高的互连密度、更低的电迁移率。另一方面，为了降低严重影响键合可靠性的天然氧化物形成的趋势，Cu-Cu 键合通常在高温（约 400℃）、高压、长时间（60～120min）下完成，这对提高产率（不考虑冷却时间）和保障器件可靠性是不利的。

图 8.1 所示为键合温度对临界界面结合能的影响（临界界面结合能也被称为界面上的临界能量释放率，如果最大的界面结合能大于临界界面结合能，那么就会发生界面分层）。可以看出，键合温度越高，临界界面结合能（G_c）越高，也就是说连接越可靠。另外，从图 8.1 可以看出，温度越高，由于两个界面层间的互扩散被激活，界面和原始键合界面之间的接缝就越少，趋于消失。这就是 Cu-Cu 键合需要高温条件的主要原因[1]。

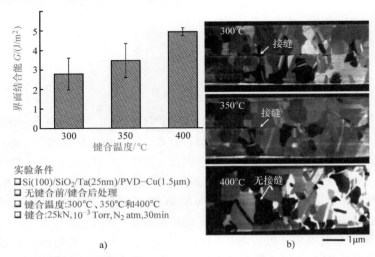

图 8.1　键合温度对键合界面特性的影响：a）界面结合能；b）微观结构的 SEM 图像

降低键合温度并获得高质量键合（互连）可以采用退火。图 8.2 所示为各种退火温度对临界界面结合能 G_c 的影响。可以看出，对于 25kN、300℃温度键合 30min 的 8in 晶圆，在大气压氮气环境下以 300℃温度退火 60min 后，G_c 从 2.8J/m² （未退火）增大到 12.2J/m²。甚至在 250℃的退火温度下保温 60min，G_c 也能增大到 8.9J/m²。然而，太低的退火温度不会有很大助益，如图 8.2 所示的 200℃退火。

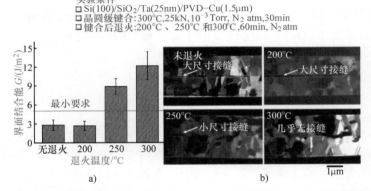

图 8.2　Cu-Cu 键合后退火温度对键合界面特性的影响：a）界面结合能；b）微观结构的 SEM 图像

8.2.2　IBM/RPI 的 Cu-Cu TCB

图 8.3a 所示为 IBM/ 伦斯勒理工学院（Rensselaer Polytechnic Institute，RPI）的两个器件层以面对面方式键合的结构示意图，图 8.3b 所示为以面对背方式键合的结构示意图[2-4]。图 8.4 为一个典型的 Cu-Cu 互连结构，其键合界面质量很高。在键合之前，用标准的后道工序（back end of line，BEOL）大马士革工艺制造铜互连（焊盘），然后用氧化物化学机械抛光（chemical-mechanical polishing，CMP）工艺（氧化物修整）将氧化物层处理至比铜表面低 40nm 的高度，键合温度提高到 400℃。

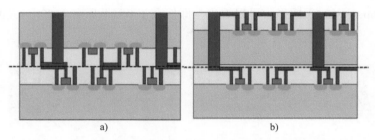

图 8.3　a）面对面型 WoW 键合；b）面对背型 WoW 键合[2]

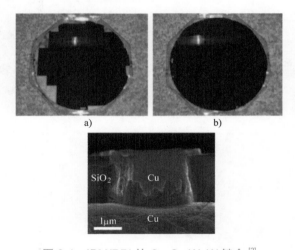

图 8.4　IBM/RPI 的 Cu-Cu WoW 键合[2]

8.3　室温 Cu-Cu TCB

8.3.1　室温 Cu-Cu TCB 的一些基本原理

室温下的 Cu-Cu 键合能带来最高的产率和最少的器件可靠性问题，同时成本非常低。然而，室温键合的缺点是对以下方面有严格要求：

1)焊盘/布线/晶圆的平整性。

2)表面处理,以确保光滑的亲水表面能实现高质量键合。

8.3.2　NIMS/AIST/东芝/东京大学的室温 Cu-Cu TCB

图 8.5a 所示为日本物质材料研究机构(National Institute for Material Science,NIMS)/日本产业技术综合研究所(National Institute of Advanced Industrial Science and Technology,AIST)/东芝(Toshiba)/东京大学(University of Tokyo)在室温下两个器件层的键合结构[5-11]。图 8.5b 为高温存储试验后无凸点电极(焊盘)界面间的典型截面扫描电子显微镜(scanning electron microscope,SEM)图像。可以看出,即使在 150℃下保温 1000h,界面仍保持着紧密的结合[5-11]。

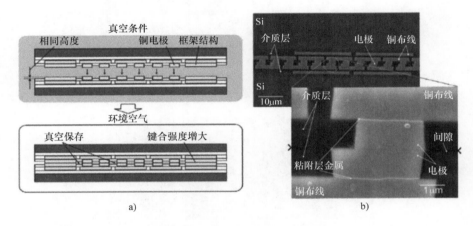

图 8.5　NIMS/AIST/东芝/东京大学的室温 Cu-Cu 键合[8]

8.4　SiO$_2$–SiO$_2$ TCB

8.4.1　SiO$_2$-SiO$_2$ TCB 的一些基本原理

SiO$_2$-SiO$_2$ 键合通常需要三个步骤,即预键合、键合和键合后处理。预键合是在室温下操作的,这就消除了晶圆对位过程中产生的跑偏误差,从而使键合后的对准精度更高。为了获得共价键(互连),键合温度需要非常高(约 400℃)。为了在较低的退火(键合后处理)温度(200~400℃)下实现强化学键(互连),必须通过等离子体活化来改变表面的化学特性。图 8.6 所示为退火温度对临界结合能的影响。正如所料,退火温度越高,临界结合能越大。考虑到大多数器件的最大允许温度,400℃是常用的退火工艺温度。图 8.7 所示为在 300℃退火温度下,退火时间对临界表面能的影响,可以看出:

1)退火时间越长,临界表面能越大。

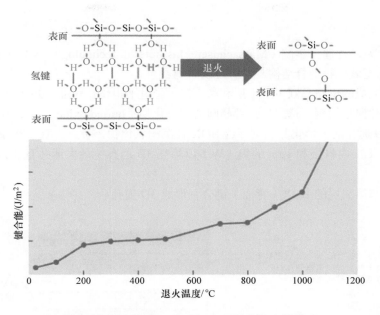

图 8.6 键合能（SiO_2-SiO_2）与退火温度的关系

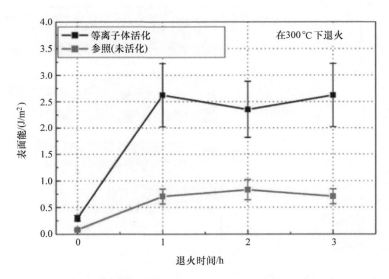

- Si到SiO_2晶圆级键合
- 表面能与300℃退火时间

图 8.7 表面能与300℃下退火时间的关系（SiO_2-SiO_2）

2）1h 的退火时间是绰绰有余的。
3）键合前对表面的化学特性的等离子体活化对临界表面能有很大影响。

8.4.2 麻省理工学院的 SiO_2-SiO_2 TCB

图 8.8 所示为麻省理工学院（Massachusetts Institute of Technology，MIT）在 275℃ 下三层器件层的氧化物至氧化物键合结构[38-44]。可以看到：

1）两个已完成的电路晶圆（第 1 层和第 2 层）经过平坦化、对准，并面对面键合。

2）支撑硅片湿法蚀刻，暴露出顶部晶圆的埋氧化物（buried oxide，BOX）结构。

3）完成 3D 通孔的图形化并蚀刻穿透 BOX 及沉积氧化物，暴露出双层器件的金属触点。

4）沉积 Ti/TiN 内衬层和 1μm 的钨（W）以填充 3D 通孔（最大通孔直径为 1.5μm），将两层电连接起来。

5）第三层与第二层的 BOX（背面）键合，形成 3D 通孔。

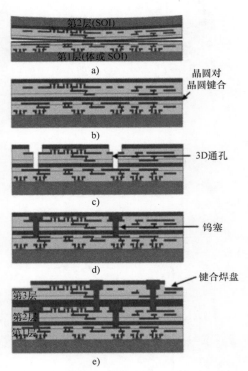

图 8.8 麻省理工学院的 SiO_2-SiO_2 WoW 键合组装工艺[38]

图 8.9 为三层 3D（环形振荡器）结构的典型截面图。可以看出：

1）各层键合在一起，用钨塞实现互连。

2）传统的层间互连分布在底部两层。

3）3D 通孔位于晶体管之间的隔离（场）区。

一些功能性的 3D 结构/电路已经被开发并在参考文献 [38-44] 中阐述。

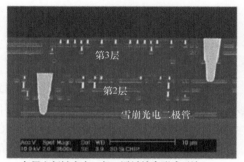
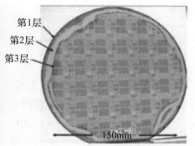

各层之间键合在一起，通过钨塞形成互连；传统的层间连接见于第2层和第3层，它们是FDSOI层。请注意，3D通孔位于晶体管之间的隔离(场)区域

通过BOX可以看到第3层的金属图案。在晶圆的边缘可以看到第2层和第1层，由于非平面表面的原因，这两层没有被键合在一起

图 8.9　麻省理工学院的 SiO_2-SiO_2 WoW 键合结果[38]

8.4.3　Leti/ 飞思卡尔 / 意法半导体的 SiO_2-SiO_2 TCB

图 8.10 所示为法国电子信息技术研究所（Leti）/ 飞思卡尔（Freescale）/ 意法半导体（STMicroelectronics）的两个器件层在 <400℃下介质层 - 介质层键合结构[45-47]。可以看到：

1）首先，在 200mm 的晶圆和 SOI 晶圆上形成金属层；接下来，将这些晶圆面对面键合，并将 SOI 晶圆的体硅去除至 BOX 层。

2）形成层间过孔（interstrata via，ISV），实现从上层到下层的接触。

3）在 SOI 晶圆的背面顶部形成金属层。

ISV 的典型截面如图 8.10 [45-47] 所示，可以看出该 ISV（约 1.5μm）形成了良好接触。

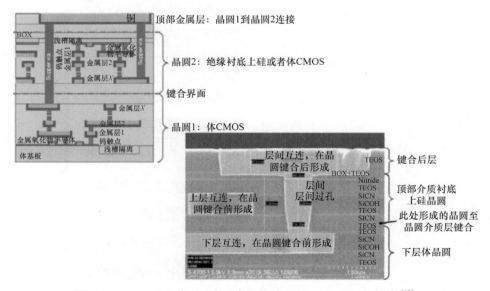

图 8.10　Leti/ 飞思卡尔 / 意法半导体的 SiO_2-SiO_2 WoW 键合[45]

8.5 低温 DBI

8.5.1 低温 DBI 的一些基本原理

图 8.11 所示为低温 DBI 的关键工艺步骤[12-37]。首先，控制纳米级表面形貌对于 DBI 技术非常重要。在活化和键合之前，介质层表面需要非常平坦和光滑。如图 8.11a 所示，采用化学机械抛光（chemical-mechanical polishing，CMP）实现非常低的介质表面粗糙度（<0.5nm RMS）并保证金属区域相对于介质层有一定幅度的凹陷。如图 8.11b 所示，干法等离子体活化的介质表面在室温下接触时瞬间键合在一起（如图 8.12[18] 所示，在非常低的温度下可以获得很高的结合能）。如图 8.11c 所示，碟形间隙可以通过加热来闭合（此步骤是选择性的，因为该间隙也可以通过后续的退火步骤来闭合）。金属 - 金属键合发生在随后的退火过程中。金属的热膨胀系数通常远大于介质，金属膨胀从而填充间隙，然后产生内部压力，如图 8.11d 所示。正是在这种内部压力和退火温度的作用下，金属原子穿过界面进行扩散，进而形成良好的金属 - 金属键合，实现电连接[18]。对于这种类型的键合，外加压力不是必需的。在这种情况下，键合中铜氧化现象最小。因为铜互连周围的键合氧化物层能够保护互连结构在退火炉中免受氧化，从而最大限度地减少退火中的铜氧化。键合的氧化物表面还实现了在操作期间对铜互连的密封。

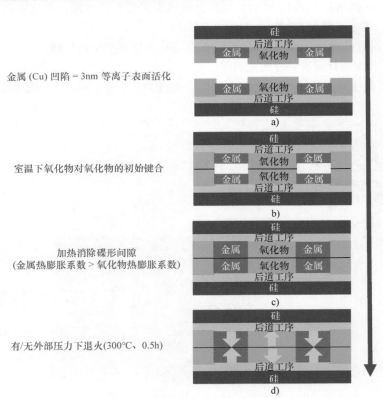

图 8.11 低温 DBI 的关键工艺步骤

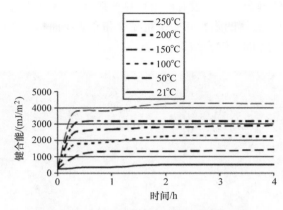

图 8.12 ZiBond™ 键合能和时间的关系

CMP 对 DBI 的影响如图 8.13 所示[18]。图 8.13a 所示为 CMP 优化前的键合，可以看到存在大尺寸接缝（未键合的 SiO_2-SiO_2 区域），且它们靠近 Cu 键合区域。图 8.13b 所示为 CMP 和

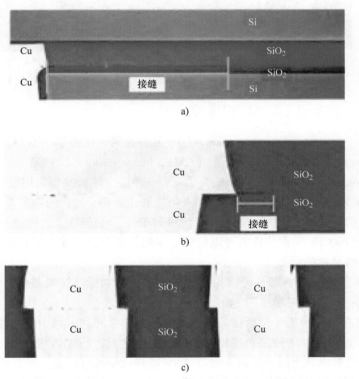

图 8.13 DBI 的截面 SEM 图像：a）铜焊盘附近明显未键合的 SiO_2 区域（接缝）；
b）键合过程中最小的接缝区域；c）没有可见接缝的 CMP 优化后的 DBI[18]

DBI 设计后形成的非常平坦氧化物的键合界面,从而最小化了接缝的产生。图 8.13c 所示为 CMP 和 DBI 设计优化后的键合界面,没有产生可见的接缝。CMP 优化是为 DBI [18] 获得合适的表面特征参数的关键,包括金属凹陷、介质层粗糙度和介质层曲率等。图 8.14 所示为具有 4μm 节距和 2μm 直径焊盘的优化 DBI 键合结果。

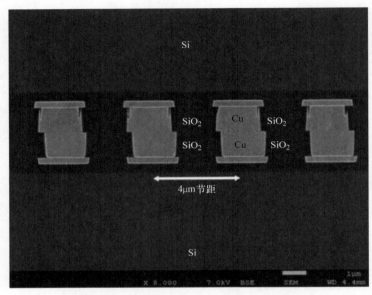

图 8.14　CMP 优化后的 DBI 的 SEM 图像

8.5.2　有 TSV 的索尼 CMOS 图像传感器

索尼(Sony)的三维(three-dimension,3D)互补金属氧化物半导体(Complementary Metal-Oxide-Semiconductor,CMOS)图像传感器(CMOS image sensor,CIS)至少有两种类型,一种是使用硅通孔(through silicon via,TSV)的类型,另一种是没有 TSV 但采用混合键合的类型。

图 8.15 所示为有 TSV 的索尼 CIS 产品。可以看出,CIS 包含两颗芯片,即 CIS 像素芯片和逻辑芯片,它们通过图 8.15 所示的分布于边缘的 TSV 垂直连接。这种设计的优点是

1) 相同尺寸的 CIS 像素芯片上可以集成更多的像素单元(或者说在更小尺寸的芯片上可以集成相同数量的像素单元)。

2) CIS 像素芯片和逻辑芯片可以用不同的工艺分别制造。结果是,CIS 芯片尺寸减小了 30%,逻辑芯片则从 50 万门增加到 240 万门[48]。

TSV 的数量在数千量级,包括信号、电源和地,像素阵列区域中不包含 TSV。列 TSV 被放置在像素 CIS 芯片上的比较器和逻辑芯片的计数器之间,行 TSV 放置在 CIS 芯片的行驱动器和逻辑芯片的行解码器之间(见图 8.15),TSV 这样的安排可以减少噪声的影响,并且使 CIS 芯片的制造变得更加简单。例如,为了减少噪声的影响,比较器被放置在使用索尼成熟的工艺技术制造的 CIS 像素芯片上,而非逻辑芯片上。

图 8.15 有 TSV 的索尼 3D CIS 像素芯片和逻辑芯片的集成[48]

CIS 像素芯片采用索尼传统的 1P4M BI-CIS（90nm）工艺制造，逻辑芯片采用成熟的 65nm 1P7M 逻辑工艺制造。像素芯片和逻辑芯片的尺寸差不多。CIS 晶圆的 CIS 硅介质层与逻辑晶圆的逻辑硅介质层键合在一起（SiO_2–SiO_2 晶圆至晶圆 ZiBond）。然后在晶圆键合后制备 TSV 并填充铜。图 8.16 和图 8.17 所示为 3D CIS 像素芯片和逻辑芯片集成结构截面的 SEM 图像。可以看到：

1）顶部是 CIS 像素芯片。
2）底部是逻辑芯片。
3）CIS 晶圆和逻辑晶圆是绝缘体至绝缘体（晶圆至晶圆）键合（见图 8.16）。
4）CIS 芯片通过 TSV 连接到逻辑芯片（见图 8.17）。

8.5.3 无 TSV（混合键合）的索尼 CMOS 图像传感器

图 8.18、图 8.19 和图 8.20 所示为无 TSV 而采用混合键合的索尼 CIS。索尼是首家在大批量制造中使用低温 Cu-Cu DBI 的公司[12, 13]。索尼为三星 2016 年推出的 Galaxy S7 手机生产了 IMX260 背照式 CMOS 图像传感器（backside illuminated CMOS image sensor，BI-CIS）。电气测试结果表明，它们坚固的 Cu-Cu 直接混合键合实现了卓越的连接性和可靠性，图像传感器的性能也非常出色。IMX260 BI-CIS 的俯视图和截面图分别如图 8.18、图 8.19 和图 8.20 所示。可以看出，与参考文献 [48] 中索尼的 ISX014 堆叠相机传感器不同，IMX260 产品中消除了 TSV，BI-CIS 芯片和处理器芯片之间的互连由 Cu-Cu DBI 实现。信号从封装基板通过键合引线与处理器芯片边缘连接在一起。

330 半导体先进封装技术

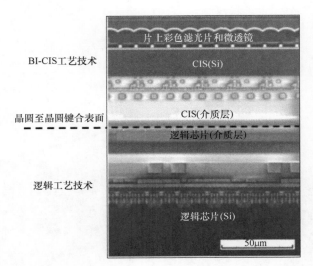

图 8.16 索尼 CIS（介质层）晶圆至逻辑芯片（介质层）晶圆键合[48]

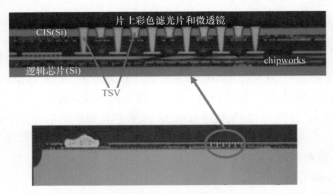

图 8.17 连接 CIS 像素芯片和逻辑芯片的 TSV[48]

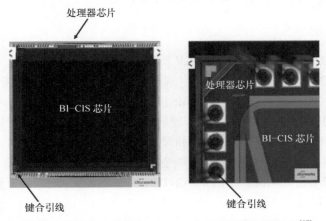

图 8.18 无 TSV 的索尼 3D CIS 和处理器 IC 的集成[12]

第 8 章 混合键合

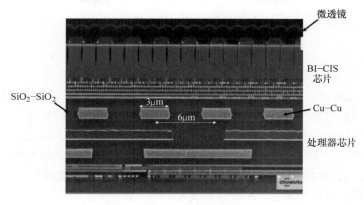

图 8.19 索尼 3D CIS 和处理器 IC 混合键合的 SEM 图像[12]

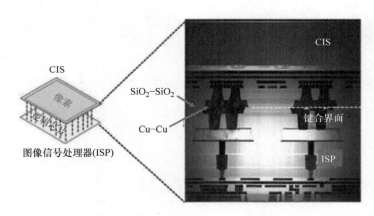

图 8.20 截面示意图和图片表明 Cu 焊盘之间的空隙在退火之后闭合[12]

通常，晶圆至晶圆的键合适用于两个晶圆上相同尺寸的芯片。在索尼的实例中，处理器芯片比像素芯片略大。为了完成晶圆至晶圆的键合，像素芯片晶圆的一些面积不得不浪费。此外，由于两个芯片中都没有 TSV，因此通过处理器芯片上的键合引线实现信号向下一互连层的传递。

Cu-Cu DBI 的组装过程从晶圆的表面清洁、金属氧化物去除和 SiO_2/SiN 的活化（通过湿法清洁和干法等离子体活化）以提高键合强度开始。然后，使用光学对准将晶圆在典型洁净室环境中在室温下相接触。第一次退火（100～150℃）的目的是加强晶圆 SiO_2/SiN 表面之间的结合力，同时最大限度地减少由于 Si、Cu 和 SiO_2/SiN 之间的热膨胀失配而导致的界面应力。然后，施加更高的温度和压力（300℃、25kN、10^{-3} Torr、N_2 atm）进行 30min 退火，以促进界面间的 Cu 扩散和界面上的晶粒生长。键合后退火是 N_2 气氛中 300℃持续处理 60min。该过程可以同时带来 Cu 和 SiO_2/SiN 的无缝键合（见图 8.19 和图 8.20）。

8.6 低温混合键合的近期发展

8.6.1 IME 混合键合的热机械性能

参考文献 [19, 20] 研究了与热机械键合性能相关的通用设计和工艺参数,包括碟形坑深度、退火温度和保温时间、TSV 节距和深度等。热机械仿真的结构如图 8.21 所示,重要的尺寸见表 8.1,有限元模型如图 8.22 所示。从图 8.22a 看到构造了一个仅包括 2×2 Cu 焊盘阵列的局部四分之一对称有限元模型,网格剖分细节如图 8.22b、c 所示。退火温度曲线如图 8.23 所示,表 8.2 列出了用于定义退火温度曲线的所有参数和退火前表面处理的碟形坑深度,仿真使用的材料特性见表 8.3,铜材料的弹塑性蠕变材料特性见表 8.4[20]。

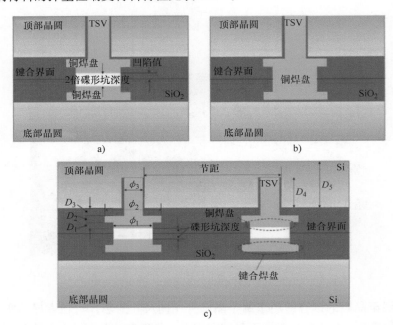

图 8.21 铜焊盘间空隙的截面示意图:a)退火前;b)退火后;c)尺寸

表 8.1 WoW 混合键合仿真中使用的设计参数 [20]

设计参数	数值 /μm
铜焊盘 / 键合焊盘 /TSV 节距	6,9,12
铜焊盘直径 ϕ_1	3
键合焊盘直径 ϕ_2	4
TSV 直径 ϕ_3	2
铜焊盘厚度 D_1	0.6
键合焊盘厚度 D_2	0.6
SiO_2 层厚度 D_3	0.2
TSV 深度 D_4	5,10,15
硅厚度 D_5	30

第 8 章 混合键合 333

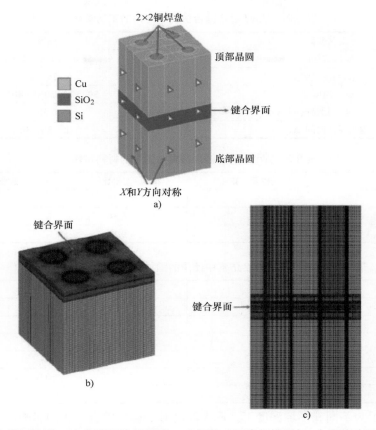

图 8.22 有限元模型：a) 有限元分析模型；b) 网格（下半部模型）；c) 网格（正视图截面）

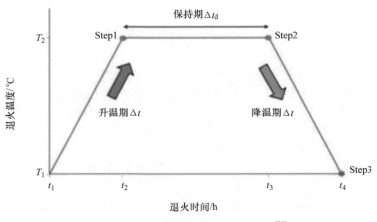

图 8.23 混合键合退火温度曲线[20]

表 8.2 WoW 混合键合仿真中使用的工艺参数[20]

工艺参数	数值
碟形坑深度	5nm，10nm，15nm
初始/结束温度 T_1	25℃
退火温度 T_2	300℃，350℃，400℃
升温/降温保持时间 Δt	0.5h
退火保持时间 Δt_d	1h，2h，3h

表 8.3 WoW 混合键合仿真中使用的材料特性[20]

材料	杨氏模量/GPa	泊松比	热膨胀系数/(10^{-6}/℃)
硅	131	0.28	2.6
铜	91.8	0.34	17.6
介质（SiO_2）	73	0.17	0.5

表 8.4 WoW 混合键合仿真中使用的铜材料弹塑性蠕变材料特性[20]

材料特性	数值
屈服强度/MPa	321
正切模量/MPa	2000
蠕变常数 1	1.43×10^{10}
蠕变常数 2	2.5
蠕变常数 3	−0.9
蠕变常数 4	23695

图 8.24 所示为不同碟形坑深度下，退火温度对 Cu-Cu 键合面积的影响。可以看到，退火温度和碟形坑深度对于获得良好的 Cu-Cu 键合非常重要。退火温度越高（400℃），碟形坑深度越小（5nm），Cu-Cu 键合越好（≥97% 铜焊盘面积）。从图 8.24 还可以看出，在低退火温度（300℃）下，碟形坑深度比在高退火温度（如 ≥350℃[20]）的影响更为明显。

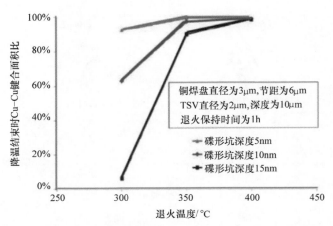

图 8.24 退火温度和碟形坑深度对 Cu-Cu 键合面积的影响[20]

对铜界面上的峰值剥离应力的观察也得到了类似的结果,如图 8.25[20] 所示。退火温度越高、碟形坑深度越小,Cu-Cu 键合界面的峰值应力越小。这符合 Cu-Cu 键合区的形成趋势。而介质键合界面的剥离应力则与之相反,如图 8.26 所示。可以看出,退火温度越高、碟形坑深度越小,介质界面的剥离应力峰值越高,导致介质层发生分层甚至断裂的可能性越大。

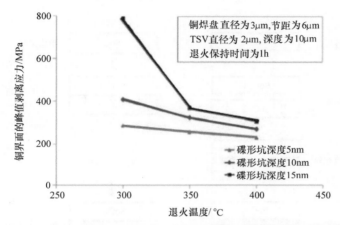

图 8.25　退火温度和碟形坑深度对 Cu 界面峰值剥离应力的影响[20]

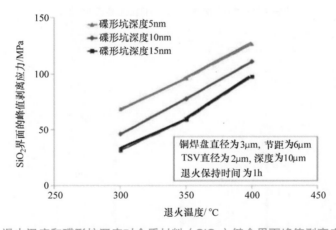

图 8.26　退火温度和碟形坑深度对介质材料(SiO_2)键合界面峰值剥离应力的影响

8.6.2　台积电的混合键合

图 8.27 所示为台积电的前端集成片上系统(system on integrated chip,SoIC)技术以及采用倒装技术的传统 3D IC 集成[21, 22]。可以看出,SoIC 与 3D IC 集成的关键区别在于 SoIC 是无凸点的,且芯粒(chiplet)之间的互连方式是 Cu-Cu 混合键合。SoIC 的组装过程既可以是晶圆 - 晶圆(wafer-on-wafer,WoW),也可以是芯片 - 晶圆(chip-on-wafer,CoW)或芯片 - 芯片

（chip-on-chip，CoC）混合键合。图 8.28 所示为不同键合组装技术如倒装芯片、2.5D/3D、SoIC 和 SoIC+ 的凸点密度。可以看到，SoIC 可以实现超窄节距、超高密度互连。SoIC 的另一个优点是不存在窄节距倒装芯片组装中的芯片 - 封装 - 交互（chip-package-interaction，CPI）可靠性问题。

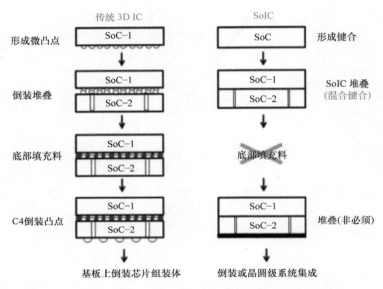

图 8.27 传统 3D IC 集成与 SoIC 集成的流程对比 [21]

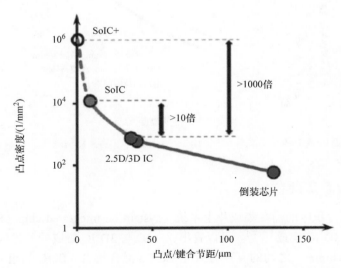

图 8.28 倒装芯片、2.5D/3D、SoIC 和 SoIC+ 的凸点密度与凸点 / 键合节距之间的关系 [21]

SoIC 技术具有比倒装芯片技术更好的电性能，如图 8.29 所示（SoIC 中芯粒是垂直混合键合的，倒装芯片是 2D 平面并排组装的）。可以看出，SoIC 技术的插入损耗几乎为零，远小于倒装芯片技术[21, 22]（见图 8.29）。

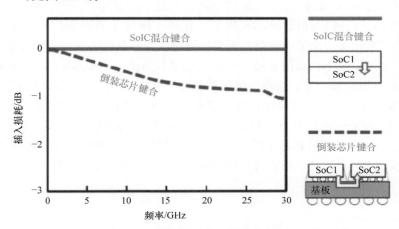

图 8.29　SoIC 和倒装芯片在不同频率下的插入损耗[21]

图 8.30a 所示为台积电后端集成 SoIC 的 CoWoS 技术，用于超高密度和高性能领域，如高性能计算（high-performance computing，HPC）等应用。图 8.30b 所示为台积电后端集成 SoIC 的集成扇出型封装堆叠（integrated fan-out package-on-package，InFO_PoP），主要应用在移动通信领域。在这两种应用场景中，SoIC 就相当于传统的片上系统（system-on-chip，SoC）。

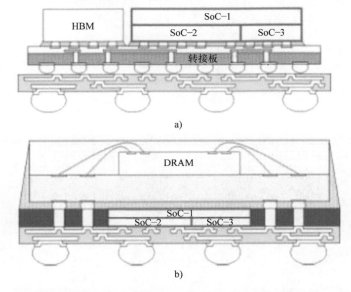

图 8.30　a）用于高性能应用场景，集成有 SoIC 的 CoWoS 技术；b）用于移动通信场景，集成有 SoIC 的 InFO_PoP 技术[21]

8.6.3 IMEC 的混合键合

图 8.31 所示为欧洲微电子研究中心（Interuniversity Microelectronics Centre，IMEC）在一个大小为 $240 \times 240 \mu m^2$ 的单元集成有 TSV 的混合键合测试结构[23-26]。该单元在 4.32mm × 4.32mm 的方形芯片内被排列成 16×16 阵列。各种不同尺寸的测试芯片均可以从晶圆上以 4.32mm 的步长切割制备。为了实现面对面混合键合，两片晶圆还进行了额外的表面处理，即 500nm SiO_2 和 120nm SiCN。如图 8.32 中的透射电子显微镜（transmission electron microscopy，TEM）图像所示，混合键合界面由嵌入顶部晶圆电介质的 $0.54 \mu m$ 方形焊盘和镀在底部晶圆上的 $1.17 \mu m$ 方形焊盘组成。对准和键合均在室温下完成，然后在 250℃下进行 2h 的键合后退火[23-26]。混合键合堆叠的截面如图 8.33 所示。顶部晶圆被减薄至 $50 \mu m$，以便从背面露出 TSV（via-middle 工艺）。RDL 和倒装芯片铜柱在顶部晶圆背面进行制作，以便进行电气测量和倒装芯片组装。有关混合键合的热、机械可靠性的更多信息，请阅读参考文献 [23-26]。

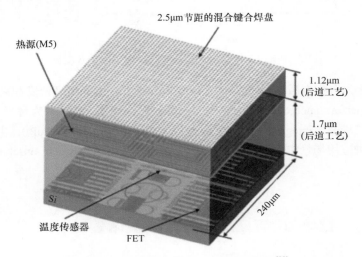

图 8.31 测试芯片的单元结构示意图[23]

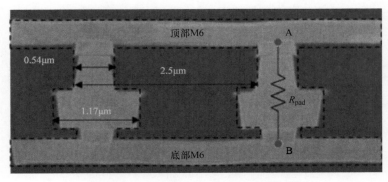

图 8.32 混合键合界面的 TEM 图像[23]

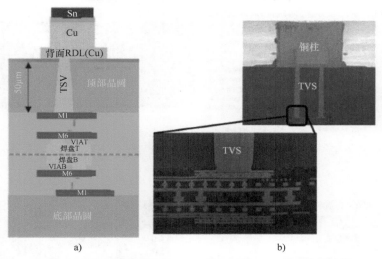

图 8.33 混合键合堆叠：a）截面示意图；b）SEM 图像[23]

8.6.4 格罗方德的混合键合

图 8.34 所示为格罗方德（Globafoundries）带有 TSV 的混合键合[27]。图 8.34a 所示为晶圆代工厂中集成 TSV 的顶部晶圆和未集成 TSV 的分立晶圆。两片晶圆的混合键合焊盘（hybrid bond terminal，HBT）金属需要匹配好。图 8.34b 所示为进行面对面键合的顶部和底部晶圆。表面处理对于混合晶圆键合（hybrid wafer bonding，HWB）工序非常重要，表面平整度是避免产生键合孔洞的关键，键合孔洞会导致较差的键合强度和导电性。形状校正在铜焊盘的对齐中也起着至关重要的作用[27]。这时两片晶圆的厚度都是完整的。图 8.34c 所示为在背面焊盘金属化工艺之后的最终晶圆堆叠的截面图，暴露出 TSV（减薄后）的一侧就是凸点/焊盘一侧。图 8.35 和图 8.36 所示为混合键合界面附近测试结构的设计和截面图。图 8.35 所示为通孔链宏观和晶圆混合键合界面处的通孔链截面图，图 8.36 所示为单触点宏观设计和混合晶圆键合界面处的单触点截面图。相关可靠性试验结果，请阅读参考文献 [27]。

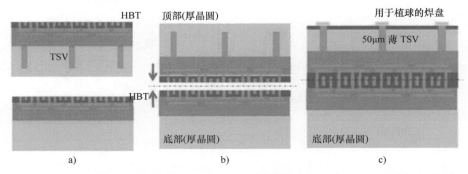

图 8.34 a）具有 TSV 的顶部晶圆和不带 TSV 的底部晶圆；b）进行面对面键合的顶部和底部晶圆；c）暴露出的 TSV（减薄后）侧是凸点/焊盘侧[27]

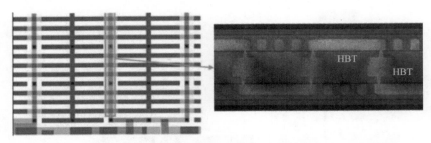

图 8.35 通孔链宏观设计及晶圆混合键合界面处的通孔链截面图[27]

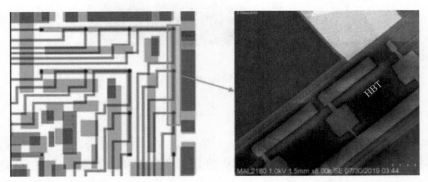

图 8.36 单触点宏观设计及混合晶圆键合界面处的单触点截面图[27]

8.6.5 三菱的混合键合

三菱（Mitsubishi）混合键合的关键工艺步骤如图 8.37 所示[28]。CMP 处理前晶圆的铜电极、SiO_2、Ti 和铝线如图 8.37a 所示，图 8.37b 所示为 CMP 处理之后的结构。铜电极和 SiO_2 之间的高度差为 10～20nm。图 8.37c 所示为使用 Si 薄膜键合后的晶圆。键合是在 6in 测试元件组（test-element-group，TEG）晶圆和 8in TEG 晶圆之间进行的。晶圆表面采用 Ar 快原子

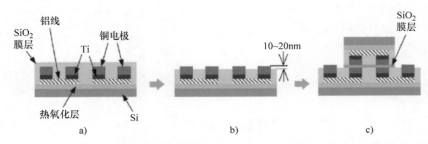

图 8.37 Cu/SiO_2 混合键合的关键工艺步骤：a）CMP 工艺前；b）CMP 工艺后，Cu 电极和 SiO_2 之间的水平高差为 10～20nm；c）使用 Si 薄膜进行键合[28]

束（fast atom beam，FAB）实现活化。在底部 6in 晶圆表面以 0.4nm/min 的速率沉积了一层 Si 薄膜，键合后的薄膜总厚度估计约为 4nm。8in 和 6in TEG 晶圆在以非常高的精度对齐后，施加 10000kgf 的键合压力完成接触。所有 Ar-FAB 辐照过程均在低于 5×10^{-6}Pa [28] 的本底真空下进行。图 8.38 所示为混合键合的结果。图 8.38a 是键合后 Cu/SiO_2 混合界面的 SEM 截面图像，可以看到没有产生明显的接缝或间隙，且对准误差约为 1μm。图 8.38b 是键合后 Cu/Cu 电极界面的 TEM 截面图，图 8.38c 是键合后 SiO_2/SiO_2 界面的 TEM 截面图。在两个键合界面，没有观察到微孔洞的出现。但是观察到有一个厚度约为 5nm 的中间层，该层可能是非晶硅层。

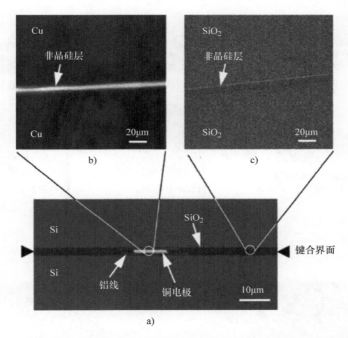

图 8.38 a）Cu/SiO_2 混合键合界面的 SEM 截面图；b）Cu/Cu 电极键合界面的 TEM 截面图；c）SiO_2/SiO_2 键合界面的 TEM 截面图 [28]

8.6.6 Leti 的混合键合

图 8.39 所示为法国电子信息技术研究所（Leti）混合键合的关键工艺步骤 [29]。他们的 Ti/SiO_2 混合键合是在 200mm 非功能性非全制程晶圆上开发的。图 8.39（1）所示为金属图形的制备流程。首先沉积 TiN 作为阻挡层，然后溅射沉积 Ti 层（500nm），再用薄 TiN 覆盖 Ti 层。图 8.39（2）所示为金属堆叠的图形化。图 8.39（3）所示为介质层的沉积。图形化之后，焊盘被正硅酸四乙酯（tetraethylorthosilicate，TEOS）氧化物包封。在 TEOS 氧化物沉积之后进行退火，原因在于：①介质层致密化和②去除介质层中吸收的水分子，以防止后续因脱气而形成键合孔洞。图 8.39（4）所示为用于晶圆平坦化和表面处理的 CMP 工艺。对于 Ti/SiO_2 平坦化，CMP 工艺从去除 SiO_2 开始，最终停止在 Ti/SiO_2 混合表面。表面形貌测量通过白光干涉

法（white-light interferometry，WLI）完成，氧化物表面粗糙度通过原子力显微镜（atomic force microscopy，AFM）测定，结果表明其具有优异的表面平整度和粗糙度[29]。图 8.39（5）所示为顶部和底部晶圆的键合。图 8.39（6）所示为金属互连的退火过程。最后，键合晶圆在 400℃下退火 2h，以确保正确的金属/金属连接。键合界面的质量采用聚焦离子束扫描电子显微镜（focused ion beam scanning electron microscopy，FIB-SEM）和 TEM 进行原位评估。图 8.40 所示为 TEM 图像，展示了出色的 Ti/Ti 连接，且金属和介质界面没有观察到明显接缝[29]。

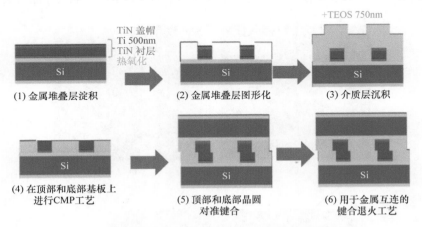

图 8.39　Ti/TEOS 混合 3D 互连工艺流程[29]

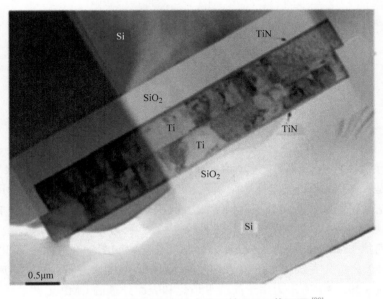

图 8.40　Ti/SiO$_2$ 混合键合界面的 TEM 截面图[29]

8.6.7 英特尔的混合键合

在英特尔架构日（2020 年 8 月 13 日）期间，英特尔展示了具有混合键合技术及传统微凸点倒装芯片技术的 FOVEROS，如图 8.41 所示。可以看出，使用混合键合技术，焊盘节距可以降至 10μm，每平方毫米内可制备 10000 个无凸点互连结构，这比采用 50μm 节距微凸点的倒装芯片技术高出许多倍。

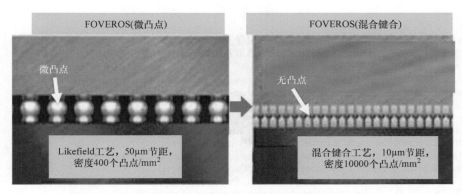

图 8.41 英特尔的 FOVEROS 混合键合

8.7 总结和建议

一些重要的结论和建议总结如下。
1）介绍了一些低温 DBI（混合键合）的基本原理。
2）简要介绍了混合键合技术的一些近期进展。
3）受不同芯片尺寸和晶圆良率的影响，芯片 - 晶圆（CoW）混合键合技术是另一种提高组装良率的方法。
4）迄今为止，索尼的 BI-CIS 是唯一使用 WoW 无凸点混合键合的大规模量产产品。
5）为了让更多的大规模量产产品采用无凸点混合键合技术，应在以下领域投入更多的研发精力：
① 降低成本；
② 纳米级形貌（CMP）；
③ 薄晶圆处理；
④ 设计参数优化；
⑤ 工艺参数优化；
⑥ 键合环境；
⑦ CoW 和 WoW 键合对准；
⑧ 晶圆变形和翘曲（控制）；
⑨ 检查和测试；

⑩ 触点完整性；

⑪ 触点质量和可靠性；

⑫ 制造良率；

⑬ 生产量；

⑭ 热管理。

参 考 文 献

1. Kim, B., T. Matthias, M. Wimplinger, P. Kettner, and P. Lindner, "Comparison of Enabling Wafer Bonding Techniques for TSV Integration", *ASME Paper No. IMECE2010-400002*.
2. Chen, K., S. Lee, P. Andry, C. Tsang, A. Topop, Y. Lin, Y., J. Lu, A. Young, M., Ieong, and W. Haensch, W., "Structure, Design and Process Control for Cu Bonded Interconnects in 3D Integrated Circuits", *IEEE Proceedings of International Electron Devices Meeting,* (IEDM 2006), San Francisco, CA, December 11–13, 2006, pp. 367-370.
3. Liu, F., Yu, R., Young, A., Doyle, J., Wang, X., Shi, L., Chen, K., Li, X., Dipaola, D., Brown, D., Ryan, C., Hagan, J., Wong, K., Lu, M., Gu, X., Klymko, N., Perfecto, E., Merryman, A., Kelly K., Purushothaman, S., Koester, S., Wisnieff, R., and Haensch, W., "A 300- Wafer-Level Three-Dimensional Integration Scheme Using Tungsten Through-Silicon Via and Hyprid Cu-Adhesive Bonding", *IEEE Proceedings of IEDM*, December 2008, pp. 1–4.
4. Yu, R., Liu, F., Polastre, R., Chen, K., Liu, X., Shi, L., Perfecto, E., Klymko, N., Chace, M., Shaw, T., Dimilia, D., Kinser, E., Young, A., Purushothaman, S., Koester, S., and Haensch W., "Reliability of a 300-mm-compatible 3DI Technology Base on Hybrid Cu-adhesive Wafer Bonding", *Proceedings of Symposium on VLSI Technology Digest of Technical Papers*, 2009, pp. 170–171.
5. Shigetou, A. Itoh, T., Sawada, K., and Suga, T., "Bumpless Interconnect of 6-um pitch Cu Electrodes at Room Temperature", In *IEEE Proceedings of ECTC*, Lake Buena Vista, FL, May 27–30, 2008, pp. 1405-1409.
6. Tsukamoto, K., E. Higurashi, and T. Suga, "Evaluation of Surface Microroughness for Surface Activated Bonding", *Proceedings of IEEE CPMT Symposium Japan*, August 2010,, pp. 147–150.
7. Kondou, R., C. Wang, and T. Suga, "Room-temperature Si-Si and Si-SiN wafer bonding", *Proceedings of IEEE CPMT Symposium Japan*, August 2010,, pp. 161–164.
8. Shigetou, A. Itoh, T., Matsuo, M., Hayasaka, N., Okumura, K., and Suga, T., "Bumpless Interconnect Through Ultrafine Cu Electrodes by Mans of Surface-Activated Bonding (SAB) Method", *IEEE Transaction on Advanced Packaging*, Vol. 29, No. 2, May 2006, pp. 226.
9. Wang, C., and Suga, T., "A Novel Moire Fringe Assisted Method for Nanoprecision Alignment in Wafer Bonding", In *IEEE Proceedings of ECTC*, San Diego, CA, May 25–29, 2009, pp. 872-878.
10. Wang, C., and Suga, T., "Moire Method for Nanoprecision Wafer-to-Wafer Alignment: Theory, Simulation and Application", *IEEE Proceedings of Int. Conference on Electronic Packaging Technology & High Density Packaging*, August 2009, pp. 219–224.
11. Higurashi, E., Chino, D., Suga, T., and Sawada, R., "Au-Au Surface-Activated Bonding and Its Application to Optical Microsensors with 3-D Structure", *IEEE Journal of Selected Topic in Quantum Electronics*, Vol. 15, No. 5 September/October 2009, pp. 1500–1505.
12. Kagawa, Y., N. Fujii, K. Aoyagi, Y. Kobayashi, S. Nishi, N. Todaka, et al., "Novel stacked CMOS image sensor with advanced Cu2Cu hybrid bonding," *Proceedings of IEEE/IEDM*, Dec. 2016, pp. 8.4.1–4.
13. Kagawa, Y., N. Fujii, K. Aoyagi, Y. Kobayashi. S. Nishi, N. Todaka, S. Takeshita, J. Taura, H. Takahashi, Y. Nishimura, et al., "An Advanced CuCu Hybrid Bonding for Novel Stack CMOS Image Sensor", *IEEE/EDTM Proceedings*, March 2018, pp. 1–3.
14. Gao, G., L. Mirkarimi, G. Fountain, T. Workman, J. Theil, and G. Guevara, C. Uzoh, D. Suwito, B. Lee, K. Bang, and R. Katkar, "Die to Wafer Stacking with Low Temperature Hybrid Bonding", *IEEE/ECTC Proceedings*, May 2020, pp. 589–594.

15. Gao, G., L. Mirkarimi, T. Workman, G. Fountain, J. Theil, G. Guevara, P. Liu, B. Lee, P. Mrozek, M. Huynh, C. Rudolph, T. Werner, and A. Hanisch, "Low Temperature Cu Interconnect with Chip to Wafer Hybrid Bonding", *IEEE/ECTC Proceedings*, May 2019, pp. 628–635.
16. Gao, G., T. Workman, L. Mirkarimi, G. Fountain, J. Theil, G. Guevara, C. Uzoh, B. Lee, P. Liu, and P. Mrozek, "Chip to Wafer Hybrid Bonding with Cu Interconnect: High Volume Manufacturing Process Compatibility Study", *IWLPC Proceedings*, October 2019, pp. 1–9.
17. Gao, G., L. Mirkarimi, T. Workman, G. Guevara, J. Theil, C. Uzoh, G. Fountain, B. Lee, P. Mrozek, M. Huynh, and R. Katkar, "Development of Low Temperture Direct Bond Interconnect Technology for Die-to-Wafer and Die-to-Die Applications – Stacking, Yield Improvement, Reliability Assessment", *IWLPC Proceedings,* October 2018, pp. 1–7.
18. Lee, B., P. Mrozek, G. Fountain, J. Posthill, J. Theil, G. Gao, R. Katkar, and L. Mirkarimi, "Nanoscale Topography Characterization for Direct Bond Interconnect", *IEEE/ECTC Proceedings*, May 2019, pp. 1041–1046.
19. Ji, L., F. Che, H. Ji, H. Li, and M. Kawano, "Modelling and characterization on wafer to wafer hybrid bonding technology for 3D IC packaging", *IEEE/EPTC Proceedings*, December 2019, pp. 87–94.
20. Ji, L., F. Che, H. Ji, H. Li, and M. Kawano, "Bonding integrity enhancement in wafer to wafer fine pitch hybrid bonding by advanced numerical modeling", *IEEE/ECTC Proceedings,* May 2020, pp. 568–575.
21. Chen, M. F., C. S. Lin, E. B. Liao, W. C. Chiou, C. C. Kuo, C. C. Hu, C. H. Tsai, C. T. Wang and D. Yu, "SoIC for Low-Temperature, Multi-Layer 3D Memory Integration", *IEEE/ECTC Proceedings,* May 2020, pp. 855–860.
22. Chen, F., M. Chen, W. Chiou, D. Yu, "System on Integrated Chips (SoIC^{TM}) for 3D Heterogeneous Integration", *IEEE/ECTC Proceedings,* May 2019, pp. 594–599.
23. Cherman, V., S. Van Huylenbroeck, M. Lofrano, X. Chang, H. Oprins, M. Gonzalez, G. Van der Plas, G. Beyer, K. Rebibis, and E. Beyne, "Thermal, Mechanical and Reliability assessment of Hybrid bonded wafers, bonded at 2.5 μm pitch", *IEEE/ECTC Proceedings,* May 2020, pp. 548–553.
24. Kennes, K., A. Phommahaxay, A. Guerrero, O. Bauder, S. Suhard, P. Bex, S. Iacovo, X. Liu, T. Schmidt, G. Beyer, and E. Beyne, "Introduction of a New Carrier System for Collective Die-to-Wafer Hybrid Bonding and Laser-Assisted Die Transfer"*, IEEE/ECTC Proceedings,* May 2020, pp. 296–302.
25. Huylenbroeck, S., J. De Vos, Z. El-Mekki, G. Jamieson, N. Tutunjyan, K. Muga, M. Stucchi, A. Miller, G. Beyer, and E. Beyne, "A Highly Reliable 1.4 μm pitch Via-last TSV Module for Wafer-to-Wafer Hybrid Bonded 3D-SOC Systems", *IEEE/ECTC Proceedings,* May 2019, pp. 1035–1040.
26. Suhard, S., A. Phommahaxay, K. Kennes, P. Bex, F. Fodor, M. Liebens, J. Slabbekoorn, A. Miller, G. Beyer, and E. Beyne, "Demonstration of a collective hybrid die-to-wafer integration", *IEEE/ECTC Proceedings,* May 2020, pp. 1315–1321.
27. Fisher, D., S. Knickerbocker, D. Smith, R. Katz, J. Garant, J. Lubguban, V. Soler, and N. Robson, "Face to Face Hybrid Wafer Bonding for Fine Pitch Applications", *IEEE/ECTC Proceedings,* May 2019, pp. 595–600.
28. Utsumi, J., K. Ide, and Y. Ichiyanagi, "Cu/SiO2 hybrid bonding obtained by surface-activated bonding method at room temperature using Si ultrathin films", *Micro and Nano Engineering*, February 2019, pp. 1–6.
29. Jouve, A., E. Lagoutte1, R. Crochemore1, G. Mauguen1, T. Flahaut1, C. Dubarry1, V. Balan, F. Fournel, E. Bourjot,, F. Servant1, M. Scannell1, K. Rohracher, T. Bodner, A. Faes, and J. Hofrichter, "A reliable copper-free wafer level hybrid bonding technology for high-performance medical imaging sensors", *IEEE/ECTC Proceedings,* May 2020, pp. 201–209.
30. Jani, I., D. Lattard, P. Vivet, L. Arnaud, S. Cheramy, E. Beigné, Alexis Farcy, Joris Jourdon, Y. Henrion, E. Deloffre, and H. Bilgen, "Characterization of fine pitch Hybrid Bonding pads using electrical misalignment test vehicle", *IEEE/ECTC Proceedings,* May 2019, pp. 1926–1932.
31. Chong, S., X. Ling, H. Li, and S. Lim, "Development of Multi-Die Stacking with Cu–Cu interconnects using Gang Bonding Approach", *IEEE/ECTC Proceedings,* May 2020, pp. 188–193.
32. Chong, S., and S. Lim, "Comprehensive Study of Copper Nano-paste for Cu–Cu Bonding", *IEEE/ECTC Proceedings,* May 2019, pp. 191–196.

33. Araki, N., S. Maetani, Y. Kim, S. Kodama, and T. Ohba, "Development of Resins for Bumpless Interconnects and Wafer-On-Wafer (WOW) Integration", *IEEE/ECTC Proceedings,* May 2019, pp. 1002–1008.
34. Fujino, M., K.Takahashi, Y. Araga, and K. Kikuchi, "300 mm wafer-level hybrid bonding for Cu/interlayer dielectric bonding in vacuum", *Japanese J. Appl. Phys.* Vol. 59, February 2020, pp. 1–8.
35. Kim, S., P. Kang, T. Kim, K. Lee, J. Jang, K. Moon, H. Na, S. Hyun, and K. Hwang, "Cu Microstructure of High Density Cu Hybrid Bonding Interconnection", *IEEE/ECTC Proceedings,* May 2019, pp. 636–641.
36. Tong, Q., G. Fountain, and P. Enquist, "Method for Low Temperature Bonding and Bonded Structure", *US 6,902,987 B1*, filed on February 16, 2000, granted on June 7, 2005.
37. Tong, Q., G. Fountain, and P. Enquist, "Method for Low Temperature Bonding and Bonded Structure", *US 7,387,944 B2*, filed on August 14, 2005, granted on June 17, 2008.
38. Burns, J., Aull, B., Keast, C., Chen, C., Chen, C. Keast, C., Knecht, J., Suntharalingam, V., Warner, K., Wyatt, P., and Yost, D., "A Wafer-Scale 3-D Circuit Integration Technology", *IEEE Transactions on Electron Devices*, Vol. 53, No. 10, October 2006, pp. 2507–2516.
39. Chen, C., Warner, K., Yost, D., Knecht, J., Suntharalingam, V., Chen, C., Burns, J., and Keast, C., "Sealing Three-Dimensional SOI Integrated-Circuit Technology", *IEEE Proceedings of Int. SOI Conference*, 2007, pp. 87–88.
40. Chen, C., Chen, C., Yost, D., Knecht, J., Wyatt, P., Burns, J., Warner, K., Gouker, P., Healey, P., Wheeler, B., and Keast, C., "Three-dimensional integration of silicon-on-insulator RF amplifier", *Electronics Letters*, Vol. 44, No. 12, June 2008, pp. 1–2.
41. Chen, C., Chen, C., Yost, D., Knecht, J., Wyatt, P., Burns, J., Warner, K., Gouker, P., Healey, P., Wheeler, B., and Keast, C., "Wafer-Scale 3D Integration of Silicon-on-Insulator RF Amplifiers", *IEEE Proceedings of Silicon Monolithic IC in RF Systems*, 2009, pp. 1–4.
42. Chen, C., Chen, C., Wyatt, P., Gouker, P., Burns, J., Knecht, J., Yost, D., Healey, P., and Keast, C, "Effects of Through-BOX Vias on SOI MOSFETs", *IEEE Proceedings of VLSI Technology, Systems and Applications*, 2008, pp. 1–2.
43. Chen, C., Chen, C., Burns, J., Yost, D., Warner, K., Knecht, J., Shibles, D., and Keast, C, "Thermal Effects of Three Dimensional Integrated Circuit Stacks", *IEEE Proceedings of Int. SOI Conference*, 2007, pp. 91–92.
44. Aull, B., Burns, J., Chen, C., Felton, B., Hanson, H., Keast, C., Knecht, J., Loomis, A., Renzi, M., Soares, A., Suntharalingam, V., Warner, K., Wolfson, D., Yost, D., and Young, D., "Laser Radar Imager Based on 3D Integration of Geiger-Mode Avalanche Photodiodes with Two SOI Timing Circuit Layers", *IEEE Proceedings of Int. Solid-State Circuits Conference,* 2006, pp. 16.9.
45. Chatterjee, R., Fayolle, M., Leduc, P., Pozder, S., Jones, B., Acosta, E., Charlet, B., Enot, T., Heitzmann, M., Zussy, M., Roman, A., Louveau, O., Maitreqean, S., Louis, D., Kernevez, N., Sillon, N., Passemard, G., Pol, V., Mathew, V., Garcia, S., Sparks, T., and Huang, Z., "Three dimensional chip stacking using a wafer-to-wafer integration", *IEEE Proceedings of IITC*, 2007, pp. 81–83.
46. Ledus, P., Crecy, F., Fayolle, M., Fayolle, M., Charlet, B., Enot, T., Zussy, M., Jones, B., Barbe, J., Kernevez, N., Sillon, N., Maitreqean, S., Louis, D., and Passemard, G., "Challenges for 3D IC integration: bonding quality and thermal management", *IEEE Proceedings of IITC*, 2007, pp. 210–212.
47. Poupon, G., Sillon, N., Henry, D., Gillot, C., Mathewson, A., Cioccio, L., Charlet, B., Leduc, P., Vinet, M., and Batude, P., "System on Wafer: A New Silicon Concept in Sip", *Proceedings of the IEEE*, Vol. 97, No. 1, January 2009, pp. 60–69.
48. Sukegawa, S., T. Umebayashi, T. Nakajima, H. Kawanobe, K. Koseki, I. Hirota, T. Haruta, et al., "A 1/4-inch 8Mpixel Back-Illuminated Stacked CMOS Image Sensor," *Proceedings of IEEE/ISSCC*, San Francisco, CA, February 2013, pp. 484–486.

第 9 章

芯粒异质集成

9.1 引言

近些年,芯粒的异质集成(或者异质芯粒集成)受到了广泛关注[1-18]。如 AMD 的 EPYZ 和英特尔的 Lakefield 等大批量生产的微处理器均采用芯粒设计和异质集成封装技术,本章将对其进行介绍。芯粒异质集成的定义和优缺点也会在本章介绍。首先简要介绍美国国防部高级研究计划局(defense advanced research projects agency,DARPA)在芯粒异质集成方面的工作,以及片上系统(SoC)与芯粒异质集成的对比。

9.2 DARPA 在芯粒异质集成方面的工作

DARPA 在过去超过 15 年的时间里与 30 多家一流的公司和高校(如英特尔、美光、Cadence、Synopsys、洛克希德·马丁、Northrop Grumman、密歇根大学和佐治亚理工学院)在异质集成方面取得了很好的进展,本节将简要介绍他们在异质集成方面的关键项目。DARPA 在异质集成方面的第一项工作是硅上化合物半导体材料(compound semiconductor materials on silicon,COSMOS)项目[19],该项目于 2007 年 5 月启动。COSMOS 开发了三种独特的技术方法来实现磷化铟异质结双极晶体管与深亚微米 Si 互补金属氧化物半导体(complementary metal-oxide semiconductor,CMOS)的异质集成。COSMOS 目前是多样化可用异质集成(diverse accessible heterogenous integration,DAHI)项目[20]中的关键。DAHI 项目正在面向以下关键技术挑战进行开发:

1) 异质集成工艺开发。
2) 高良率制造和晶圆代工厂建设。
3) 电路设计和架构创新。

DARPA 于 2017 年启动了通用异质集成和 IP(知识产权)复用策略 [common heterogeneous integration and IP(intellectual Property)reuse strategies,CHIPS] 项目[21]。CHIPS 项目的目标是用芯粒制造模块化计算机。CHIPS 项目目前正在解决集成标准、IP 块和设计工具的问题。英特尔向 CHIPS 项目参与者提供其先进接口总线(advanced interface bus,AIB)技术的免费使用许可。

2019年年中，美国海军提出了一项现阶段最先进的异质集成封装（state-of-the-art heterogeneous integrated packaging，SHIP）项目[22]。SHIP项目的主要目标是充分利用产业界专业知识，实现安全、可评估和经济高效的现阶段最先进的集成、设计、组装和测试技术。其设计还必须遵守DARPA的CHIPS项目下制定的接口标准，以确保最终产品的正确嵌入和可测试性。

9.3 SoC（片上系统）

如第2章所述，SoC将具有不同功能的集成电路，如中央处理器（central processing unit，CPU）、图形处理器（graphic processing unit，GPU）、存储器等集成进单个芯片以构成系统或子系统。最著名的SoC是苹果的应用处理器（application processor，AP），其A10到A14应用处理器简单展示于图9.1中。图9.2所示为不同特征尺寸（工艺技术）的芯片晶体管数量随年份的变化关系，从中可以看到摩尔定律的影响，它通过减小特征尺寸来增加晶体管数量和功能复杂性。但不幸的是，摩尔定律的终结正在迅速临近，要想减小特征尺寸（继续微缩）以制作SoC越来越困难而且成本高昂。根据国际商业策略（International Business Strategies）公司的调研，图9.3所示为先进芯片设计成本随特征尺寸减小（直至5nm）的变化关系。可以看出，仅完成5nm特征尺寸芯片的设计就需要5亿多美元，5nm工艺技术的开发还需要10亿美元。

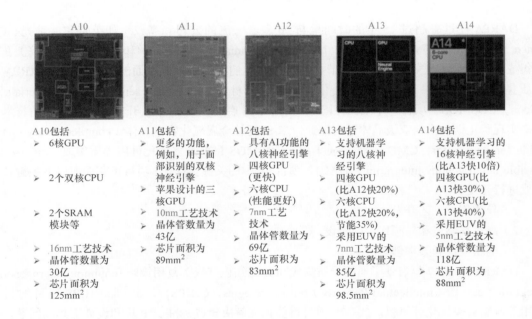

图9.1 苹果的应用处理器（SoC）

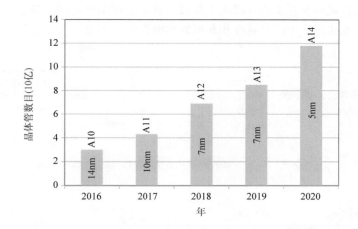

图 9.2　不同特征尺寸苹果应用处理器芯片晶体管数目随年份的变化

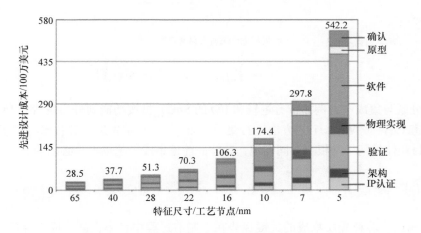

图 9.3　不同特征尺寸先进芯片设计成本

9.4　芯粒异质集成

芯粒异质集成是与 SoC 相对的。芯粒异质集成将 SoC 重新设计为更小的芯粒，然后利用封装技术将不同材料制作的，具有不同功能的，由不同设计公司和代工厂制作的，具有不同晶圆尺寸、特征尺寸的芯粒集成到一个系统或子系统中[23, 24]。其中一颗芯粒就是一种由可复用 IP（知识产权）块组成的功能集成电路（IC）模块。

如图 9.4 所示，目前至少有两种不同的芯粒异质集成方法，即芯片切分与集成（由成本和良率驱动）和芯片分区与集成（由成本和技术优化驱动）。在芯片切分与集成中，逻辑（logic）等 SoC 被切分为更小的芯粒，如 logic1、logic2 和 logic3。这些芯粒可以通过前道芯片 - 晶圆（chip-on-wafer，CoW）或晶圆 - 晶圆（wafer-on-wafer，WoW）工艺完成堆叠（集成)[7-9]，然后采用先进封装技术将其组装（集成）在单个封装体的同一基板上。应该强调的是，前道工艺芯粒集成能获得更小的封装面积和更好的电性能，不过这不是必须的。

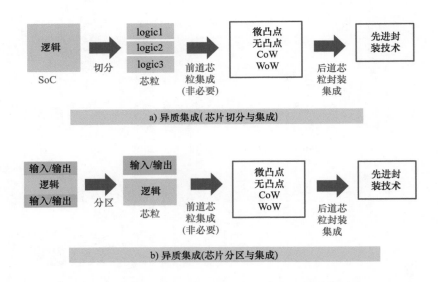

图 9.4　芯粒：a）芯片切分与集成；b）芯片分区与集成

在芯片分区与集成中，例如带有逻辑和 I/O 的 SoC，被按功能划分为逻辑和 I/O 芯粒模块，然后通过前道 CoW 或 WoW 工艺方法进行集成（堆叠）。再用先进封装技术将逻辑和 I/O 芯粒组装在单个封装体的同一基板上。同样地，芯粒的前道集成工艺也不是必须的。

9.5　芯粒异质集成的优缺点

与 SoC 相比，芯粒异质集成的关键优势在于制造过程中良率提高（成本降低）、设计过程中的上市时间短和成本降低。图 9.5 所示为单片设计和 2、3、4 芯粒设计所对应的每片晶圆良率（合格芯片百分比）与芯片尺寸的关系图[25]。可以看出，$360mm^2$ 单片芯片的良率是 15%，而 4 芯粒设计（每片 $99mm^2$）的良率将增加一倍以上，达到 37%。4 芯粒设计方式的总芯片面积会带来约 10% 的面积损失（$396mm^2$ 中用于各芯粒互连的硅面积为 $36mm^2$），但良率的显著提高会直接转化为成本的降低。同时，芯片分区也将会缩短上市时间。特别是 AMD 凭借其非常成功的 EPYC CPU 产品，已经证实，使用芯粒设计的 CPU 核可以将 32 核 CPU 的设计和制造成本降低 40%[26]。最后由于芯粒分散在整个封装体中，也会对热性能有所优化。

芯粒异质集成的缺点是
1) 接口和重复逻辑需要一些额外的面积开销。
2) 更高的封装成本。
3) 更高的复杂度和设计工作量。
4) 过去的设计方法学并不太适用于芯粒。

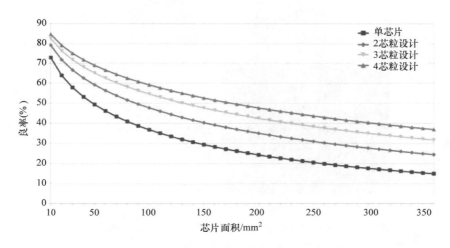

图 9.5　SoC 和不同芯粒设计的良率与芯片尺寸的关系

9.6　应用于芯粒异质集成的先进封装

9.6.1　有机基板上的 2D 芯粒异质集成

图 9.6 所示为有机基板上的 2D 芯粒异质集成示意图，这是正在和将要使用最多的结构。可以看出，这些芯粒通过焊料凸点和底部填充倒装于积层封装基板上，然后通过焊球安装在 PCB 上，组装方法主要有表面安装技术（surface mount technology，SMT）和倒装芯片技术。最有名的有机基板 2D 芯粒异质集成是 AMD EPYC 处理器[1-3]，如图 9.7 所示，将在本章 9.7 节中详细阐述。可以看到在一个较大的 I/O 芯片（采用 14nm 工艺技术制造）的两侧有 4 对芯粒（采用 7nm 工艺技术制造），它们通过键合和底部填充紧密安装在一个有机基板上。如果采用图 9.8 所示扇出型封装技术[27-30]，芯粒通过扇出 RDL 和焊球连接到 PCB 上，可以省去倒装芯片的凸点成型、积层封装基板和底部填充等。

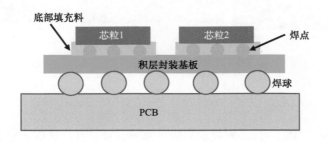

图 9.6　有机基板上的 2D 芯粒异质集成

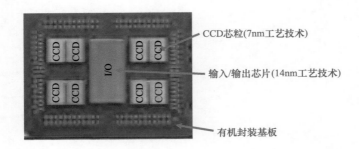

图 9.7 有机基板上的 AMD 第二代 EPYC 2D 芯粒异质集成 [1]

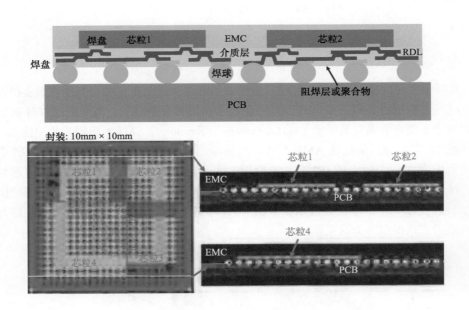

图 9.8 有机基板上的 2D 芯粒扇出型异质集成

9.6.2 有机基板上的 2.1D 芯粒异质集成

图 9.9 所示为有机基板上的 2.1D 芯粒异质集成示意图。可以看到这些芯粒通过焊料凸点倒装在具有薄膜层的积层封装基板上。最著名的 2.1D 基板是 Shinko 的 i-THOP（集成薄膜高密度有机封装）[31, 32]，如图 9.10 所示。薄膜层的金属线宽和线距为 2μm。

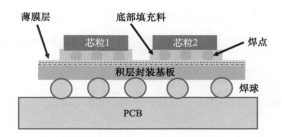

图 9.9　有机基板上的 2.1D 芯粒异质集成

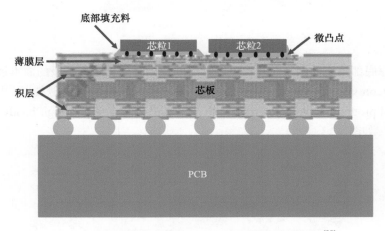

图 9.10　有机基板上的 Shinko 2.1D 芯粒异质集成[32]

9.6.3　有机基板上的 2.3D 芯粒异质集成

图 9.11 所示为有机基板上的 2.3D 芯粒异质集成。可以看到芯粒通过焊料凸点倒装在无芯板有机转接板上，最著名的例子由 Shinko 在 2012 年提出的（见图 9.12）。他们提出采用无芯板

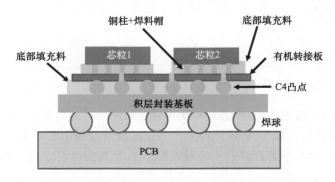

图 9.11　有机基板上的 2.3D 芯粒异质集成

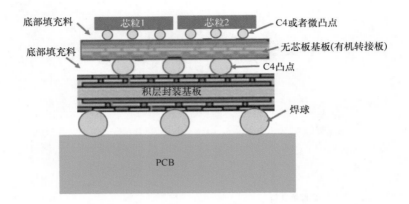

图 9.12　无芯板基板上的 Shinko 2.3D 芯粒异质集成

封装基板来支撑底部填充的倒装芯片。可以肯定的是，制作无芯板基板的成本远低于制作硅通孔（through-silicon via，TSV）转接板的成本。2016 年，思科使用类似技术[33]支持现场可编程门阵列（field programmable gate array，FPGA）和高带宽存储器（high bandwidth memories，HBM）的集成，如图 9.13 所示。

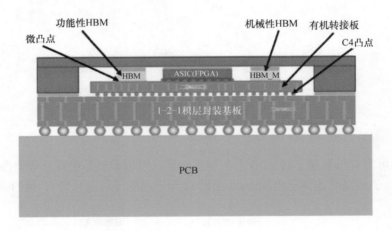

图 9.13　有机转接板上的思科 2.3D 芯粒异质集成[33]

9.6.4　硅基板（无源 TSV 转接板）上的 2.5D 芯粒异质集成

图 9.14 所示为硅基板上的 2.5D 芯粒异质集成。从图中可以看到，这些芯粒是通过微凸点（铜柱 + 焊料帽）倒装在无源 TSV 转接板上[34-41]，这块转接板上没有 CMOS 器件。最著名的例子是图 9.15 所示的台积电为其客户（如赛灵思[42-45]和英伟达[46]）提供的基板上晶圆上芯片（chip on wafer on substrate，CoWoS）技术。TSV 转接板通常有 4 层 RDL，最小布线间距为 0.4μm，用于支持 SoC 和 HBM，它对于高密度和高性能应用非常有用。

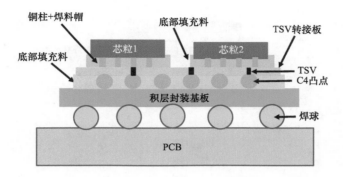

图 9.14　无源 TSV 转接板上的 2.5D 芯粒异质集成

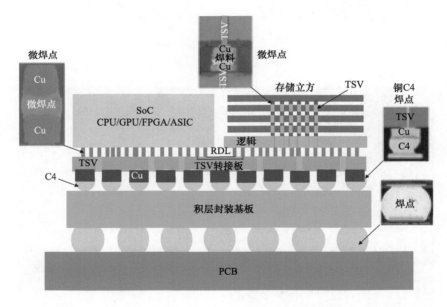

图 9.15　无源 TSV 转接板上的 2.5D（CoWoS-2）芯粒异质集成[41]

9.6.5　硅基板（有源 TSV 转接板）上的 3D 芯粒异质集成

图 9.16 所示为硅基板上的 3D 芯粒异质集成。可以看到芯粒通过微凸点（铜柱＋焊料帽）倒装在有源 TSV 转接板上，有源转接板上有 CMOS 器件。最著名的也是目前唯一大批量生产的一款产品是英特尔的 FOVEROS，如图 9.17 所示[4-6]。可以看到芯粒（计算芯片和存储芯片等）面对面地键合到有源 TSV 转接板上，然后该模块再键合到有机封装基板上。请阅读 7.3.3 节了解更多详细信息。

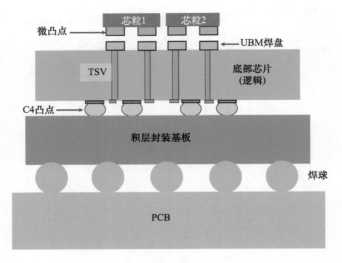

图 9.16　3D 芯粒异质集成

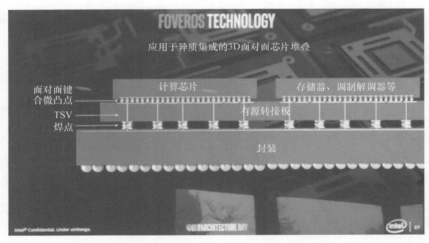

图 9.17　英特尔的 3D 芯粒异质集成（FOVEROS）

9.6.6　带互连桥的有机基板上的芯粒异质集成

图 9.18 所示为带互连桥的有机基板上的芯粒异质集成。芯粒同时通过 C4 凸点和 C2 凸点倒装键合在有机积层封装基板上，基板中埋入了带 RDL 的互连桥[47, 48]。最著名的互连桥是英特尔的埋入式多芯片互连桥（embedded multi-die interconnect bridge，EMIB），如图 9.19 所示，在英特尔的 FPGA（Agilex）中采用。

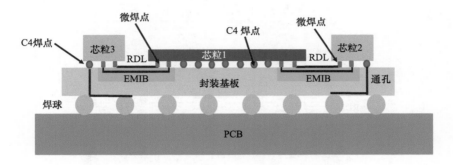

图 9.18 带互连桥的有机基板上的芯粒异质集成

图 9.19 带互连桥的有机基板上的英特尔芯粒异质集成（Agilex FPGA）[47]

9.6.7 PoP 芯粒异质集成

图 9.20 所示为封装堆叠（package-on-package，PoP）芯粒异质集成技术。可以看到，应用处理器芯粒采用台积电的集成扇出型（integrated fan-out，InFO）晶圆级系统集成（wafer level system integration，WLSI）技术进行封装（底部封装），而存储芯粒则采用引线键合到无芯板有机基板的技术进行封装（顶部封装）。iPhone XS 中应用处理器芯片组的截面如图 9.21 所示。

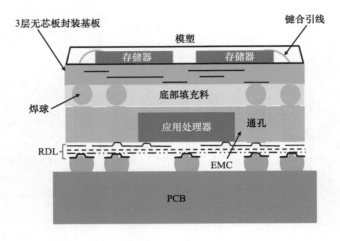

图 9.20　PoP 芯粒异质集成

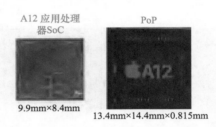

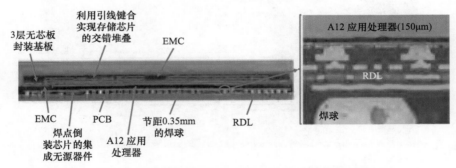

图 9.21　苹果 iPhone 手机的 PoP InFO 芯粒异质集成

9.6.8　扇出型 RDL 基板上的芯粒异质集成

图 9.22 所示为 RDL 基板上的芯粒异质集成。可以看到这些芯粒通过 C2 凸点（铜柱 + 焊料帽）倒装互连在扇出型后上晶（先 RDL）基板上，然后用 C4 凸点键合连接在积层有机封装基板上（形为 2.3D IC 集成），最后整个模块用焊球组装在 PCB 上[15, 49-55]。这种封装技术对于高密度、高性能和大型芯粒应用很有意义，仅次于 9.6.4 节所示的 CoWoS。对于较小的芯粒，如 5mm × 5mm，则使用先上晶扇出型基板（成本较低）[56-60] 就足够了。

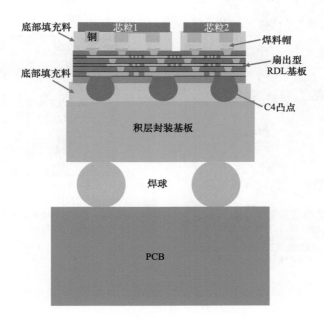

图 9.22　扇出型（先 RDL）基板上的芯粒异质集成

9.7　AMD 的芯粒异质集成

2019 年年中，AMD 推出了代号为 Rome 的第二代 EPYC 7002 系列处理器，它的核心数量翻了一番，达到 64 个。第二代 EPYC 是新一代服务器处理器，为数据中心设定了更高的标准。参考文献 [1-3] 介绍，Rome 服务器产品使用 9-2-9 封装基板完成了信号连接，在封装基板芯板上有 4 层布线用于信号互连。图 9.23 所示为其中一个信号布线层（其他层与之类似），以及 CCD（CPU 计算芯片）、IOD（IO 芯片）、主外部动态随机存取存储器（dynamic random-access memory，DRAM）和 SerDes 接口的物理位置。

图 9.24 所示为 AMD 芯粒架构的演进（开发）过程[1-3]。对于高性能服务器和桌面处理器，I/O 非常繁多。模拟器件和 I/O 的凸点节距并不能从先进节点工艺中获得很多益处，反而成本非常高昂。这一问题的解决方案之一是将 SoC 划分为多个芯粒，为 CPU 核保留成本高昂的先进硅工艺，同时对 I/O 和内存接口则采用次一代的硅工艺[1-3]。由于 AMD 致力于保持 EPYC 封装尺寸和输出引脚不变，因此随着芯片数量从第一代 EPYC 的 4 个增加到第二代 EPYC 的 9 个，需要更加紧密的芯片/封装协同设计。

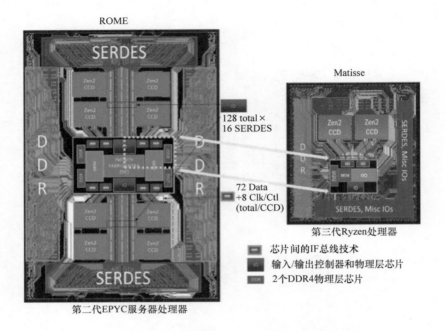

图 9.23　AMD 基于芯粒的第二代 EPYC 服务器和桌面处理器[1]

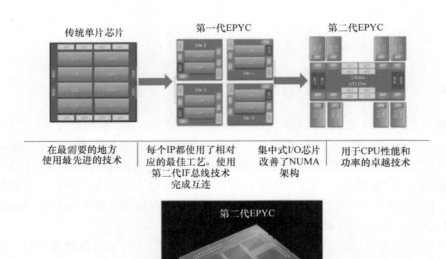

图 9.24　AMD 芯粒架构的演进[1]

第二代 EPYC 芯粒的性能与成本如图 9.25 所示[1-3]。AMD 透露，采用台积电的 7nm 工艺技术制造 16 核 CPU 单芯片的成本是多芯粒 CPU 的两倍多。从图 9.25 可以看出：

1）与同等核心数的单片设计相比，采用芯粒设计的芯片核心数量越多（少），所节省的成本就越多（少）。

2）采用芯粒设计可以突破单片设计最高核心数和性能的限制。

3）采用芯粒设计有助于降低产品线中所有核心数/性能节点的整体成本。

4）采用芯粒设计的芯片成本及性能随着芯粒数目的减少按比例下降。

5）采用 14nm 工艺制备 IO 芯片有助于降低固定成本。

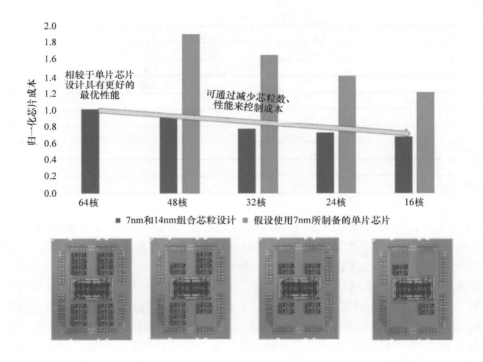

图 9.25　AMD 芯片成本比较：芯粒（7nm+14nm）与单片芯片（7nm）[1]

AMD 还通过使用更小的芯粒来优化成本结构和提高芯片良率。AMD 在核心缓存芯片上采用了台积电昂贵的 7nm 工艺技术，而将 DRAM 和 Pie 逻辑转移到了由格罗方德制造的 14nm I/O 芯片上。

第二代 EPYC 是一种 2D 芯粒集成技术，即所有芯粒并排放在单个封装的同一基板上。AMD 未来的芯粒异质集成技术[3]将是 3D 芯粒集成架构，如图 9.26 所示，一些芯粒堆叠在另一些芯粒（如逻辑芯片）的顶部，称之为有源 TSV 转接板。

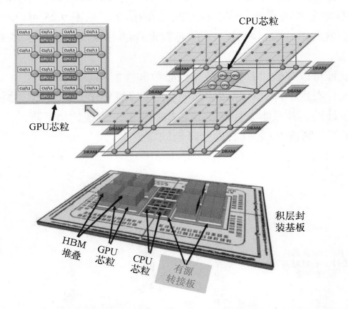

图 9.26　AMD 未来的芯粒技术——3D 芯粒集成（有源 TSV 转接板）[3]

9.8　英特尔的芯粒异质集成

2020 年 7 月，英特尔发布了基于 FOVEROS 技术（2019 年 7 月发布的全向互连技术类型 3）的移动终端（笔记本电脑）处理器 "Lakefield"。SoC 芯片被分区（如 CPU、GPU、LP-DDR4 等）并切分（例如，CPU 被切分为一颗大 CPU 和 4 颗小 CPU）为芯粒，如图 9.27 所示。然后采用 CoW 工艺将这些芯粒面对面键合（堆叠）到有源 TSV 转接板（一颗大的 22FFL 基底芯片）上，如图 9.28 和图 9.29 所示[4-6]。芯粒和逻辑基底芯片之间的互连通过微凸点（铜柱 +SnAg 焊料帽）实现，如图 9.28 和图 9.29 所示。基底芯片与封装基板之间的互连为 C4 凸点，封装基板与 PCB 之间通过焊球互连。最终封装形式为一个封装堆叠（12mm × 12mm × 1mm），如图 9.27 所示。芯粒异质集成结构在底部封装体中，顶部封装体是采用引线键合技术实现的存储芯片封装。

芯粒的制造采用英特尔的 10nm 工艺技术，基底芯片为 22nm 工艺制造。由于芯粒的尺寸更小，而且并非所有芯片都使用 10nm 工艺技术，因此整体良率必然很高，从而降低成本。

应该注意到，这是 3D 芯粒集成的第一个批量化应用。同时，这也是第一款采用 3D IC 集成并用于移动产品（如笔记本电脑）处理器的大批量制造应用。

在英特尔架构日（2020 年 8 月 13 日）期间，他们发布了采用 Cu-Cu 混合键合的 FOVEROS 技术。他们演示了采用无凸点混合键合，节距可以降到 10μm，而不再是图 9.30 所示的像 Lakefield 的 50μm。

第 9 章 芯粒异质集成　363

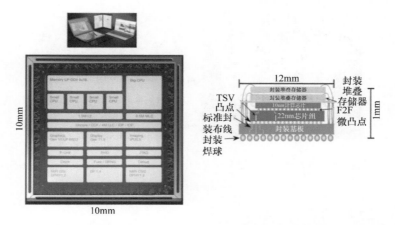

图 9.27　采用 FOVEROS 技术的英特尔 Lakefield 移动终端（笔记本电脑）处理器[5]

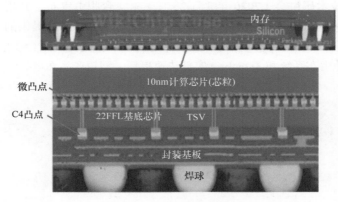

图 9.28　Lakefield 处理器截面的 SEM 照片[5]

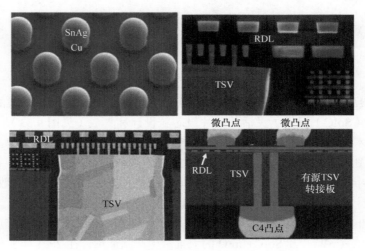

图 9.29　Lakefield 处理器截面细节的 SEM 照片[5]

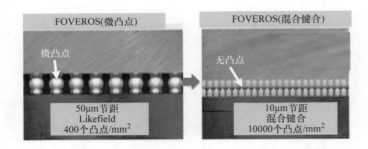

图 9.30　具有微凸点（50μm 节距）和无凸点（10μm 节距）的英特尔 FOVEROS[5]

9.9　台积电的芯粒异质集成

在台积电年度技术研讨会（2020 年 8 月 25 日）上，台积电宣布了其用于移动端、高性能计算、汽车和物联网应用的 3DFabric（3D 制造）技术[7-9]。3DFabric 提供的芯粒异质集成技术是从前道到后道的全集成，如图 9.31 所示。该特定应用平台充分利用台积电的先进晶圆技术、开放式创新平台设计生态系统和 3DFabric 实现快速更新、缩短产品面市时间。

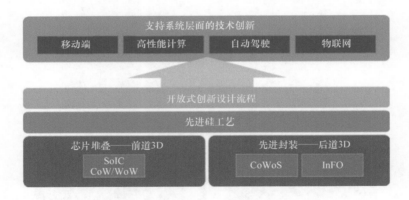

图 9.31　台积电的 3DFabric 集成

台积电的芯粒异质集成路线图如图 9.32 所示[8]。采用 CoW 和 WoW 形式的前道 3D 混合键合（堆叠）技术（SoIC）提供了灵活的芯片级芯粒设计和集成（见图 9.33）。与传统的芯粒微凸点倒装芯片技术相比，混合键合集成片上系统（SoIC）具有许多优点，例如更好的热性能和更小的能耗，如图 9.34 所示[9]。

在 3D 后道封装集成中，CoWoS 所增加和丰富的技术内涵，为满足云、数据中心和高端服务器上的高性能计算需求提供了极高的计算性能和内存带宽，如图 9.35a 所示。在另一类 3D 后道封装集成中，InFO 系列技术提供了内存到逻辑、逻辑到逻辑、封装堆叠等应用，如图 9.35b 所示。SoIC、SoIC+CoWoS 和 SoIC+InFO 的大批量生产制造预计在 2021 年实现。

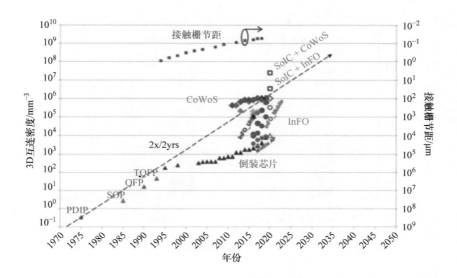

图 9.32　台积电的 3DFabric 路线图 [9]

来源：Doug Yu, IEEE IEDM 2019, Panelist presentation, Dec 2019。

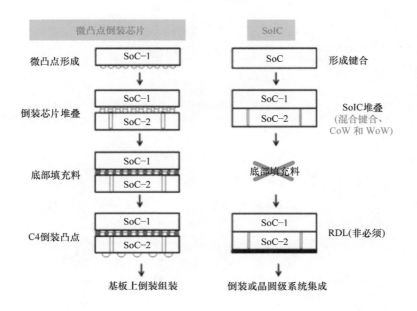

图 9.33　台积电混合键合 SoIC 与传统微凸点倒装技术的对比 [9]

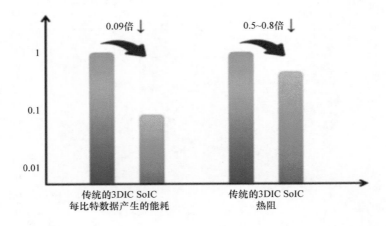

图 9.34 热性能和能耗：SoIC 与传统 3D IC 的对比[9]

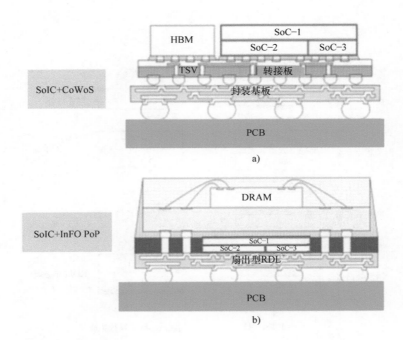

图 9.35 台积电的后道集成：a）SoIC+ CoWoS；b）SoIC+InFO PoP[9]

9.10 总结和建议

一些重要的结论和建议总结如下。

1）工艺微缩的 SoC 将继续存在。然而只有少数几家公司，如苹果、三星、华为、谷歌，能够负担起更小的特征尺寸（先进节点）。通常他们采用这种方式是有原因的，以苹果为例，至少有三个原因：

① 2008 年 4 月 23 日，自苹果收购 Palo Alto Semiconductor 起，便一直在用大量 IP 构建芯片，并且与其软件开发进行紧密耦合（集成）。

② 因为额外的芯片间互连和通信开销会带来更多的问题，将其 SoC 设计分解为芯粒并非那么有吸引力。

③ 世界排名第一的代工厂（台积电）是苹果的忠实合作伙伴，他们致力于完成苹果的产品，例如，应用处理器（A16）计划于 2022 年下半年采用台积电的 3nm 工艺技术制造。

2）芯粒提供了 SoC 的替代品（非必须），尤其是对于大多数公司无法负担的先进节点。

3）为了促进/普及芯粒异质集成，急需构建相关标准！DARPA CHIPS 项目正朝着正确的方向进展。

4）对于复杂的芯粒异质集成，迫切需要电子设计自动化（EDA）工具来自动化地进行拆解、分类和设计。

5）有关芯粒异质集成的更多信息，请阅读参考文献 [23]。

参 考 文 献

1. Naffziger, S., K. Lepak, M. Paraschour, and M. Subramony, "AMD Chiplet Architecture for High-Performance Server and Desktop Products", *IEEE/ISSCC Proceedings,* February 2020, pp. 44–45.
2. Naffziger, S., "Chiplet Meets the Real World: Benefits and Limits of Chiplet Designs", *Symposia on VLSI Technology and Circuits,* June 2020, pp. 1–39.
3. Stow, D., Y. Xie, T. Siddiqua, and G. Loh, "Cost-Effective Design of Scalable High-Performance Systems Using Active and Passive Interposers", *IEEE/ICCAD Proceedings*, November 2017, pp. 1–8.
4. Ingerly, D., S. Amin, L. Aryasomayajula, A. Balankutty, D. Borst, A. Chandra, K. Cheemalapati, C. Cook, R. Criss, K. Enamul1, W. Gomes, D. Jones, K. Kolluru, A. Kandas, G.. Kim, H. Ma, D. Pantuso, C. Petersburg, M. Phen-givoni, A. Pillai, A. Sairam, P. Shekhar, P. Sinha, P. Stover, A. Telang, and Z. Zell, "Foveros: 3D Integration and the use of Face-to-Face Chip Stacking for Logic Devices", *IEEE/IEDM Proceedings,* December 2019, pp. 19.6.1–19.6.4.
5. Gomes, W., S. Khushu, D. Ingerly, P. Stover, N. Chowdhury, F. O'Mahony, etc., "Lakefield and Mobility Computer: A 3D Stacked 10 nm and 2FFL Hybrid Processor System in $12\times12\ mm^2$, 1 mm Package-on-Package", *IEEE/ISSCC Proceedings,* February 2020, pp. 40–41.
6. WikiChip, "A Look at Intel Lakefield: A 3D-Stacked Single-ISA Heterogeneous Penta-Core SoC", https://en.wikichip.org/wiki/chiplet, May 27, 2020.
7. Chen, M. F., C. S. Lin, E. B. Liao, W. C. Chiou, C. C. Kuo, C. C. Hu, C. H. Tsai, C. T. Wang and D. Yu, "SoIC for Low-Temperature, Multi-Layer 3D Memory Integration", *IEEE/ECTC Proceedings,* May 2020, pp. 855–860.
8. Chen, Y. H., C. A. Yang, C. C. Kuo, M. F. Chen, C. H. Tung, W. C. Chiou, and D. Yu, "Ultra High Density SoIC with Sub-micron Bond Pitch", *IEEE/ECTC Proceedings,* May 2020, pp. 576–581.

9. Chen, F., M. Chen, W. Chiou, D. Yu, "System on Integrated Chips (SoIC™) for 3D Heterogeneous Integration", *IEEE/ECTC Proceedings,* May 2019, pp. 594–599.
10. Lin, J., C. Chung, C. Lin, A. Liao, Y. Lu, J.g Chen, and D. Ng, "Scalable Chiplet package using Fan-Out Embedded Bridge", *IEEE/ECTC Proceedings*, May 2020, pp. 14–18.
11. Coudrain, P., J. Charbonnier, A. Garnier, P. Vivet, R. Vélard, A. Vinci, F. Ponthenier, A. Farcy, R. Segaud, P. Chausse, L. Arnaud, D. Lattard, E. Guthmuller, G. Romano, A. Gueugnot, F. Berger, J. Beltritti, T. Mourier, M. Gottardi, S. Minoret, C. Ribière, G. Romero, P.-E. Philip, Y. Exbrayat, D. Scevola, D. Campos, M. Argoud, N. Allouti, R. Eleouet, C. Fuguet Tortolero, C. Aumont, D. Dutoit, C. Legalland, J. Michailos, S. Chéramy, and G. Simon, "Active interposer technology for chiplet-based advanced 3D system architectures", *EEE/ECTC Proceedings*, May 2019, pp. 569–578.
12. Jo, P., T. Zheng, and M. Bakir, "Polylithic Integration of 2.5D and 3D Chiplets using Interconnect Stitching", *IEEE/ECTC Proceedings*, May 2019, pp. 1803–1808.
13. Ko, T., H. P. Pu, Y. Chiang, H. J. Kuo, C. T. Wang, C. S. Liu, and D. C. Yu, "Applications and Reliability Study of InFO_UHD (Ultra-High-Density) Technology", *IEEE/ECTC Proceedings*, May 2020, pp. 1120–1125.
14. Huang, Y., C. Chung, C. Lin, G. Lu, C. Tseng, H. Chang, C. Hsu, and B. Xu, "Challenges of large Fan out multi-chip module and fine Cu line space", *IEEE/ECTC Proceedings*, May 2020, pp. 1140–1145.
15. Chuang, P., M. Lin, S. Hung, Y. Wu, D. Wong, M. Yew, C. Hsu, L. Liao, P. Lai, P. Tsai, S. Chen, S. Cheng, and S. Jeng, "Hybrid Fan-out Package for Vertical Heterogeneous Integration", *IEEE/ECTC Proceedings*, May 2020, pp. 333–338.
16. Rupp, B., A. Plochowietz, L. Crawford, M. Shreve, S. Raychaudhuri, S. Butylkov, Y. Wang, P. Mei, Q. Wang, J. Kalb, Y. Wang, E. Chow, and J. Lu, "Chiplet Micro-Assembly Printer", *IEEE/ECTC Proceedings*, May 2019, pp. 1312–1315.
17. Fish, M., P. McCluskey, and A. Bar-Cohen, "Thermal Isolation Within High-Power 2.5D Heterogenously Integrated Electronic Packages", *IEEE/ECTC Proceedings*, May 2016, pp. 1847–1855.
18. Gomez, D., K. Ghosal, M. Meitl, S. Bonafede, C. Prevatte, T. Moore, B. Raymond, D. Kneeburg, A. Fecioru, A. Trindade, and C. Bower1, "Process Capability and Elastomer Stamp Lifetime in Micro Transfer Printing", *IEEE/ECTC Proceedings*, May 2016, pp. 680–687.
19. https://www.darpa.mil/program/compound-semiconductor-materials-on-silicon
20. https://www.darpa.mil/program/dahi-compound-semiconductor-materials-on-silicon.
21. https://www.darpa.mil/program/common-heterogeneous-integration-and-ip-reuse-strategies.
22. https://nstxl.org/opportunity/state-of-the-art-heterogeneous-integrated-packaging-ship-prototype-project/.
23. Lau, J. H., *Heterogeneous Integration, Springer*, New York, 2019.
24. Lau, J. H., "Recent Advances and Trends in Heterogeneous Integrations", *IMAPS Transactions, Journal of Microelectronics and Electronic Packaging*, Vol. 16, April 2019, pp. 45–77.
25. https://en.wikichip.org/wiki/chiplet, March 27, 2020.
26. https://www.netronome.com/blog/its-time-disaggregated-silicon/, March 12, 2020.
27. Lau, J. H., M. Li, M. Li, T. Chen, I. Xu, X. Qing, Z. Cheng, N. Fan, E. Kuah, Z. Li, K. Tan, Y. Cheung, E. Ng, P. Lo, K. Wu, J. Hao, S. Koh, R. Jiang, X. Cao, R. Beica, S. Lim, N. Lee, C. Ko, H. Yang, Y. Chen, M. Tao, J. Lo, and R. Lee, "Fan-Out Wafer-Level Packaging for Heterogeneous Integration", *IEEE Transactions on CPMT*, 2018, September 2018, pp. 1544–1560.
28. Lau, J. H., M. Li, Y. Lei, M. Li, I. Xu, T. Chen, Q. Yong, Z. Cheng, K. Wu, P. Lo, Z. Li, K. Tan, Y. Cheung, N. Fan, E. Kuah, C. Xi, J. Ran, R. Beica, S. Lim, N. Lee, C. Ko, H. Yang, Y. Chen, M. Tao, J. Lo, and R. Lee, "Reliability of Fan-Out Wafer-Level Heterogeneous Integration", *IMAPS Transactions, Journal of Microelectronics and Electronic Packaging*, Vol. 15, Issue: 4, October 2018, pp. 148–162.
29. Ko, CT, H. Yang, J. H. Lau, M. Li, M. Li, C. Lin, J. W. Lin, T. Chen, I. Xu, C. Chang, J. Pan, H. Wu, Q. Yong, N. Fan, E. Kuah, Z. Li, K. Tan, Y. Cheung, E. Ng, K. Wu, J. Hao, R. Beica, M. Lin, Y. Chen, Z. Cheng, S. Koh, R. Jiang, X. Cao, S. Lim, N. Lee, M. Tao, J. Lo, and R. Lee, "Chip-First Fan-Out Panel-Level Packaging for Heterogeneous Integration", *IEEE Transactions on CPMT*, September 2018, pp. 1561–1572.

30. Ko, C. T., H. Yang, J. H. Lau, M. Li, M. Li, C. Lin, J. Lin, C. Chang, J. Pan, H. Wu, Y. Chen, T. Chen, I. Xu, P. Lo, N. Fan, E. Kuah, Z. Li, K. Tan, C. Lin, R. Beica, M. Lin, C. Xi, S. Lim, N. Lee, M. Tao, J. Lo, and R. Lee, "Design, Materials, Process, and Fabrication of Fan-Out Panel-Level Heterogeneous Integration", *IMAPS Transactions, Journal of Microelectronics and Electronic Packaging*, Vol. 15, Issue: 4, October 2018, pp. 141–147.
31. Shimizu, N., Kaneda, W., Arisaka, H., Koizumi, N., Sunohara, S., Rokugawa, A., and Koyama, T., "Development of Organic Multi Chip Package for High Performance Application," *IMAPS International Symposium on Microelectronics*, Orlando, FL, Sep. 30–Oct. 3, 2013, pp. 414–419.
32. Oi, K., Otake, S., Shimizu, N., Watanabe, S., Kunimoto, Y., Kurihara, T., Koyama, T., Tanaka, M., Aryasomayajula, L., and Kutlu, Z., "Development of New 2.5D Package With Novel Integrated Organic Interposer Substrate With Ultra-Fine Wiring and High Density Bumps," *IEEE/ECTC Proceedings,* May 2014, pp. 348–353.
33. Li, L., P. Chia, P. Ton, M. Nagar, S. Patil, J. Xue, J. DeLaCruz, M. Voicu, J. Hellings, B. Isaacson, M. Coor, and R. Havens, "3D SiP with organic interposer of ASIC and memory integration," *IEEE/ECTC Proceedings*, May 2016, pp. 1445–1450.
34. Souriau, J., O. Lignier, M. Charrier, and G. Poupon, "Wafer Level Processing Of 3D System in Package for RF and Data Applications", *IEEE/ECTC Proceedings*, 2005, pp. 356–361.
35. Henry, D., D. Belhachemi, J-C. Souriau, C. Brunet-Manquat, C. Puget, G. Ponthenier, J. Vallejo, C. Lecouvey, and N. Sillon, "Low Electrical Resistance Silicon Through Vias: Technology and Characterization", *IEEE/ECTC Proceedings*, 2006, pp. 1360–1366.
36. Shao, S., Y. Niu, J. Wang, R. Liu, S. Park, H. Lee, G. Refai-Ahmed, and L. Yip, "Comprehensive Study on 2.5D Package Design for Board-Level Reliability in Thermal Cycling and Power Cycling", *Proceedings of IEEE/ECTC*, May 2018, pp. 1662–1669.
37. McCann, S., H. Lee, G. Refai-Ahmed, T. Lee, and S. Ramalingam, "Warpage and Reliability Challenges for Stacked Silicon Interconnect Technology in Large Packages", *Proceedings of IEEE/ECTC*, May 2018, pp. 2339–2344.
38. Lai, C., H. Li, S. Peng, T. Lu, and S. Chen, "Warpage Study of Large 2.5D IC Chip Module", *Proceedings of IEEE/ECTC*, May 2017, pp. 1263–1268.
39. Hong, J., K. Choi, D. Oh, S. Park, S. Shao, H. Wang, Y. Niu, and V. Pham, "Design Guideline of 2.5D Package with Emphasis on Warpage Control and Thermal Management", *Proceedings of IEEE/ECTC*, May 2018, pp. 682–692.
40. Lee, J., C. Lee, C. Kim, and S. Kalchuri, "Micro Bump System for 2nd Generation Silicon Interposer with GPU and High Bandwidth Memory (HBM) Concurrent Integration", *Proceedings of IEEE/ECTC*, May 2018, pp. 607–612.
41. Hou, S., W. Chen, C. Hu, C. Chiu, K. Ting, T. Lin, W. Wei, W. Chiou, V. Lin, V. Chang, C. Wang, C. Wu, and D. Yu, "Wafer-Level Integration of an Advanced Logic-Memory System Through the Second-Generation CoWoS Technology", *IEEE Transactions on Electron Devices*, October 2017, pp. 4071–4077.
42. Chaware, R., K. Nagarajan, and S. Ramalingam, "Assembly and reliability challenges in 3D integration of 28 nm FPGA die on a large high-density 65 nm passive interposer," *IEEE/ECTC Proceedings*, May 2012, pp. 279–283.
43. Banijamali, B., S. Ramalingam, K. Nagarajan, and R. Chaware, "Advanced reliability study of TSV interposers and interconnects for the 28 nm technology FPGA," *IEEE/ECTC Proceedings*, May 2011, pp. 285–290.
44. Banijamali, B., S. Ramalingam, H. Liu, and M. Kim, "Outstanding and innovative reliability study of 3D TSV interposer and fine-pitch solder micro-bumps," *IEEE/ECTC Proceedings*, May 2012, pp. 309–314.
45. Banijamali, B., C. Chiu, C. Hsieh, T. Lin, C. Hu, S. Hou, et al., "Reliability evaluation of a CoWoS-enabled 3D IC package," *IEEE/ECTC Proceedings*, May 2013, pp. 35–40.
46. Lee, J., C. Lee, C. Kim, and S. Kalchuri, "Micro Bump System for 2nd Generation Silicon Interposer with GPU and High Bandwidth Memory (HBM) Concurrent Integration", *IEEE/ECTC Proceedings*, May 2018, pp. 607–612.
47. Mahajan, R., R. Sankman, N. Patel, D. Kim, K. Aygun, Z. Qian, et al., "Embedded multi-die interconnect bridge (EMIB) – a high-density, high-bandwidth packaging interconnect," *IEEE/ECTC Proceedings*, May 2016, pp. 557–565.

48. Podpod, A., J. Slabbekoorn, A. Phommahaxay, F. Duval, A. Salahouedlhadj, M. Gonzalez, K. Rebibis, R.A. Miller, G. Beyer, and E. Beyne, "A Novel Fan-Out Concept for Ultra-High Chip-to-Chip Interconnect Density with 20-μm Pitch", *IEEE/ECTC Proceedings*, May 2018, pp. 370–378.
49. Suk, K., S. Lee, J. Kim, S. Lee, H. Kim, S. Lee, H. Kim, S. Lee, P. Kim, D. Kim, D. Oh, and J. Byun, "Low-cost Si-less RDL interposer package for high-performance computing applications," *IEEE/ECTC Proceedings*, May 2018, pp. 64–69.
50. Miki, S., H. Taneda, N. Kobayashi, K. Oi, K. Nagai, T. Koyama, "Development of 2.3D High Density Organic Package using Low Temperature Bonding Process with Sn-Bi Solder", *IEEE/ECTC Proceedings*, May 2019, pp., 1599–1604.
51. Murayama, K., S. Miki, H. Sugahara, and K. Oi, "Electro-migration evaluation between organic interposer and build-up substrate on 2.3D organic package", *IEEE/ECTC Proceedings*, May 2020, pp. 716–722.
52. Chang, K., C. Huang, H. Kuo, M. Jhong, T. Hsieh, M. Hung, and C. Wang, "Ultra High Density IO Fan-Out Design Optimization with Signal Integrity and Power Integrity", *IEEE/ECTC Proceedings*, May 2019, pp. 41–46.
53. Lin, Y., M. Yew, M. Liu, S. Chen, T. Lai, P. Kavle, C. Lin, T. Fang, C. Chen, C. Yu, K. Lee, C. Hsu, P. Lin, F. Hsu and S. Jeng, "Multilayer RDL Interposer for Heterogeneous Device and Module Integration", *IEEE/ECTC Proceedings*, May 2019, pp. 931–936.
54. Lau, J. H., C. Ko, T. Peng, K. Yang, T. Xia, P. Lin, J. Chen, P. Huang, T. Tseng, E. Lin, L. Chang, C. Lin, and W. Lu, "Chip-Last (RDL-First) Fan-Out Panel-Level Packaging (FOPLP) for Heterogeneous Integration", *IMAPS Transactions, Journal of Microelectronics and Electronic Packaging*, Vol. 17, No. 3, October 2020, pp. 89–98.
55. Lau, J. H., C. Ko, K. Yang, C. Peng, T. Xia, P. Lin, J. Chen, P. Huang, H. Liu, T. Tseng, E. Lin, and L. Chang, "Panel-Level Fan-Out RDL-first Packaging for Heterogeneous Integration", *IEEE Transactions on CPMT*, Vol. 10, No. 7, July 2020, pp. 1125–1137.
56. Pendse, R., "Semiconductor device and method of forming extended semiconductor device with fan-out interconnect structure to reduce complexity of substrate," filed on Dec. 23, 2011, *US 2013/0161833 A1*, pub. date: June 27, 2013.
57. Yoon, S. W., P. Tang, R. Emigh, Y. Lin, P. C. Marimuthu, and R. Pendse, "Fan-out flip-chip eWLB (embedded wafer-level ball grid array) technology as 2.5D packaging solutions," *IEEE/ECTC Proceedings*, May 2013, pp. 1855–1860.
58. Lin, Y., W. Lai, C. Kao, J. Lou, P. Yang, C. Wang, et al., "Wafer warpage experiments and simulation for fan-out chip-on-substrate," *IEEE/ECTC Proceedings*, May 2016, pp. 13–18.
59. Chen, N. C., T. Hsieh, J. Jinn, P. Chang, F. Huang, J. Xiao, A. Chou, and B. Lin, "A Novel System in Package with Fan-out WLP for high speed SERDES application", *IEEE/ECTC Proceedings*, May 2016, pp. 1496–1501.
60. Lee, Y., W. Lai, I. Hu, M. Shih, C. Kao, D. Tarng, and C. Hung, "Fan-Out Chip on Substrate Device Interconnection Reliability Analysis", *IEEE/ECTC Proceedings*, May 2017, pp. 22–27.

第 10 章

低损耗介电材料

10.1 引言

半导体行业公认有五个主要的增长引擎（应用）[1, 2]：

1）移动设备，如智能手机、智能手表、笔记本电脑、可穿戴设备、平板电脑等。

2）高性能计算（high-performance computing，HPC），也称为超级计算，能够在超级计算机上高速处理数据并执行复杂计算。

3）自动驾驶汽车。

4）物联网（internet of things，IoT），如智能工厂和智能健康。

5）大数据（用于云计算）和即时数据（用于边缘计算）。

宽带蜂窝网络的第 5 代技术标准（5th generation technology standard for broadband cellular networks，5G）等系统技术正在驱动这 5 种半导体应用快速增长。根据美国联邦通信委员会的规定：

1）中频频谱（也称为 Sub-6GHz 5G）定义为 900MHz＜频率＜6GHz、数据速度≤1Gbit/s。

2）高频频谱（也称为 5G 毫米波）定义为 24GHz≤频率≤100GHz、1Gbit/s＜数据速度≤10Gbit/s。

Sub-6GHz 5G 和 LTE（4G）的应用并存，天线和多模射频收发器之间的距离较大。28/39GHz 应用于 5G 移动场景下的天线等，60GHz 应用于高速无线数据链路等，77GHz 应用于汽车雷达等，94GHz 应用于雷达成像等。

为了满足提高信号传输速度 / 速率和管理海量数据的要求，半导体、封装和材料等方面的不断发展是非常必要的。就绝缘材料的电性能而言，低损耗因子或损耗角正切（dissipation factor or loss tangent，D_f）和低介电常数（dielectric constant or permittivity，D_k）材料在 5G 应用中是优先选择的[3-18]。本章系统地介绍了近三年来参考文献 [3-18] 中用于高速和高频场景下材料的 D_f 和 D_k 特性。首先简单介绍 5G 应用中需要使用低 D_f 和 D_k 特性以及低热膨胀系数的介电材料的原因。

10.2 为什么需要低 Dk 和 Df 的介电材料

图 10.1 所示为传输损耗的计算，它等于导体损耗和介电损耗之和。介电损耗与频率、Df、Dk 的二次方根成正比。因此，为了具有较低的传输损耗，需要采用较低的 Df 和 Dk 值。

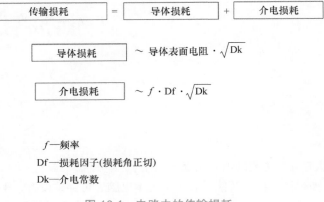

f——频率
Df——损耗因子(损耗角正切)
Dk——介电常数

图 10.1 电路中的传输损耗

导体损耗与导体表面电阻、Dk 的二次方根成正比。通常，频率越高，电流信号越靠近导体表面流动（趋肤效应）。对于粗糙的导体表面，则可以假定电流信号在表面上传播的距离更长，进而导致更大的传输损耗。因此，使用具有较低表面粗糙度的铜可以降低导体表层电阻。关于导体损耗的内容不在本书中讨论。

10.3 为什么需要低热膨胀系数的介电材料

对于多层基板或再布线层（redistribution-layer，RDL），绝缘膜（介电材料）被用作导体层之间的层间粘接剂。由于大多数导体层由热膨胀系数（coefficient of thermal expansion，CTE）约为 $17.5 \times 10^{-6}/℃$ 的电镀铜制成，因此优选低 CTE [≤ $(20 \sim 30) \times 10^{-6}/℃$] 的介电材料。Cu 导体层与介质层之间的低热膨胀失配具有如下优点：

1）更低的基板翘曲。
2）更少的层间分层。
3）更好的可靠性。

除了低 Df、Dk 和 CTE 外，下一代介电材料还需要具有低吸湿性、良好的机械和热性能，以抵抗基板和 RDL 中的内部应力。此外，新兴的介电材料必须能够低温固化、易于制造并克服复杂的组装过程。

10.4 NAMICS 材料的 Dk 和 Df

通过适当的填料选择和树脂组合，NAMICS 开发了一种稳定的介电材料，具有低 Df、低

Dk 以及良好的附着力[3]。图 10.2 所示为薄膜材料的外观,该薄膜保持足够的柔韧性,同时由于添加了球形 SiO_2 颗粒填料而具有低 CTE。他们的测试结构是采用优化后的树脂组合物通过真空热模压工艺(在 200℃、1MPa 下处理 1h)制备。测试结果如图 10.2 的表所示。可以看出:

1)在 2GHz 时,Dk = 3 和 Df = 0.0025。
2)CTE = 25×10^{-6}/℃。
3)抗拉强度为 630MPa。
4)铜粗糙面(Ra = 1.8μm)的剥离强度为 7N/cm,铜光洁面(Ra = 0.25μm)剥离强度为 5.5N/cm。

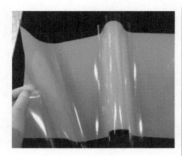

指标	膜层
Dk (2GHz)	3
Df (2GHz)	0.0025
抗拉强度/MPa	630
铜剥离强度/(N/cm) 铜粗糙面	7
铜剥离强度/(N/cm) 铜光洁面	5.5
CTE/(10^{-6}/℃)	25
Tg/℃ (DMA)	150

图 10.2 薄膜外观及材料特性[3]

所有这些值都优于预期。图 10.3 和图 10.4 所示为介电膜在不同频率下的 Df 和 Dk 测试结果。可以看出,介电膜材料特性保持在低 Dk 和低 Df 值,且与频率无关。图 10.5 所示为介电膜的 Df 和 Dk 在 85℃/85%RH 条件下长达 1000h 的测试结果。可以看出:

1)Dk 始终保持不变。
2)Df 略有增加(从 0.0030 到 0.0034)。

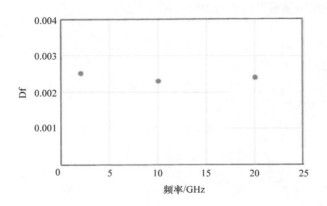

图 10.3 Df 与频率的关系[3]

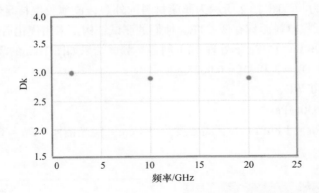

图 10.4　Dk 与频率的关系[3]

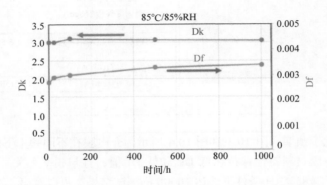

图 10.5　在 85℃/85%RH 条件下 Df 和 Dk 与时间的关系[3]

图 10.6 和图 10.7 所示分别为 125℃温度和 85℃/85%RH 条件下介电膜粘附强度的变化。可以看出，介电膜的粘附强度可以长达 1000h 保持不变。

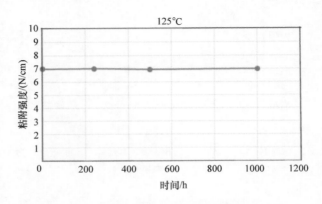

图 10.6　125℃下的粘附强度与时间的关系[3]

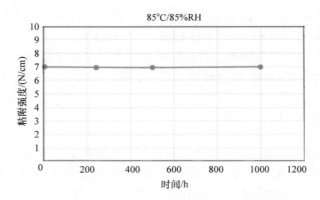

图 10.7　85℃/85%RH 条件下的粘附强度与时间的关系[3]

10.5　Arakawa 材料的 Dk 和 Df

Arakawa[4] 通过优化聚酰亚胺主链中脂肪族、脂环族和芳香族基团的组成比，开发了具有良好耐热性、低介电常数（Dk）和损耗因子（Df）特性的可溶性聚酰亚胺。图 10.8 所示为聚酰亚胺特性（Dk、Df 与软化点温度的关系）的可调范围，图中左纵轴是 Df，右纵轴是 Dk，横轴表示聚酰亚胺的软化点（softening point，Tg）。方形点为 Dk 值，菱形点为 Df 值。他们能够将聚酰亚胺的 Df 控制在 0.0035 以下，将 Dk 控制在 2.8 以下，并将 Tg 控制在 80~180℃范围之内。

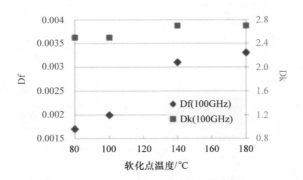

指标	单位	值
剥离强度	N/mm	0.8
断裂伸长率	%	100
Dk(10GHz)Tg=80℃	—	2.5
Df(10GHz)Tg=80℃	—	0.0017
弹性模量	GPa	0.136

图 10.8　Df、Dk 与软化点温度的关系

将聚酰亚胺（25μm）夹在粘接层（12~13μm）之间，随后再夹在轧制铜层（12μm）之间构成了柔性覆铜板（flexible copper clad laminate，FCCL）。对这一5层结构进行剥离试验，结果如图10.8的表所示。可以看出，聚酰亚胺薄膜与薄型铜箔之间达到了良好的粘附强度（0.8N/mm）。此外，在10GHz和Tg = 80℃条件下，Dk = 2.5，Df = 0.0017。

10.6 杜邦材料的Dk和Df

基于更低固化温度的新型芳烷基热固性聚合物，杜邦开发了5G感光介电应用的XP配方A和XP配方B材料。XP配方A和XP配方B的整体性能总结见表10.1[5, 6]。XP配方A使用丙烯酸网络来实现光刻中的负胶特性。丙烯酸方案获得了一些理想特性，例如良好的玻璃化转变温度（Tg = 220℃）、良好的耐化学性和光刻能力。XP配方B使用一种新型的非丙烯酸酯的感光体系，以实现高对比度（深宽比超过1:1）的光刻图形，同时获得Df = 0.0022（28GHz）、Df = 0.0029（77GHz）和Dk = 2.5（10~40GHz）。利用XP配方B所制备的通孔的扫描电子显微镜（scanning electron microscope，SEM）图像的截面如图10.9所示，图10.9a为厚度5μm膜层中实现了4.44μm的CD尺寸，图10.9b为厚度10μm膜层中实现了7.74μm的CD尺寸，图10.9c为厚度15μm膜层中实现了10.14μm的CD尺寸。

表10.1 配方A和配方B的材料特性

	XP配方A	XP配方B
薄膜厚度/μm	5~25	5~15
拉伸强度/MPa	91	85
断裂延伸率（%）	12	18（最大30%）
拉伸模量/GPa	2.1	1.6
介电常数（10~40GHz）	2.6	2.5
耗散系数（28GHz，39GHz，60GHz，77GHz）	0.0041，0.0044，未检出，未检出	0.0022，0.0024，0.0028，0.0029
耗散系数，后HAST	无变化	无变化
分解温度/℃	300	320
Tg/℃，DMA	220	170
吸水率（23℃/45%RH）	0.16%	0.13%
深宽比	2.5:1	1:1
分辨率	5μm通孔	5μm通孔
显影剂/漂洗	PGMEA/PGMEA	PGMEA/PGMEA
单层固化条件	170℃，1h	170℃，1h
硬固化条件	200℃，1h	200℃，1h
保质期（20℃）	>3周	>3周

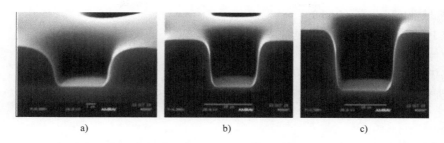

图 10.9 用 XP 配方 B 制造的通孔 SEM 图像：a）厚度 5μm 膜层中实现了 4.44μm 的 CD 尺寸；b）厚度 10μm 膜层中实现了 7.74μm 的 CD 尺寸；c）厚度 15μm 膜层中实现了 10.14μm 的 CD 尺寸[6]

10.7 日立/杜邦微系统材料的 Dk 和 Df

　　日立/杜邦微系统（Hitachi/Dupont MicroSystems）开发了具有优异的电气和机械性能的非光敏和光敏聚酰亚胺（PI）材料[7]。首先他们重新设计了聚合物主链以获得低 Dk 和 Df 特性。结果表明，新型非光敏 PI 在 20GHz 时分别达到了 Dk=2.9 和 Df=0.003，如图 10.10a、b 所示。随后他们仔细选择了新型 PI 的感光体系，以同时保持低 Dk/Df 值和高分辨率。然后调整了光引发剂含量、交联剂含量等，使其兼具优异的光刻性能和电性能。结果表明，在 320℃ 条件下固化的新型光敏 PI 在 20GHz 时的 Dk 为 3.0，在 20GHz 时的 Df 为 0.006，如图 10.10c、d 所示。此外，若将该材料在 200℃ 下固化，则 20GHz 下 Dk=3.0，Df=0.009。他们还证实了新型 PI 具

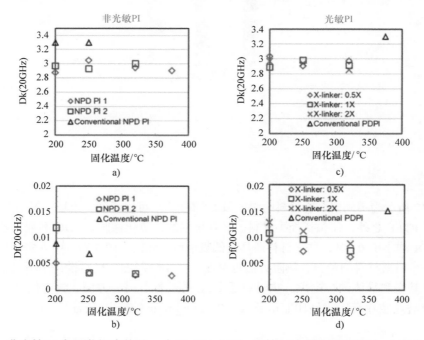

图 10.10 非光敏 PI 与固化温度关系：a）Dk 和 b）Df。光敏 PI 与固化温度关系：c）Dk 和 d）Df

有优异的机械性能（见图 10.11）：

1）该 PI 在 >250℃ 下固化的延伸率约为 70%。
2）拉伸强度大于 100MPa。
3）杨氏模量大于 2GPa。

从光刻角度来看，新型 PI 在厚度 10μm 下可以实现 15μm 的线宽 / 线距。

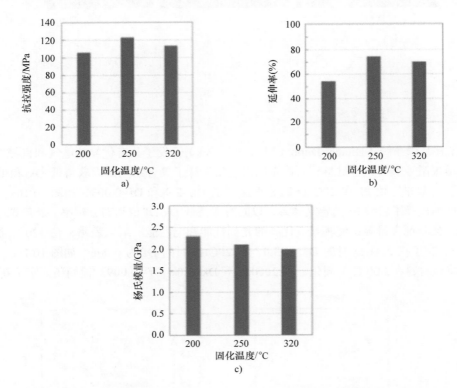

图 10.11 新型光敏 PI 与固化温度的关系：a）抗拉强度；b）延伸率；c）杨氏模量

10.8 JSR 材料的 Dk 和 Df

图 10.12 所示为不同聚合物材料的 Dk 和 Df 值[8, 9]。可以看到，市场上低 Dk 和低 Df 的常规材料除了 PTFE 之外，还包括烃类材料。烃基聚烯烃、氢化橡胶、COC 和聚苯乙烯的 Dk 均小于 2.5、Df 均小于 0.001，这些材料还具有低吸水性。然而，图 10.12 中许多材料的粘附性都很弱。此外，根据材料的种类，热阻方面也存在一些实际问题。材料设计中不仅要考虑低 Df，还要考虑对基板和导体的粘附性、热阻和低 CTE 等特性。另一方面，具有优异热阻特性和机械强度的聚酰亚胺在高频柔性印制电路板（flexible printed circuit board，FPC）应用中实现了低 Df。

第 10 章 低损耗介电材料

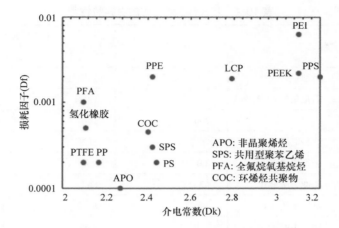

图 10.12　10GHz 条件下不同聚合物的 Df 和 Dk 值[9]

图 10.13 所示为 JSR 采用其独特的聚合物合成技术基于芳香族聚醚的 HC 聚合物。表 10.2（第 3 列）为 JSR HC 聚合物的材料特性。可以看出，在 10GHz 时 Dk = 2.46、Df = 0.0027、杨氏模量为 3GPa、抗拉强度为 62MPa、延伸率为 34%、Tg 为 206℃。选择合适的单体结构，就可以对 HC 聚合物进行改性，从而调控 Df 和 Dk 的特性。例如，通过使用各种不同单体，在 150℃ 左右的 Tg 下，Df 可以降低到 0.002，如图 10.13 所示。

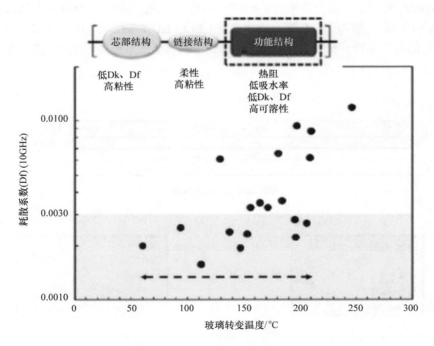

图 10.13　10GHz 条件下 HC 聚合物 Df 与 Tg 关系[9]

表 10.2 JSR HC 聚合物的材料特性[9]

类别名称		HC	HC-F 复合物 (TPE)	
			HC-F	新型 HC-F
基本信息	类型		热固性	热固性
	固化条件/(℃/h)		185/1	185/1
热特性	Tg/℃	206	168	138
	焊料电阻		良好	良好
机械特性	杨氏模量/GPa	3.0	3.2	1.1
	拉伸强度/MPa	62	86	48
	延伸率(%)	34	6	60
电特性	Dk(10GHz)	2.46	2.67	2.49
	Df(10GHz)	0.0027	0.0059	0.0016

基于这些 HC 聚合物，针对 FPC 应用开发了热固性 HC-F[8, 9]。可以对交联剂、添加剂和其他化学成分进行宽范围调整，进而改变热固性 HC-F 基础聚合物的相关特性，如热性能、机械性能和 Dk 值（见图 10.14）。表 10.2（第 4 列和第 5 列）为两种类型的热固性聚醚 HC-F 复合材料的材料特性，其固化条件是 185℃下 1h。HC 聚合物的 Tg 最高（206℃），而新型 HC-F 聚合物的 Tg 最低（138℃）。在 10GHz 下进行测试，新型 HC-F 聚合物 Df 最低（0.0016），HC 聚合物次之（0.0027），HC-F 聚合物最高（0.0059）。另一方面，同样在 10GHz 下测试，HC 聚合物的 Dk 最低（2.46），新型 HC-F 聚合物次之（2.49），HC-F 聚合物第三（2.67）。因此，从低损耗介电材料的角度来看，新型 HC-F 和 HC 聚合物适用于高速和高频应用，尤其是新型 HC-F 聚合物。

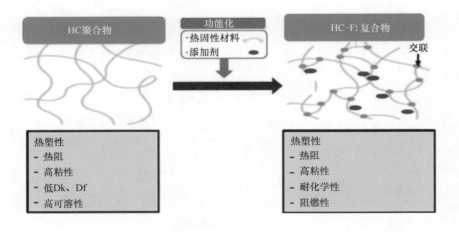

图 10.14 HC 聚合物的 HC-F 复合物[9]

10.9 Toray 材料的 Dk 和 Df

通过研究在聚合物主链中含有各种不同迁移率和不同极性分子的聚酰亚胺材料，Toray 得以获知如何设计得到低 Dk 和 Df 的聚酰亚胺材料[10-13]。他们发现 −150～−50℃ 下的分子迁移率对应于 10～100GHz 频段内的 Df。为了在 GHz 量级降低 Df，很重要的一点是需要在 −50℃ 左右的低温下降低分子迁移率。此外，降低聚酰亚胺主链的极性和柔韧性对于获得低 Df 和 Dk 的聚酰亚胺也很重要。基于这些数据，Toray 提出了两种不同的分子设计方法，一种是控制聚酰亚胺链中的分子移动性，另一种是减小分子极性。随后，他们开发了三种类型的低 Dk 和 Df 聚酰亚胺材料（PI-A、PI-B 和 PI-C），见表 10.3。可以看出，PI-B 聚酰亚胺可以达到 Dk = 2.7 和 Df = 0.002（在 20GHz 时），是最低的。PI-C 聚酰亚胺的固化温度最低（200℃）。PI-A 聚酰亚胺的拉伸强度（100MPa）和杨氏模量（2GPa）最高。

表 10.3 聚酰亚胺的材料特性[11]

图形化方法	PI-A	PI-B	PI-C
	湿法刻蚀激光	激光	光刻激光
Dk（20GHz）	3.0	2.7	2.7
Df（20GHz）	0.003	0.002（0.001①）	0.007
Tg/℃	145	175	120
热膨胀系数/(10^{-6}/℃)	70	65	70
杨氏模量/GPa	2	1.9	1.7
拉伸强度/MPa	100	95	65
延伸率（%）	150	40	15
低吸水性（%）	0.6	0.6	0.6
固化温度/℃	220	220	200

① 1GHz 条件下的结果。

10.10 富士通材料的 Dk 和 Df

富士通（Fujitsu）对厚度为 0.356mm 的低损耗玻璃布基板中瑞翁株式会社（Zeon Corporation）的环烯烃聚合物材料在 10～95GHz 频段内于不同温度条件下（−30℃、25℃ 和 130℃）进行了性能测试，获得了图 10.15 和图 10.16[14] 所示的 Dk 和 Df 值。可以看到

1）Dk 关于频率近似常数。
2）Dk 关于温度的变化很小。
3）Df 是频率的函数，频率越高，Df 越高。
4）Df 具有温度依赖性，温度越高，Df 越高。

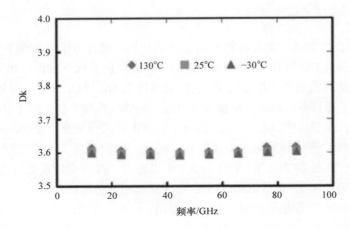

图 10.15 在 −30℃、25℃、130℃条件下 Dk 与频率关系的测试数据[14]

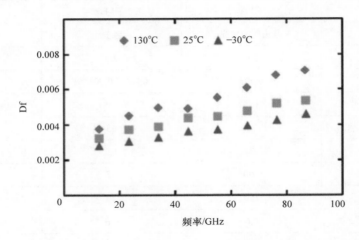

图 10.16 在 −30℃、25℃、130℃条件下 Df 与频率关系的测试数据[14]

10.11 Kayaku 材料的 Dk 和 Df

Kayaku Advanced Materials 和 Nippon Kayaku 联合合成了一种新型嵌段共聚物和对光固化敏感的配方，这种特殊材料被称为 PRL-29 [15]。PRL-29 是一种负性胶，可用于光刻图形化，固化后厚度可达 15μm，具有低于 225℃ 的低温固化能力。PRL-29 的材料特性见表 10.4。可以看出，在 20～85GHz 频段内，其具有 Dk = 2.5 和 Df = 0.004（平均值）（见图 10.17）。

如图 10.18 所示，玻璃化转变温度（glass transition temperature，Tg）为 220℃。PRL-29 的延伸率和杨氏模量分别为 35% 和 1.8GPa（线性应力 - 应变曲线的斜率），如图 10.19 所示。PRL-29 的杨氏模量与温度的关系（见图 10.20）与其他材料（如环氧树脂和 KMRD）一样，即温度越高，杨氏模量越低。

表 10.4　PRL-29 的材料特性[15]

项目	PRL-29
固化温度 /℃	200
深宽比	1:1
Tg/℃	220
杨氏模量 /GPa	1.8
拉伸强度 /MPa	60
断裂延伸率（最大）（%）	35
热膨胀系数（<Tg）/(10^{-6}/℃)	62
剪切附着力 /MPa	35
介电常数（1~85GHz）	2.5
耗散系数（1~85GHz）	0.004
吸水率（%）	0.3
导热性能 /[W/(m·K)]	0.23
PCT（121℃/100%RH, 48h）	通过
Bias HAST（85/85RH, 3.3V, 168h）	通过
耐化学性	优异

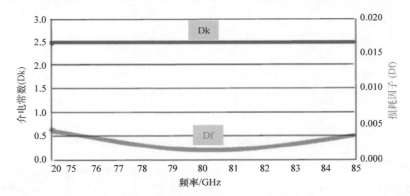

图 10.17　Df 和 Dk 与频率的关系[15]

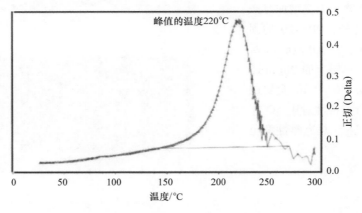

图 10.18　PRL-29 的玻璃化转变温度（Tg=220℃）[15]

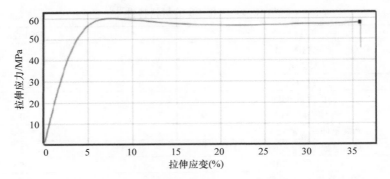

图 10.19　PRL-29 的应力 - 应变曲线（延伸率为 35%）[15]

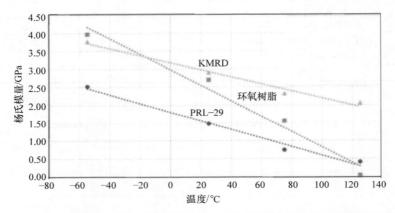

图 10.20　PRL-29、KMRD 和环氧树脂材料的杨氏模量和温度关系曲线 [15]

值得注意的是，PRL-29 的杨氏模量小于环氧树脂和 KMRD。这意味着 PRL-29 更柔软，应变增加时产生的应力更小。此外，PRL-29 的固化温度约为 200℃。图 10.21a、b 所示分别为 15μm 厚膜层上的 20μm 通孔和 15μm 厚膜层上的 15μm 金属线宽和线距。上述所有特性对于 5G 应用下的电性能和机械 / 热完整性都至关重要。

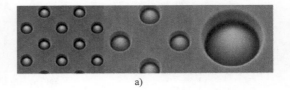

图 10.21　采用 PRL-29 材料：a）15μm 厚膜层上的 20μm 通孔；b）15μm 厚膜层上的 15μm 金属线宽和线距 [15]

10.12 三菱材料的 Dk 和 Df

在参考文献 [16] 中，三菱（Mitsubishi）基于双马来酰亚胺三嗪（bismaleimide triazine，BT）树脂开发了一种新的附加低极性组分，并通过新开发的聚合物共混技术与其中的主要组分相结合。为了确定材料配方中低极性组分的最佳比例，三菱在制造层压板之前对多种 BT 复合材料进行了可行性研究。图 10.22 中的表所示为 BT 复合材料中低极性组分的含量。例如，样品 1 中不含任何极性组分，样品 6 中低极性组分含量高达 80%。样品的截面如图 10.22 所示（浅灰色区域为主要组分，深灰色区域为低极性组分），可以发现[16]，最适合 BT 树脂进行交联的组分比例为 25%（样品 3）。图 10.23 所示为 BT 复合材料的电学特性。可以看出，Dk 和 Df 都随着低极性组分含量的增加而减小。85℃/85%RH 和温度（130℃）对 Df 的影响分别如图 10.24 和 图 10.25 所示。可以看到：

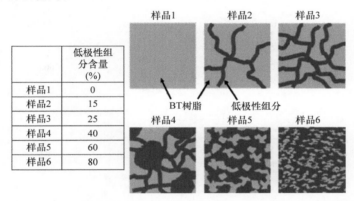

图 10.22 具有低极性组分的 BT 复合材料的截面图像。深灰色区域表示低极性组分，浅灰色区域表示主要组分[16]

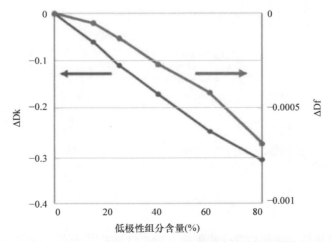

图 10.23 Df 和 Dk 与低极性组分含量的关系[16]

1）Df 在 10GHz、85℃/85%RH 条件下随时间延长而增加。
2）Df 在 10GHz、130℃ 条件下随时间延长基本保持不变。
此外，根据图 10.24 和图 10.25 可知，在 10GHz 下 Df = 0.0025。

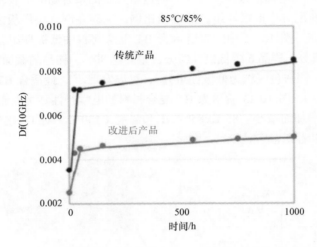

图 10.24　85℃/85%RH 条件下 Df 与时间的关系[16]

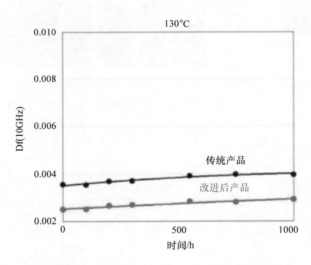

图 10.25　130℃条件下 Df 与时间的关系[16]

10.13　TAITO INK 材料的 Dk 和 Df

通过优化环氧树脂和二氧化硅混合物配比，TAITO INK[17] 开发了一种新型的干膜积层材料，以实现下一代基板的低损耗传输。图 10.26 的左侧为常规材料设计中具有交联点的分子结构。

新的设计（材料 A）采用了一种新型固化体系，该体系具有较低极性的分子结构和较小数量的交联点，其中大部分是二氧化硅填料（见图 10.26 右侧）。图 10.27 所示为以商用卷对卷形式制备的干膜材料 A。

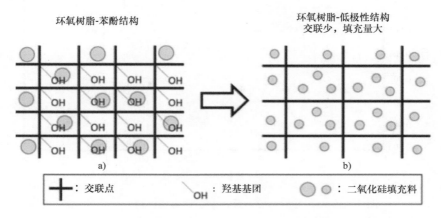

图 10.26　a）传统的环氧基材料；b）新设计的低极性材料[17]

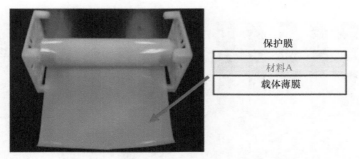

图 10.27　低 Df 干膜堆积材料（材料 A）[17]

材料 A 的特性见表 10.5。采用分离柱电介质谐振腔（split post dielectric resonator，SPDR）测试，10GHz 时 Df = 0.0025、Dk = 3.2，适用于 10~80 GHz 频段，如图 10.28 所示。图 10.29 所示为材料 A 和常规材料所制备的金属线结构 SEM 图。可以看到，材料 A 制备的金属线表面更加光滑，这是由于使用了更小尺寸的纳米级二氧化硅，且树脂和二氧化硅的分散性提高所致。

表 10.5　材料 A 的材料特性[17]

固化后的材料特性	材料 A	常规材料
Dk（10GHz，SPDR）	3.2	3.4
Df（10GHz，SPDR）	0.0025	0.0045
频率	10~80GHz	10GHz
CTE（30~100℃）	18×10^{-6}/K	17×10^{-6}/K
Tg（TMA）	160℃	170℃
杨氏模量	10GPa	12GPa

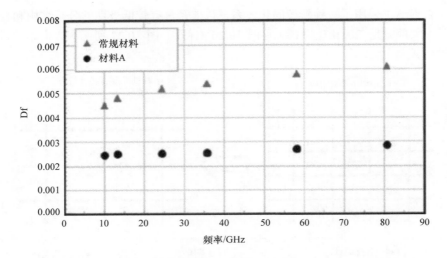

图 10.28 Df 与频率的关系，常规材料与材料 A 的比较[17]

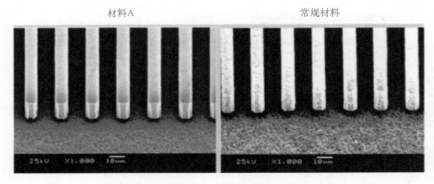

图 10.29 闪蚀后表面状态的 SEM 图像：材料 A 与常规材料[17]

10.14 浙江大学材料的 Dk 和 Df

在参考文献 [18] 中，浙江大学提供了 P-CCL（PCB- 覆铜板）和 LT-CCL（Li_2TiO_3- 覆铜板）的电学材料特性参数值（Dk 和 Df）。用树脂和无碱玻璃纤维布制备的复合层压板简称为 P-CCL。类似地，填充有 LT 陶瓷粉末的复合层压板（MPPE：SEBS = 5：1）简称为 LT-CCL。图 10.30 为各类改性聚苯醚（MPPE）和苯乙烯 - 乙烯 / 丁烯 - 苯乙烯（SEBS）不同质量比所对应的 P-CCL 的 Dk 和 Df 值。可以看出，P-CCL 的 Dk 维持在 3.1 附近。图 10.30 所示的 P-CCL 的介电损耗在 10GHz 时保持在 0.0027 ~ 0.0029 之间，在 CCL 的制造和应用中具有相当大的竞争力。

图 10.31 所示为当 MPPE 与 SEBS 的质量比为 5：1 时，LT 陶瓷粉末含量对 LT-CCL 的 Dk 和 Df 的影响。可以看出，Dk 是 LT 陶瓷粉末含量的函数；LT 含量越高，Dk 越高。另一方面，

添加 LT 陶瓷粉末后，Df 变低并在 10GHz 时保持在 0.0026～0.0028 之间。此外，如图 10.31 所示，随着质量比增加到 1，LT-CCL 的 Df 略有上升。

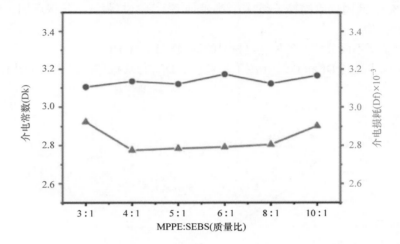

图 10.30　P-CCL 材料 Df 和 Dk 与 MPPE：SEBS（质量比）的关系

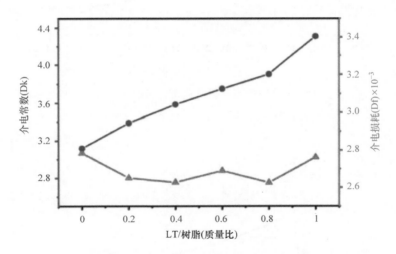

图 10.31　LT-CCL 材料 Df 和 Dk 与 LT/树脂（质量比）的关系

10.15　总结和建议

一些重要的结论和建议总结如下。

1）高速和高频电路中最重要的任务是减少传输损耗，它等于导体损耗和介电损耗之和。导体损耗的解决方案是对表面粗糙度极低的铜箔使用高附着力技术，介电损耗的解决方案是在宽频率、温度、湿度等范围内使用具有优异介电性能和稳定的低 Dk 及 Df 材料。

2）对于 5G 等高速、高频应用，介电材料不仅应具有低损耗（值），以及在变化的湿度条件下具有稳定的 Dk 和 Df 值，而且还应具有低 CTE、低固化温度、低杨氏模量（<2GPa）、低吸湿率（<0.3%）、低固化收缩率（<5%）、高延伸率、高抗拉强度、长保质期、易于制造、适合组装等特点。

3）系统介绍了不同公司的各种介电材料的低损耗 Dk 和 Df。

4）基于文献报道的 Df 和 Dk，预测了未来五年 Df 和 Dk 的路线图分别如图 10.32 和图 10.33 所示。

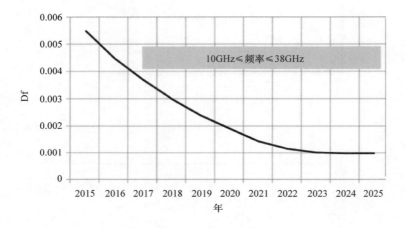

图 10.32　Df 未来五年路线图

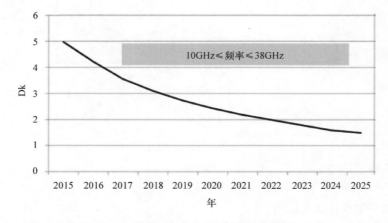

图 10.33　Dk 未来五年路线图

参 考 文 献

1. Lau, J. H., *Fan-Out Wafer-Level Packaging*, Springer, New York, 2018.
2. Lau, J. H., *Heterogeneous Integration*, Springer, New York, 2019.
3. Sato, J., S. Teraki, M. Yoshida, and H. Kondo, "High Performance Insulating Adhesive Film for High-Frequency Applications", *Proceedings of IEEE/ECTC*, May 2017, pp. 1322–1327.
4. Tasaki, T., "Low Transmission Loss Flexible Substrates using Low Dk/Df Polyimide Adhesives", *TechConnect Briefs*, V4, May 2018, pp. 75–78.
5. Hayes, C., K. Wang, R. Bell, C. Calabrese, J. Kong, J. Paik, L. Wei, K. Thompson, M. Gallagher, and R. Barr, "Low Loss Photodielectric Materials for 5G HS/HF Applications", *Proceeding of International Symposium on Microelectronics*, October 2019, pp. 1–5.
6. Hayes, C., K. Wang, R. Bell, C. Calabrese, M. Gallagher, K. Thompson, and R. Barr, "High Aspect Ratio, High Resolution, and Broad Process Window Description of a Low Loss Photodielectric for 5G HS/HF Applications Using High and Low Numerical Aperture Photolithography Tools", *Proceedings of IEEE/ECTC*, May 2020, pp. 623–628.
7. Matsukawa, D., N. Nagami, K. Mizuno, N. Saito, T. Enomoto, and T. Motobe, "Development of Low Dk and Df Polyimides for 5G Application", *Proceeding of International Symposium on Microelectronics*, October 2019, pp. 1–4.
8. Ito, H., K. Kanno, A. Watanabe, R. Tsuyuki, R. Tatara, M. Raj, and R. Tummala, "Advanced Low-Loss and High-Density Photosensitive Dielectric Material for RF/Millimeter-Wave Applications" *Proceedings of International Wafer Level Packaging Conference*, October 2019, pp. 1–6.
9. Nishimura, I., S. Fujitomi, Y. Yamashita, N. Kawashima, and N. Miyaki, "Development of new dielectric material to reduce transmission loss", *Proceedings of IEEE/ECTC*, May 2020, pp. 641–646.
10. Araki, H., Y. Kiuchi, A. Shimada, H. Ogasawara, M. Jukei, and M. Tomikawa, "Low Df Polyimide with Photosenditivity for High Frequency Applications", *Journal of Photopolymer Science and Technology*, V33, 2020, pp. 165–170.
11. Araki, H., Y. Kiuchi, A. Shimada, H. Ogasawara, M. Jukei, and M. Tomikawa, "Low Permittivity and Dielectric Loss Polyimide with Patternability for High Frequency Applications", *Proceedings of IEEE/ECTC*, May 2020, pp. 635–640.
12. Tomikawa, M., H. Araki, M. Jukei, H. Ogasawarai, and A. Shimada, "Low Temperature Curable Low Df Photosensitive Polyimide", *Proceeding of International Symposium on Microelectronics*, October 2019, pp. 1–5.
13. Tomikawa, M., H. Araki, M. Jukei, H. Ogasawarai, and A. Shimada, "Hsigh Frequency Dielectric Properties of Low Dk, Df Polyimides", *Proceeding of International Symposium on Microelectronics*, October 2020, pp. 1–5.
14. Takahashi, K., S. Kikuchi, A. Matsui, M. Abe and K. Chouraku, "Complex Permittivity Measurements in a Wide Temperature Range for Printed Circuit Board Material Used in Millimeter Wave Band", *Proceedings of IEEE/ECTC*, May 2020, pp. 938–945.
15. Han, K., Y. Akatsuka, J. Cordero, S. Inagaki, and D. Nawrocki, "Novel Low Temperature Curable Photo-Patternable Low Dk/Df for Wafer Level Packaging (WLP)", *Proceedings of IEEE/ECTC*, May 2020, pp. 83–88.
16. Yamamoto, K, S. Koga, S. Seino, K. Higashita, K. Hasebe, E. Shiga, T. Kida, and S. Yoshida, "Low Loss BT resin for substrates in 5G communication module", *Proceedings of IEEE/ECTC*, May 2020, pp. 1795–1800.
17. Kakutani, T., D. Okamoto, Z. Guan, Y. Suzuki, M. Ali, A. Watanabe, M. Kathaperumal, and M. Swaminathan, "Advanced Low Loss Dielectric Material Reliability and Filter Characteristics at High Frequency for mmWave Applications", *Proceedings of IEEE/ECTC*, May 2020, pp. 1795–1800.
18. Guo, J., H. Wang, C. Zhang, Q. Zhang, and H. Yang, "MPPE/SEBS Composites with Low Dielectric Loss for High-Frequency Copper Clad Laminates Applications", *Polymers*, V12, August 2020, pp. 1875–1887.

第 11 章

先进封装未来趋势

11.1 引言

本章将对先进封装技术的发展趋势进行介绍。组装工艺如表面安装技术（surface mount technology，SMT），以及引线键合技术、倒装芯片技术、芯片 - 芯片（chip-on-chip，CoC）、芯片 - 晶圆（chip-on-wafer，CoW）、晶圆 - 晶圆（wafer-on-wafer，WoW）、热压键合（thermocompression bonding，TCB）、混合键合等相关技术的发展趋势也会被简要讨论。本章还会介绍片上系统（system-on-chip，SoC）及芯粒的未来趋势，并首先简要提及 COVID-19 对半导体行业的影响。

11.2 COVID-19 对半导体产业的影响

毕马威（KPMG）与全球半导体联盟（Global Semiconductor Alliance，GSA）达成合作，就 COVID-19 对半导体行业的影响展开了市场倾向性调查[1]。他们关于 COVID-19 对第五代宽带蜂窝网络技术标准（5th Generation Technology Standard for Broadband Cellular Networks，5G）、人工智能（artificial intelligence，AI）、物联网（internet of things，IoT）和自动驾驶汽车终端市场投资、增长和采购影响的主要调查结果如图 11.1 所示（评分标准：1 表示显著负面影响，5 表示显著积极影响）。

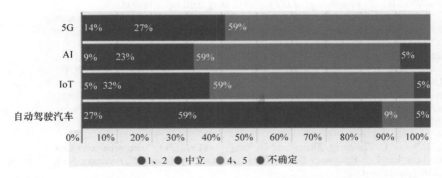

图 11.1 COVID-19 对 5G、AI、IoT 和自动驾驶汽车投资、增长和采购的影响
（评分标准：1 表示显著负面影响，5 表示显著积极影响）

11.3 COVID-19 对晶圆代工行业的影响

工艺技术向先进节点迁移,如 7nm、5nm 以及极紫外(extreme ultraviolet,EUV)光刻技术,使之加速满足了 5G 智能手机、AI 高性能计算(high performance computing,HPC)和 AI/图形处理器(graphics processing unit,GPU)/现场可编程门阵列(field-programmable gate array,FPGA)在大数据(用于云计算)和实时数据(用于边缘计算)等方面的需求。

2020 年,晶圆代工行业营收达到约 820 亿美元,同比(年)增长 23%。尽管 2020 年的基数很高,但 2021 年仍将保持两位数的增长。Counterpoint 预测,2021 年的总营收将为 920 亿美元,同比(年)增长 12%,2022~2023 年将为 1000 亿美元[2]。世界上最大的晶圆代工厂(台积电)预计将继续领跑行业,在 2021 年实现 13%~16% 的年度销售增长。

11.4 COVID-19 对半导体客户的影响

2021 年,世界上只有两家公司(台积电和三星)可以实现 5nm 和 7nm 工艺。根据 Counterpoint 的报告[3],2021 年 5nm 工艺的潜在客户如图 11.2 所示,7nm 工艺的潜在客户如图 11.3 所示。

对于 5nm 工艺,苹果是 2021 年 5nm 工艺的第一大客户(所有订单都给了台积电)(见图 11.2),包括 iPhone(A14/A15)和新发布的 Apple Silicon。高通将是 5nm 的第二大客户,因为 iPhone 13 可能采用其 X60 调制解调器。台积电预计将在 2021 年从 5nm 中获得 100 亿美元的营收。三星代工厂也将从 5nm 的订单中获得强效推动力,包括其内部品牌(Exynos)的 SoC 产品及高通的订单。

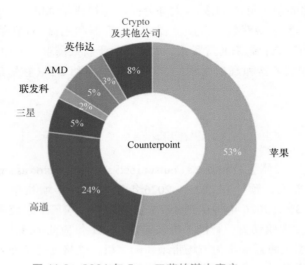

图 11.2 2021 年 5nm 工艺的潜力客户

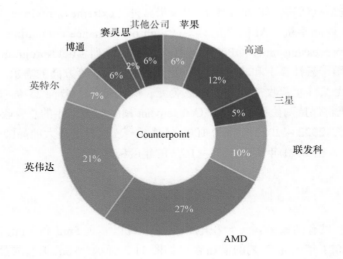

图 11.3 2021 年 7nm 工艺的潜力客户

不同于 5nm 工艺超过 80% 的晶圆被用于智能手机，7nm 工艺的应用更加多样化，包括 AI/GPU、中央处理器（central processing unit，CPU）、网络处理器和车用处理器。台积电在其 7nm 系列中拥有 7nm [仅深紫外（deep ultraviolet，DUV）]、7nm+（采用 EUV）和 6nm（采用 EUV）等多种方案，三星则推出了均采用 EUV 生产的 7nm/6nm 工艺。在这一布局中，智能手机仅消耗 35% 的晶圆（见图 11.3），大部分晶圆将提供给 AMD（占 7nm 出货量的 27%）和英伟达（占 21%）。鉴于游戏机、云服务器/AI 处理器和主流 5G 智能手机的强劲需求，整个 2021 年期间 7nm 的产能看起来非常紧张。因此，对于新兴需求，如加密货币挖矿 ASIC 和基于 ARM 的处理器（在服务器和汽车领域），芯片组供应商和 OEM 在短期内很难获得额外的产能分配。

11.5 COVID-19 对封测行业的影响

2020 年，外包半导体组装和测试（outsourced semiconductor assembly and test，OSAT）市场价值为 316.4 亿美元，预计在 2021~2026 年期间复合年增长率（compound annual growth rate，CAGR）为 7.3%，2026 年将达到 497.1 亿美元[4]。汽车子系统和连接设备的需求增加被认为是预测期内的主要驱动力。智能手机领域是 OSAT 供应商的最大客户，去年的出货量超过了 10 亿颗。5G 的引入将进一步为市场带来新的生机。连接度是工业 4.0 的核心，因此在未来，边缘计算和物联网设备的需求被认为将呈指数趋势增长。从图 11.4 可以看出 COVID-19 对 2020 年 OSAT 市场的影响[4]。

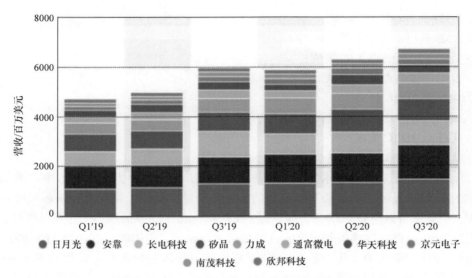

图 11.4　2019 年和 2020 年 OSAT 产业的营收

11.6　驱动端、半导体和先进封装

半导体产业有五个主要的增长引擎（应用）：
1）移动终端；
2）高性能计算；
3）自动驾驶汽车；
4）物联网；
5）用于云计算的大数据和用于边缘计算的实时数据处理。

以下 2 种系统技术驱动端在促进 5 个应用快速发展：
1）人工智能；
2）5G。

集成有半导体材料及器件的先进封装技术种类包括：
1）2D 扇出型（先上晶）IC 集成；
2）2D 倒装芯片 IC 集成；
3）封装堆叠（package-on-package，PoP）；
4）系统级封装（system-in-package，SiP）或异质集成；
5）2D 扇出型（后上晶）IC 集成；
6）2.1D 倒装芯片 IC 集成；
7）2.1D 含互连桥倒装芯片 IC 集成；
8）2.1D 含互连桥扇出型 IC 集成；
9）2.3D 扇出型（先上晶）IC 集成；

10) 2.3D 倒装芯片 IC 集成；

11) 2.3D 扇出型（后上晶）IC 集成；

12) 2.5D（焊料凸点）IC 集成；

13) 2.5D（微凸点）IC 集成；

14) 微凸点 3D IC 集成；

15) 微凸点芯粒 3D IC 集成；

16) 无凸点 3D IC 集成；

17) 无凸点芯粒 3D IC 集成。

图 11.5 是上述先进封装技术的密度和性能对应图，它们被归类于图 11.6 中。图 11.7 是未来 5 年内它们的基板尺寸和基板上引脚数的关系图。可以看出，积层封装基板尺寸可以达到 5000mm^2，引脚数可以达到 6000，金属线宽/线距 ≥ 6μm（2.1D 的积层上还有薄膜层，金属线宽/线距 ≥ 2μm）；TSV 转接板基板尺寸可以达到 3000mm^2，引脚数可以达到 >100000，金属线宽/线距 ≤ 1μm；扇出型（先上晶）RDL-基板（或转接板）衬底尺寸可以达到 600mm^2，引脚数可以达到 2500，金属线宽/线距 ≥ 5μm。扇出型（后上晶）RDL-转接板基板尺寸可以达到 2500mm^2，引脚数可以达到 5000，金属线宽/线距 ≥ 2μm；互连桥尺寸非常小（<200mm^2），引脚数很少（<2000），金属线宽/线距 ≥ 2μm。

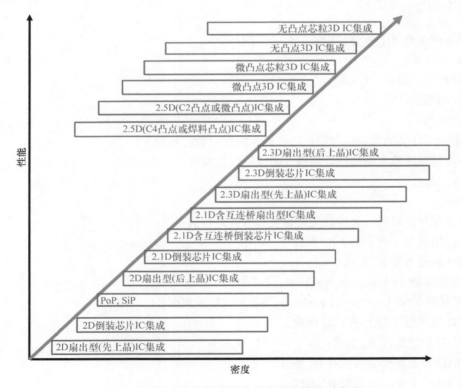

图 11.5　各种先进封装技术的密度和性能对应图

第 11 章　先进封装未来趋势　397

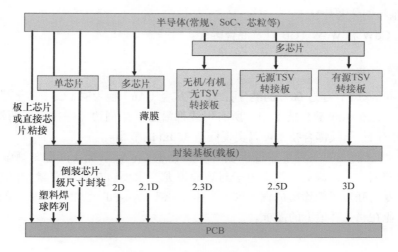

图 11.6　先进封装的分类

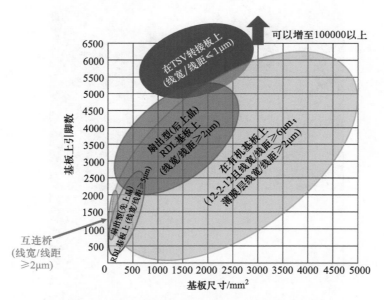

图 11.7　先进封装技术：基板尺寸和基板上引脚数

11.7　先进封装的组装工艺

先进封装技术包含许多装配工艺，本节将简要介绍以下内容：
1）引线键合技术；
2）表面安装技术；
3）晶圆凸点成型技术；

4)有机基板倒装芯片技术;

5)CoC、CoW 和 WoW TCB 以及混合键合。

11.7.1 引线键合

目前在成本驱动下,超过 50% 的键合引线已经从金(Au)换为铜(Cu)甚至部分银(Ag)[5-8]。如图 11.8[5] 所示,在引线键合技术中,所有引线都沿芯片的周边(一排或者两排)进行键合。对于一些特殊应用,引线键合技术可以实现低至 35μm 的节距。

铝焊盘上键合铜线的主要可靠性挑战之一是克服 Cu-Al 系统在高温、潮湿和偏压环境下产生腐蚀的敏感性 [6, 7],这对于汽车电子封装而言尤为重要。在参考文献 [7] 中,弗劳恩霍夫(Fraunhofer)专门开发了钯涂覆的铜(PCC)和金 - 钯(Au-Pd)涂覆(APC)的铜引线,以满足汽车电子产业在恶劣环境下的需求。

图 11.8　引线键合技术

第 11 章 先进封装未来趋势

近期 Ag 键合引线作为一种相对较新的材料出现在商业化电子产品中,因为它具有适中的硬度、高延展性、金属中最好的导热性和导电性以及低生长速度的金属间化合物(intermetallic compound,IMC)[8]。铝焊盘会产生 Ag_2Al 和 Ag_3Al 等界面 IMC。然而 Ag_2Al 的软度、断裂韧度和耐腐蚀性要比 Ag_3Al 好得多。因此从长期可靠性的角度来看,Ag_3Al 及其相邻相间界面将变为薄弱区域。在参考文献 [8] 中,加利福尼亚大学尔湾分校展示了一种新的 An-10In 合金,其中 Ag_3Al 可以在互扩散过程中被消除,并被一个三元相所取代。与 Ag_3Al 相比,这种新相坚韧得多且更耐腐蚀。

11.7.2 SMT

如图 11.9 所示,在第 2 章中所讨论的 SMT 关键要素包括模板印刷、射片机、芯片拾取和放置(pick and place,P&P)以及回流。SMT 的利润率非常低,因此成本压力相对较高。近年来,随着许多电子产品向微型分立元件、高密度、超窄焊盘节距等方向发展,它们对 SMT 提出了严峻的挑战。同时,免清洗焊膏也被广泛使用。所有上述趋势都对回流工艺提出了新的要求,需要开发更先进的回流热传导技术以实现节能需求及温度均一性。

随着工业 4.0 的发展,SMT 的自动化程度越来越高,人工成本大幅降低,人均产量也随之提升。SMT 的未来趋势将会朝着高度小型化、高性能、高可靠性、高效率及环境友好的方向发展。

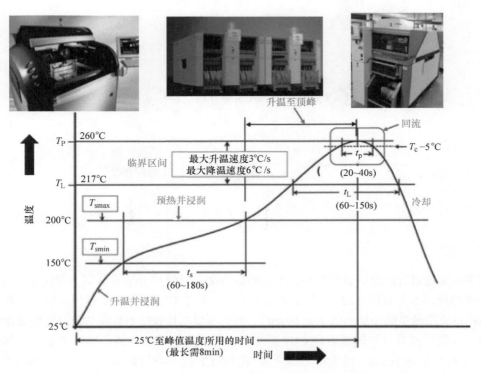

图 11.9　SMT:模板印刷、芯片贴装机、芯片拾取和放置、回流

11.7.3 倒装芯片技术的晶圆凸点成型

如第 2 章所述,晶圆凸点成型是倒装芯片技术之母。至少有两种不同的凸点可供选择,即可控塌陷芯片互连(controlled collapse chip connection,C4)凸点和芯片互连(chip connection,C2)凸点,工艺流程分别如图 11.10a、b 所示。其技术趋势将朝着更窄节距发展:在未来 5 年内,C4 凸点的最小节距将达到 50μm,C2 凸点的最小节距将达到 20μm。

图 11.10c 展示了 Amkor 的 Double-POSSUM 封装[9]。可以看出,封装实际上是由两级嵌套芯片定义的。三个子芯片倒装互连到较大的母芯片上,然后再连接到最大的祖母芯片上,最后再将祖母芯片倒装互连到封装基板上。子、母芯片之间的凸点是微凸点(带有焊料帽的铜柱),母、祖母芯片之间以及祖母芯片与封装基板之间采用的是 C4 凸点。

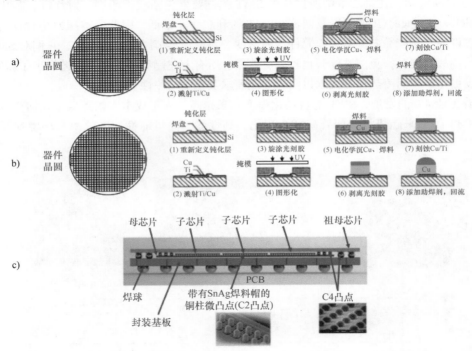

图 11.10 a)和 b)C4 及 C2 晶圆凸点成型;c)Amkor 的 Double-POSSUM 封装

11.7.4 有机基板上的倒装芯片技术

倒装芯片技术的最新进展和趋势已经在如参考文献[10]中介绍过。第 2 章讨论了有机基板上的倒装芯片技术(见图 2.33)。在未来的 5 年里,基于 CUF 的 C4 凸点的倒装芯片批量回流技术仍将是使用最多的(最小节距为 60μm)。由于薄型芯片和薄型封装基板的使用量增加,C2 凸点小压力热压键合及回流再进行 CUF 的工艺的使用将会增多(最小节距 50μm)。倒装芯片 C2 凸点(最小节距 30μm)的大压力键合及 NCP/NCF 技术将频繁用于高引脚数和高密度的应用场景。

11.7.5 CoC、CoW 和 WoW TCB 以及混合键合

如图 11.11 所示，硅到硅倒装芯片面对面或面对背键合通常通过 TCB 或者 Cu-Cu 无凸点混合键合实现，互连结构是 C2 凸点或无凸点结构。图 11.12 所示为英特尔带有 C2 凸点结构的 TCB 和无凸点结构的混合键合。

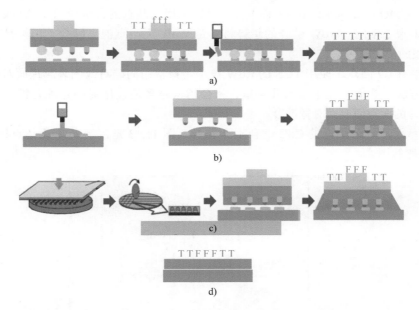

图 11.11 倒装芯片（硅到硅）组装工艺：a）基于 CUF 的 C2 凸点的小压力 TCB；b）基于 NCP 的 C2 凸点的大压力 TCB；c）基于 NCF 的 C2 凸点的大压力 TCB；d）无凸点混合键合

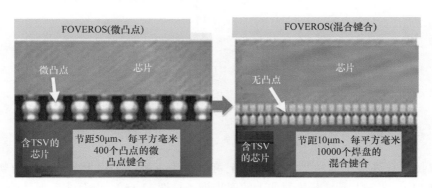

图 11.12 英特尔的微凸点键合与无凸点混合键合

WoW 键合形式能够提供最高的产率，然而其良率问题仍待解决，即可能将好芯片键合在坏芯片上。通常情况下，WoW 适用于两个晶圆的键合场景。对于四片晶圆堆叠的场景，如高带宽存储器（high bandwidth memory, HBM），良率造成的损失是不被允许的。比如，如果单片 DRAM

晶圆良率为 60%，则组装后良率为 12.96%；如果 DRAM 晶圆的良率为 80%，则组装后良率为 40.96%；如果 DRAM 晶圆的良率为 90%，则组装后良率为 65.61%。此外，芯片尺寸各异也是一个难题。

目前，微机电系统（micro electro mechanical system，MEMS）器件晶圆和盖板（或 ASIC）晶圆通过 C2 凸点的 TCB 技术已被几家公司用于大规模生产。只有索尼的处理器晶圆和 CMOS 图像传感器（CMOS image sensor，CIS）像素晶圆采用无凸点混合键合的 WoW 技术实现了大规模量产。未来 5 年内，WoW 键合的趋势将会是 WoW 无凸点混合键合（仍然是两个晶圆），并将比现阶段的使用量更大。

目前，带有 C2 凸点、使用 TCB 的 CoW 技术是最常用的硅对硅倒装芯片面对面或面对背键合的组装方法。然而，在未来的几年里，将会越来越多地使用无凸点混合键合方法来实现 CoW 倒装芯片的面对面或面对背键合。

由于产率因素，无凸点的 CoC 键合不会普及。图 11.13 是倒装芯片无凸点键合 CoW 和 WoW 技术的路线图。

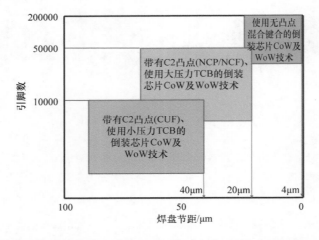

图 11.13　基于 CUF 的带有 C2 凸点、使用小压力 TCB，基于 NCP/NCF 的带有 C2 凸点、使用大压力 TCB，使用无凸点混合键合等方式实现的倒装芯片（硅对硅）CoW 及 WoW 技术

11.8　扇出型先上晶（芯片面朝上）、先上晶（芯片面朝下）以及后上晶技术

扇出型封装有许多不同种类[11]，基本上可以分为三类：①先上晶、芯片面朝下的扇出型（见图 11.14），②先上晶、芯片面朝上的扇出型（见图 11.15），以及③后上晶的扇出型（见图 11.16）。表 11.1 对这三种封装型式进行了比较。可以看出，先上晶、芯片面朝下（见图 11.14）是最简单、成本最低的型式，而后上晶或先 RDL（见图 11.16）是最复杂、成本最高的型式（后上晶需要晶圆植球、芯片-RDL-基板键合、底部填充或模塑底部填充以及封装基板）。先上晶型式中芯片面朝上（见图 11.15）比芯片面朝下需要的工艺步骤稍多一些（因此成本也稍高）。

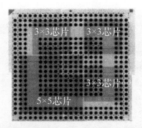

图 11.14 先上晶、芯片面朝下的扇出型封装

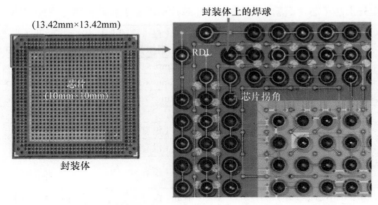

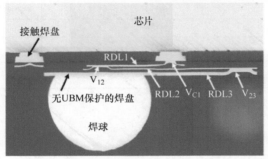

图 11.15 先上晶、芯片面朝上的扇出型封装

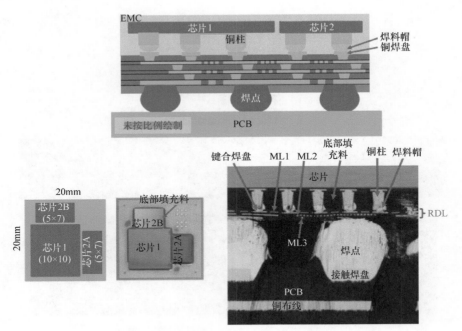

图 11.16 扇出型后上晶（或先 RDL）封装

表 11.1 扇出型先上晶（芯片面朝下）、先上晶（芯片面朝上）和后上晶（或先 RDL）的对比

	先上晶（面朝下），例如 eWLB	先上晶（面朝上），例如 In_FO	后上晶（先 RDL），例如 SiWLP
芯片尺寸	≤ 5mm × 5mm[①]	≤ 12mm × 12mm[①]	≤ 20mm × 20mm
封装尺寸	≤ 10mm × 10mm[②]	≤ 25mm × 25mm[②]	≤ 45mm × 45mm
RDL（金属线宽/线距）	≥ 10μm	≥ 5μm	≥ 2μm 或 < 1μm[③]
RDL（层数）	≤ 3	≤ 4	≤ 6
晶圆植球	不是（没有）	不是（没有）	是（有）
芯片-基板键合	不是（没有）	不是（没有）	是（有）
底部填充或模塑底部填充	不是（没有）	不是（没有）	是（有）
积层封装基板	不是（没有）	不是（没有）	是（有）
工艺步骤	简单	稍多	较多
成本	低	中等	高
性能	低	中等	高
应用	基带、微处理器、射频/模拟、电源管理集成电路等	苹果的应用处理器芯片组	高性能及高密度（非大规模量产）

[①] 受限于芯片移位。
[②] 受限于翘曲。
[③] 采用 PECVD + Cu 大马士革工艺 + CMP。

先上晶扇出型封装的应用比扇入型晶圆级封装（见第 3 章）更广。然而，也有一些事情虽然传统的塑料球栅阵列（plastic ball grid array，PBGA）封装就可以做到，但是先上晶扇出型封装却做不到，如①更大的芯片尺寸（≥12mm×12mm），②更大的封装尺寸（≥25mm×25mm）。这是由于热膨胀系数不匹配及扇出型先上晶封装的翘曲度限制所造成的。采用扇出型后上晶（先RDL）封装可以将应用范围扩展到芯片尺寸≤20mm×20mm，封装尺寸≤45mm×45mm。

扇出型先上晶型式是对用于便携式、移动和可穿戴产品的基带、射频/模拟、电源管理集成电路、应用处理器、低端ASIC、中央处理器（central processing unit，CPU）和图形处理器（graphics processing unit，GPU）半导体IC封装的合适选择。而扇出型后上晶（后RDL）型式适用于封装例如用于服务器、网络和电信产品的HBM、高端CPU、GPU、ASIC和现场可编程门阵列（field programmable grid array，FPGA）等IC。

在未来5年内，先上晶、芯片面朝下的扇出型封装仍然并将继续是使用最多的型式；先上晶、芯片面朝上的扇出型封装将是第二大型式；而后上晶的扇出型封装将是使用更少的型式。

11.9　互连桥与TSV转接板

图11.17是英特尔的在积层封装基板顶层制备的嵌入式多芯片互连桥（embedded multi-die interconnect bridge，EMIB）。EMIB是一片无功能的、带有窄金属线宽/线距RDL的硅片，用于实现芯片水平方向的互连[12, 13]。EMIB的引入可以去除TSV转接板的使用，它们的一项专利如图11.18a所示。

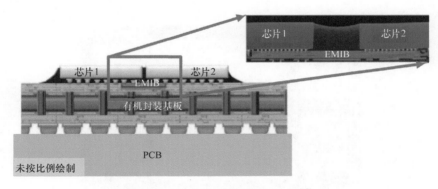

图11.17　英特尔的EMIB技术[13]

2017年12月8日，应用材料（Applied Materials）公司提出使用扇出型封装技术将互连桥嵌入到环氧树脂模塑料（epoxy molding compound，EMC）中，如图11.18b所示[14]。可以看出，芯片上的电路是用RDL扇出的，并通过互连桥实现了水平方向的连接，垂直互连是通过TMV实现的[14]。

在台积电年度技术研讨会上（2020年8月25日），台积电宣布了它的局部硅互连（local Si interconnect，LSI）技术，如图11.19所示。从图中可以看到LSI技术与应用材料公司的方案非常相似，唯一的明显不同是LSI的转接板也许不仅仅是一片无TSV的转接板，还可能还包括TSV甚至CMOS器件。

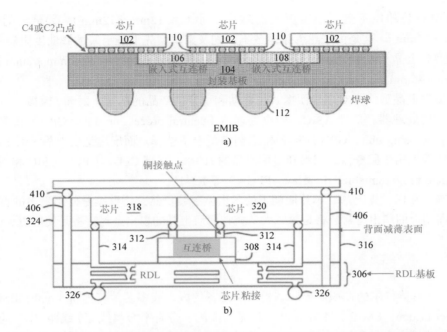

图 11.18 a）英特尔的积层封装基板顶部嵌入 EMIB 技术；b）应用材料公司的 EMC 中嵌入互连桥技术

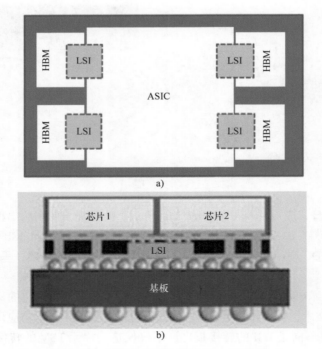

图 11.19 台积电的互连桥技术——LSI 技术

由于芯粒异质集成技术的普及,互连桥技术的使用将会增多,以实现主要在水平方向上的芯片互连,如 EMIB、LSI 和应用材料公司的技术。

11.10 SoC 与芯粒

SoC 将具有不同功能的 IC 集成到一个系统或子系统的单一芯片中。图 11.20 所示为苹果应用处理器(application processor,AP)从 A10 到 A14 的变化,图 11.21 所示为不同特征尺寸芯片的晶体管数量与年份的关系。可以看出,芯片中晶体管的数量和功能随着特征尺寸的减小而增加。不幸的是,根据国际商业策略(International Business Strategies)公司的调研,图 11.22 所示为先进工艺芯片设计成本随特征尺寸减小(直至 5nm)的变化关系。可以看出,仅完成 5nm 特征尺寸芯片的设计就需要 5 亿多美元,5nm 工艺技术的开发还需要 10 亿美元。

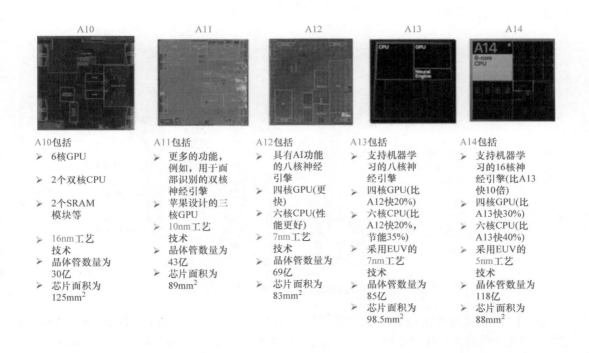

图 11.20　苹果的应用处理器(从 A10 至 A14)

芯粒异质集成与 SoC 不同。芯粒异质集成将 SoC 重新设计为更小的芯粒,然后利用封装技术将不同材料制作的,具有不同功能的,由不同设计公司和代工厂制作的,具有不同晶圆尺寸、特征尺寸的芯粒集成到一个系统或子系统中[15, 16]。其中一颗芯粒就是一种由可复用 IP(知识产权)块组成的功能集成电路(IC)模块。

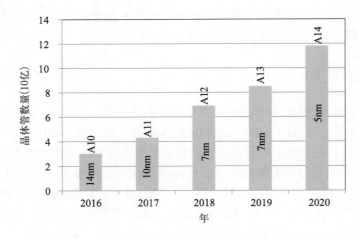

图 11.21　苹果应用处理器：晶体管数量和年份的关系

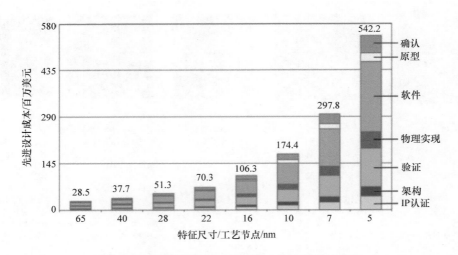

图 11.22　不同特征尺寸的先进工艺芯片设计成本

图 11.23a 所示为 AMD EPYC 处理器[17-19]。可以看到在一颗较大的 I/O 芯片（采用 14nm 工艺技术制造）的两侧有 4 对芯粒（采用 7nm 工艺技术制造），它们通过键合和底部填充紧密安装在一个有机基板上。图 11.23b 所示为英特尔移动（笔记本电脑）处理器"Lakefield"，基于其 FOVEROS 技术制造。SoC 芯片被分区（如 CPU、GPU、LPDDR4 等）并切分（例如，CPU 被切分为一颗大 CPU 和 4 颗小 CPU）为芯粒，然后采用 CoW 工艺将这些芯粒面对面键合（堆叠）到有源 TSV 转接板（一颗大的 22FFL 基底芯片）上。

如图 11.24 所示，目前至少有两种不同的芯粒异质集成方法，即芯片切分与集成（由成本和良率驱动）和芯片分区与集成（由成本和技术优化驱动）。在芯片切分与集成中，逻辑（logic）芯片等 SoC 被切分为更小的芯粒，如 logic1、logic2 和 logic3。这些芯粒可以通过前道 CoW

键合或 WoW 键合工艺完成堆叠（集成）[20-23]，然后采用先进封装技术将其组装（集成）在单个封装体的同一基板上。应该强调的是，前道工艺芯粒集成能获得更小的封装面积和更好的电性能，不过这不是必须的。在芯片分区与集成中，例如带有逻辑和 I/O 的 SoC，被按功能划分为逻辑和 I/O 芯粒模块，然后通过前道 CoW 或 WoW 工艺方法进行集成（堆叠）。再用先进封装技术将逻辑和 I/O 芯粒组装在单个封装体的同一基板上。同样地，芯粒的前道集成工艺也不是必须的。

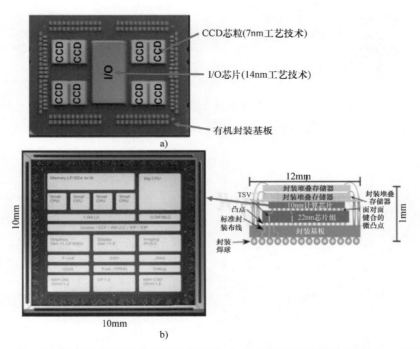

图 11.23　a）AMD 采用芯粒设计的 EPYC 处理器；b）英特尔采用芯粒设计的 Lakefield 处理器

与 SoC 相比，芯粒异质集成的关键优势在于制造过程中良率提高（成本降低）、设计过程中的上市时间短和成本降低。图 11.25 所示为单片设计和 2、3、4 芯粒设计所对应的每片晶圆良率（良好芯片百分比）与芯片尺寸的关系图[24]。

工艺微缩的 SoC 将继续存在。然而只有少数几家公司，如苹果、三星、华为、谷歌，能够负担起更小的特征尺寸（先进工艺节点）。通常他们采用这种方式是有原因的，以苹果为例，至少有三个原因：

1）2008 年 4 月 23 日，自苹果收购 Palo Alto Semiconductor 起，便一直在用大量 IP 构建芯片，并且与其软件开发进行紧密耦合（集成）。

2）因为额外的芯片间互连和通信开销会带来更多的问题，将其 SoC 设计分解为芯粒并非那么有吸引力。

3）世界排名第一的代工厂（台积电）是苹果的忠实合作伙伴，他们致力于完成苹果的产品，例如，应用处理器（A16）计划于 2022 年下半年采用台积电的 3nm 工艺技术制造。

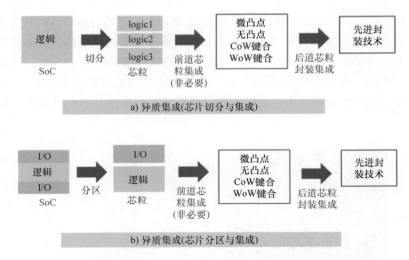

图 11.24 通过芯片切分与集成、分区与集成实现的芯粒技术

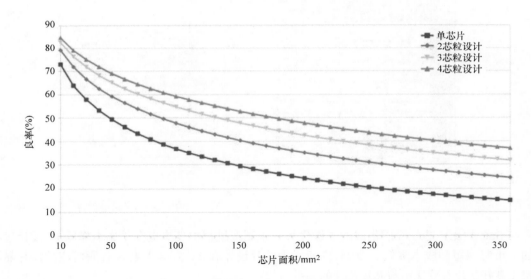

图 11.25 SoC 和不同芯粒设计的良率与芯片尺寸的关系

11.11 高速/高频器件对材料的需求

在 5G 的驱动下，为了满足提高信号传输速度/速率和管理海量数据的要求，先进封装材料的不断发展是非常必要的。就绝缘材料的电性能而言，低损耗因子或损耗角正切（dissipation factor or loss tangent，Df）和低介电常数（dielectric constant or permittivity，Dk）材料在 5G 应用中是优先选择的[25-40]，且其路线图如图 11.26、图 11.27 所示。

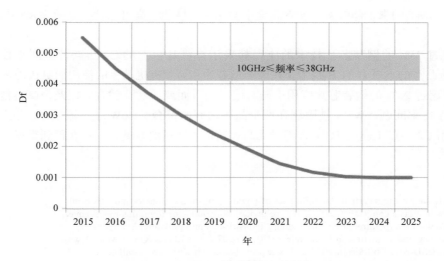

图 11.26 损耗因子或损耗角正切（Df）的路线图

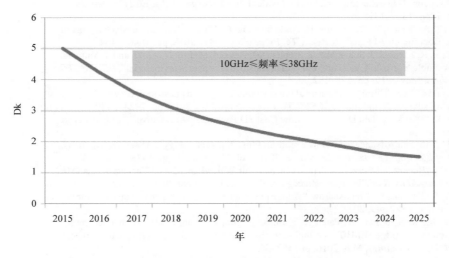

图 11.27 介电常数（Dk）的路线图

11.12 总结和建议

一些重要的结论和建议总结如下。

1）与 SoC 相比，芯粒异质集成的关键优势在于制造过程中良率提高（成本降低）、设计过程中的上市时间短和成本降低。劣势之一是增加了封装成本。

2）工艺微缩的 SoC 将继续存在。芯粒技术提供了 SoC 的另外一种选择，特别是对于大多数公司无法负担的先进工艺节点芯片。

3）随着工业 4.0 的发展，SMT 的自动化程度越来越高，人工成本大幅降低，人均产量也随之提升。SMT 的未来趋势将会朝着高度小型化、高性能、高可靠性、高效率及环境友好的方向发展。

4）超过 75% 的倒装芯片技术的应用都采用有机基板上的 C4 凸点批量回流技术。基于 CUF 的 C2 凸点的小压力 TCB 技术正在因薄型芯片和薄型有机基板的使用而更快地发展。

5）不超过 25% 的倒装芯片应用于硅对硅、面对面或面对背键合，以及极高性能和高密度场景。由于良率问题，WoW 被限制在两个晶圆之间的键合，CoW 是主流技术。互连材料/结构主要采用 C2 凸点（通过 TCB），不过无凸点结构（通过混合键合）正在得到重视。

参 考 文 献

1. https://advisory.kpmg.us/articles/2020/impact-of-covid-19-on-semiconductor-industry.html.
2. https://www.statista.com/statistics/867210/worldwide-semiconductor-foundries-by-revenue/.
3. https://www.counterpointresearch.com/foundry-industry-revenue-growth-continue-2021/.
4. https://www.globenewswire.com/news-release/2021/02/02/2167973/0/en/OSAT-Market-was-valued-at-USD-31-64-billion-in-2020-and-is-expected-to-reach-USD-49-71-billion-at-a-CAGR-of-7-3-over-the-forecast-period-2021-2026.html.
5. Qin, I., O. Yauw, G. Schulze, A. Shah, B. Chylak, and N. Wong, "Advances in Wire Bonding Technology for 3D Die Stacking and Fan Out Wafer Level Package", *IEEE/ECTC Proceedings*, May 2017, pp. 1309–1315.
6. Qin, W., T. Anderson, D. Barrientos, H. Anderson, and G. Chang, "Corrosion Mechanisms of Cu Wire Bonding on Al Pads", *IEEE/ECTC Proceedings*, May 2018, pp. 1446–1454.
7. Klengel, S., R. Klengel, J. Schischka, T. Stephan, M. Petzold, M. Eto, N. Araki, and T. Yamada, "A new reliable, corrosion resistant gold-palladium coated copper wire material", *IEEE/ECTC Proceedings*, May 2019, pp. 175–182.
8. Wua, J., and C. Lee, "Eliminating harmful intermetallic compound phase in silver wire bonding by alloying silver with indium", *IEEE/ECTC Proceedings*, May 2018, pp. 2224–2230.
9. Sutanto, J., "POSSUM Die Design as a Low Cost 3D Packaging Alternative," *3D Packaging*, 2012, pp. 16–18.
10. Lau, J. H., "Recent Advances and New Trends in Flip Chip Technology", *ASME Transactions, Journal of Electronic Packaging*, September 2016, Vol. 138, Issue 3, pp. 1–23.
11. Lau, J. H., "Recent Advances and Trends in Fan-Out Wafer/Panel-Level Packaging", *ASME Transactions, Journal of Electronic Packaging*, Vol. 141, December 2019, pp. 1–27.
12. Chiu, C., Z. Qian, and M. Manusharow, "Bridge interconnect with air gap in package assembly," *US Patent No. 8,872,349*, 2014.
13. Mahajan, R., R. Sankman, N. Patel, D. Kim, K. Aygun, Z. Qian, et al., "Embedded multi-die interconnect bridge (EMIB) – a high-density, high-bandwidth packaging interconnect," *IEEE/ECTC Proceedings*, May 2016, pp. 557–565.
14. Hsiung, C., and a. Sundarrajan, "Methods and Apparatus for Wafer-Level Die Bridge", US 10,651,126 B2, Filed on December 8, 2017, Granted on May 12, 2020.
15. Lau, J. H., *Heterogeneous Integrations*, Springer, New York, 2019.
16. Lau, J. H., "Recent Advances and Trends in Heterogeneous Integrations", *IMAPS Transactions, Journal of Microelectronics and Electronic Packaging*, Vol. 16, April 2019, pp. 45–77.
17. Naffziger, S., K. Lepak, M. Paraschour, and M. Subramony, "AMD Chiplet Architecture for High-Performance Server and Desktop Products", *IEEE/ISSCC Proceedings*, February 2020, pp. 44–45.
18. Naffziger, S., "Chiplet Meets the Real World: Benefits and Limits of Chiplet Designs", *Symposia on VLSI Technology and Circuits*, June 2020, pp. 1–39.
19. Stow, D., Y. Xie, T. Siddiqua, and G. Loh, "Cost-Effective Design of Scalable High-Performance Systems Using Active and Passive Interposers", *IEEE/ICCAD Proceedings*, November 2017, pp. 1–8.
20. Chen, M. F., C. S. Lin, E. B. Liao, W. C. Chiou, C. C. Kuo, C. C. Hu, C. H. Tsai, C. T. Wang and D. Yu, "SoIC for Low-Temperature, Multi-Layer 3D Memory Integration", *IEEE/ECTC Proceedings*, May 2020, pp. 855–860.

21. Chen, Y. H., C. A. Yang, C. C. Kuo, M. F. Chen, C. H. Tung, W. C. Chiou, and D. Yu, "Ultra High Density SoIC with Sub-micron Bond Pitch", *IEEE/ECTC Proceedings,* May 2020, pp. 576–581.
22. Chen, F., M. Chen, W. Chiou, D. Yu, "System on Integrated Chips (SoIC^{TM}) for 3D Heterogeneous Integration", *IEEE/ECTC Proceedings,* May 2019, pp. 594–599.
23. Lin, J., C. Chung, C. Lin, A. Liao, Y. Lu, J. Chen, and D. Ng, "Scalable Chiplet package using Fan-Out Embedded Bridge", *IEEE/ECTC Proceedings*, May 2020, pp. 14–18.
24. https://en.wikichip.org/wiki/chiplet, March 27, 2020.
25. Sato, J., S. Teraki, M. Yoshida, and H. Kondo, "High Performance Insulating Adhesive Film for High-Frequency Applications", *Proceedings of IEEE/ECTC*, May 2017, pp. 1322–1327.
26. Tasaki, T., "Low Transmission Loss Flexible Substrates using Low Dk/Df Polyimide Adhesives", *TechConnect Briefs*, V4, May 2018, pp. 75–78.
27. Hayes, C., K. Wang, R. Bell, C. Calabrese, J. Kong, J. Paik, L. Wei, K. Thompson, M. Gallagher, and R. Barr, "Low Loss Photodielectric Materials for 5G HS/HF Applications", *Proceeding of International Symposium on Microelectronics*, October 2019, pp. 1–5.
28. Hayes, C., K. Wang, R. Bell, C. Calabrese, M. Gallagher, K. Thompson, and R. Barr, "High Aspect Ratio, High Resolution, and Broad Process Window Description of a Low Loss Photodielectric for 5G HS/HF Applications Using High and Low Numerical Aperture Photolithography Tools", *Proceedings of IEEE/ECTC*, May 2020, pp. 623–628.
29. Matsukawa, D., N. Nagami, K. Mizuno, N. Saito, T. Enomoto, and T. Motobe, "Development of Low Dk and Df Polyimides for 5G Application", *Proceeding of International Symposium on Microelectronics*, October 2019, pp. 1–4.
30. Ito, H., K. Kanno, A. Watanabe, R. Tsuyuki, R. Tatara, M. Raj, and R. Tummala, "Advanced Low-Loss and High-Density Photosensitive Dielectric Material for RF/Millimeter-Wave Applications" *Proceedings of International Wafer Level Packaging Conference*, October 2019, pp. 1–6.
31. Nishimura, I., S. Fujitomi, Y. Yamashita, N. Kawashima, and N. Miyaki, "Development of new dielectric material to reduce transmission loss", *Proceedings of IEEE/ECTC*, May 2020, pp. 641–646.
32. Araki, H., Y. Kiuchi, A. Shimada, H. Ogasawara, M. Jukei, and M. Tomikawa, "Low Df Polyimide with Photosenditivity for High Frequency Applications", *Journal of Photopolymer Science and Technology*, V33, 2020, pp. 165–170.
33. Araki, H., Y. Kiuchi, A. Shimada, H. Ogasawara, M. Jukei, and M. Tomikawa, "Low Permittivity and Dielectric Loss Polyimide with Patternability for High Frequency Applications", *Proceedings of IEEE/ECTC*, May 2020, pp. 635–640.
34. Tomikawa, M., H. Araki, M. Jukei, H. Ogasawarai, and A. Shimada, "Low Temperature Curable Low Df Photosensitive Polyimide", *Proceeding of International Symposium on Microelectronics*, October 2019, pp. 1–5.
35. Tomikawa, M., H. Araki, M. Jukei, H. Ogasawarai, and A. Shimada, "Hsigh Frequency Dielectric Properties of Low Dk, Df Polyimides", *Proceeding of International Symposium on Microelectronics*, October 2020, pp. 1–5.
36. Takahashi, K., S. Kikuchi, A. Matsui, M. Abe and K. Chouraku, "Complex Permittivity Measurements in a Wide Temperature Range for Printed Circuit Board Material Used in Millimeter Wave Band", *Proceedings of IEEE/ECTC*, May 2020, pp. 938–945.
37. Han, K., Y. Akatsuka, J. Cordero, S. Inagaki, and D. Nawrocki, "Novel Low Temperature Curable Photo-Patternable Low Dk/Df for Wafer Level Packaging (WLP)", *Proceedings of IEEE/ECTC*, May 2020, pp. 83–88.
38. Yamamoto, K, S. Koga, S. Seino, K. Higashita, K. Hasebe, E. Shiga, T. Kida, and S. Yoshida, "Low Loss BT resin for substrates in 5G communication module", *Proceedings of IEEE/ECTC*, May 2020, pp. 1795–1800.
39. Kakutani, T., D. Okamoto, Z. Guan, Y. Suzuki, M. Ali, A. Watanabe, M. Kathaperumal, and M. Swaminathan, "Advanced Low Loss Dielectric Material Reliability and Filter Characteristics at High Frequency for mmWave Applications", *Proceedings of IEEE/ECTC*, May 2020, pp. 1795–1800.
40. Guo, J., H. Wang, C. Zhang, Q. Zhang, and H. Yang, "MPPE/SEBS Composites with Low Dielectric Loss for High-Frequency Copper Clad Laminates Applications", *Polymers*, V12, August 2020, pp. 1875–1887.